T0327744

Partial Discharges (PD)

Partial Discharges (PD)

Detection, Identification, and Localization

Norasage Pattanadech
King Mongkut Institute of Technology Ladkrabang
Bangkok, Thailand

Rainer Haller
University of West Bohemia
Plzeň, Czech Republic

Stefan Kornhuber
University of Applied Science
Zittau, Germany

Michael Muhr
Graz University of Technology
Graz, Austria

This edition first published 2023

Registered Office(s)
John Wiley & Sons, Inc., 111 River Street, Hoboken, NJ 07030, USA
John Wiley & Sons Ltd, The Atrium, Southern Gate, Chichester, West Sussex, PO19 8SQ, UK

For details of our global editorial offices, customer services, and more information about Wiley products visit us at www.wiley.com.

Wiley also publishes its books in a variety of electronic formats and by print-on-demand. Some content that appears in standard print versions of this book may not be available in other formats.

Library of Congress Cataloging-in-Publication Data applied for:

Hardback ISBN: 9781119568452

Cover Design: Wiley
Cover Image: Courtesy of Authors

Set in 9.5/12.5pt STIXTwoText by Straive, Pondicherry, India
Printed and bound by CPI Group (UK) Ltd, Croydon, CR0 4YY

C9781119568452_270723

Contents

Author Biographies

Dr. Norasage Pattanadech received his PhD in Engineering Sciences Electrical Engineering from the Institute of High Voltage Engineering and System Management, Graz University of Technology, Austria. He has more than 20 years of experience in the field of High-Voltage Testing and Analysis, especially in condition monitoring of high voltage equipment. He has served on IEC TC42 MT 23 and MT 14 committees. He received a Japanese patent in 2021 for a breaker with electrode controlling current. He is the author or co-author of more than 100 publications and four books on electrical engineering and PD measurement.

Rainer Haller received his diploma and PhD in High Voltage Engineering from the High Voltage Department at the University of Technology in Dresden (Germany). During this time, he was mainly engaged in PD measurement on high voltage insulation as well as in the development of appropriate PD measuring technique. At the same time, he participated on activities of the IEC–WG "High Voltage Test and Measurement Techniques" in cooperation with CESI (Italy). He joined a manufacturer of high voltage equipment in 1986 and was mainly responsible for the development of high voltage testing equipment as well as power transformers. In 1991, he became a professor at the University of Applied Sciences in Regensburg (Germany), where he was engaged in High Voltage Engineering and Electrical Power Engineering. In 2006, he joined the University of West Bohemia in Pilsen (Czech Republic) and is currently the head of the High Voltage Section of the Regional Innovation Center for Electrical Engineering (RICE). Dr. Haller is author and co-author of more than 100 scientific papers, numerous lectures, and one monography. Currently, he is actively engaged as lecturer and researcher on High Voltage Engineering as well as a member in international organizations like CIGRE.

Stefan Kornhuber studied Electrical Power Engineering at the Graz University of Technology. In 2005, he received his diploma degree and in 2007 his doctoral degree with the main research topic on Temperature Measurement and Uprating of OHTLs. Until 2006, he was with Test Institute for High Voltage Engineering Graz GmbH, with the main research topics in high voltage testing, simulation, and investigation of stresses of transients in electrical power networks. From 2006 to 2013 he was with Lemke Diagnostics and Doble Lemke. In 2013 he joined ABB AG Power Transformers – Engineering Solutions in Halle, Germany, as Head of Condition Management for Power Transformers and was later responsible for on-site and local high voltage test field and systems.

In 2014, Dr. Kornhuber became chair in High Voltage Engineering and Theoretical Electrical Engineering at the University of Applied Science Zittau/Görlitz. The main research topics are outer and inner electrical interfaces of polymeric materials, test and measuring methods, and methods for technical diagnostics. He is a member of working groups at CIGRE, IEC, and DKE and is a convenor of CIGRE D1.58 and IEC TC 112 WG3. In 2021, he received the CIGRE Technical Council SC D1 Award and in 2022 the IEC 1906 Award. Since the beginning of 2023 he has been convening the German CIGRE SC D1 mirror committee and the IEEE DEIS Outdoor Insulation Technical Committee.

Michael Muhr received his diploma degree in 1971, PhD in 1978, and the Habilitation 1983 from Graz University of Technology (TU Graz). In 1990, he was appointed to the head of the Institute of High Voltage Engineering at TU Graz, and in 1996 he was appointed as full professor of High Voltage Engineering.

He was managing director of the Test Institution of High Voltage Engineering of TU Graz. From 2003 to 2007, he was head of the Senate, and from 2007 to 2011 vice-rector for Academic Affairs of TU Graz.

Dr. Muhr has edited more than 190 publications and reports and supervised more than 160 diploma and 50 doctoral thesis. He has received recognitions and awards (Dr. h. c. from University in Pilsen), cooperated with many institutes in Europe and Overseas, and is a member in the national and international societies ÖVE, DKE, IEEE, IEC, and CIGRE.

Foreword

Partial discharges (PDs) igniting in the bulk dielectric of high-voltage (HV) equipment may cause irreversible insulation deterioration and hence initialize an ultimate spark breakdown. Although this has been known since the beginning of the last century, the detection of partial discharges in HV equipment became more crucial in the 1960s, when organic insulation materials were more widely introduced in the HV industry, such as epoxy resin and polyethylene. These dielectrics are very sensitive to PD events. Hence, detection, measurement, and localization of partial discharges became an indispensable tool for quality assurance tests of HV equipment. These topics were extensively addressed by F.H. Kreuger in a textbook titled *Discharge Detection in High Voltage Equipment*, first published in 1964, and updated in 1989.

Due to the fast-growing practical experiences in PD measurements as well as the development of modern PD measuring systems, the advancements achieved within the following years were summarized by D. König and Y.N. Rao and published in a monography titled *Partial Discharges in Electrical Power Apparatus,* edited in 1993. The latest edition of a textbook traces back to the year 2010, when T.S. Ramu and H.N. Nagamani published the book *Partial Discharge Based Condition Monitoring of High Voltage Equipment.*

Since that time, greater advancements have been achieved – on one hand, the further growing practical experiences and, on the other hand, the use of advanced digital signal processors and computer-aided PD measuring systems to acquire, visualize, and classify the captured PD data. However, these achievements have not yet published in an updated textbook but rather in numerous technical papers and conference papers as well as in specific chapters of various textbooks dealing with HV measuring and test techniques, and thus, are not easily accessible to technicians and engineers engaged in the field of PD measurements.

Therefore, I appreciate very much the authors' decision to write this textbook, as well as their outstanding work addressing the state-of-the art in detection, identification, and location of partial discharges. In this context, it is worth mentioning that the authors are well-acknowledged experts in the field of HV measurement and test techniques, including the specifics of PD measurements and diagnosis tests. A great deal of work on this subject has been done by them in various national and international societies and organizations, such as IEC, CIGRE, IEEE, and DKE. Moreover, one author, Michael Muhr, was appointed for a long time as chairman of the CIGRE AG HV Test Techniques and at present he is convenor of IEC TC 42 working groups: "IEC 60270 – Partial discharge measurements" and "IEC 62748 – Measurement of partial discharges by electromagnetic and acoustic methods." At this point I would also like to mention that the authors have been known to me for more than a half century, and several topics covered in this textbook have been investigated and discussed together, particularly when participating in the annual meetings of the above-mentioned organizations.

As outlined by the authors, the primary objective of the book is to present the current status of the knowledge regarding the detection and measurement of partial discharges as well as the procedures available to localize, identify, and classify harmful PD defects. The introducing chapters address the very complex physics of partial discharges in gaseous inclusions and their modeling, while the chapters that follow treat the fundamentals of PD tests and the specific aspects to be considered under on-site condition, such as the denoising of the captured PD transients. Moreover, the book provides valuable information on the opportunities and limitations of alternative approaches, such as the use of ultrasonic and electromagnetic PD detection methods. A specific chapter is dedicated to the challenges of PD measurements under direct voltages, which is especially of interest for quality assurance tests of high-voltage direct current (HVDC) equipment, increasingly used to transmit and distribute renewable energy.

The book is primarily intended for researchers, engineers, and technicians dealing with the development, design, and manufacturing and quality assurance test of HV equipment. For me, this is also the ultimate book for the maintenance staff performing PD tests to ascertain the insulation integrity of HV apparatus after manufacturing and repair. Moreover, this book would be of great interest to students educated in electrical engineering, particularly if interested in more in-depth studies of the very complex PD phenomena in gaseous inclusions. The book offers readers real-work experience, problem description, and solutions, while teaching them about the nowadays available tools for PD detection, as well as for localization, identification, and classification of the captured and acquired PD transients. Experts may use the knowledge provided by this book as they consider upgrading the current standard IEC 60270. The textbook includes full-color photos and illustrations, forms, and tables to complement the topics covered in the individual chapters. There is also an extensive reference list, supporting the readers interested in more in-depth studies of PD phenomena in dielectric bounded air gaps, as representative for gaseous inclusions embedded in the bulk dielectric of HV equipment.

Eberhard Lemke
Prof. Dr.-Ing. habil. Dr. h. c.
Dresden, Germany

Symbols and Abbreviations

Scalar quantities that change over time are written in italics, while vector quantities, like $v(t)$ and $\boldsymbol{E}(\boldsymbol{x},t)$, are written in bold and italics.

Peak values, such as $\hat{E}$ and $\hat{U}$, are denoted by an umbrella crown or caret on the letter.

Symbols

a	Random variable, coefficient
b	Random variable, coefficient
$\mathbf{d}$, d	Gap distance, distance, length (vector and magnitude)
e	e = 2.718281. . ., / Euler number
$\boldsymbol{e}$	Elementary charge
$\boldsymbol{f}$	Geometry factor
f	Frequency
$f(\ldots)$	Function of . . .
$\mathbf{g}$	Geometry factor
g	Activation function
$\boldsymbol{i}$, i	Current (vector and magnitude)
j	Imaginary unit, counting index
k	Counting index
m	Number (quantity), counting index
n	Pulse repetition rate, number (quantity), counting index
p	Gas pressure, Number (quantity)
$\boldsymbol{p}$	Dipole moment
q	Charge
r	Radius, distance
s	Response signal
t	Time
u	Voltage
v	Voltage, speed of signal

$\boldsymbol{w}$	Energy
w	Weight
x	Distance, radius, Space coordinate, charge amplitude
y	Space coordinate
z	Space coordinate
$\boldsymbol{A}$, A	Area (vector and magnitude)
A	Amplitude spectra
C	Coulomb
C	Carbon (chemical symbol)
C	Capacitance
D	Distance
$\boldsymbol{E}$, E	Electrical field strength (vector and magnitude)
H	Hydrogen (chemical symbol)
$\boldsymbol{I}$, I	Current (vector and magnitude)
$\boldsymbol{J}$	Current density
L	Length, inductance
M	Mutual inductance
N	Number, Number (quantity)
O	Oxygen (chemical symbol)
P	Power
Q	charge
R	Resistance, activation function
$\boldsymbol{S}$	Spatial vector
S	Sulfur (chemical symbol), sigmoid function
Si	Silicon (chemical symbol)
T	Time, period
U	Voltage
V	Voltage, volume
W	Energy, wavelet transform function
X	Distance, signal function
Z	Impedance
Φ	Phase angle
ψ	Wavelet basis function
α	Ionization coefficient
ε	Permittivity
η	Function
η	Field efficiency factor, coefficient
κ	Electrical conductivity
λ	Potential function
μ	Mean
σ	Surface charge density
τ	Time constant
φ	Phase angle
$\varnothing$	Magnetic flux, phase angle

Abbreviations

3CFRD	Three-frequency-related diagram
3PARD	Three-phase-relation diagram
3PTRD	Three-phase-time-relation diagram
A/D	Analog to digital
AC	Alternating electric fields
ACRF	Frequency-tuned resonant circuit for AC voltage generation
ACRL	Inductance-tuned resonant circuit for AC voltage generation
ACTC	Transformer circuit for AC voltage generation
ADC	Analogue-digital converter
AE	Acoustic emission
AI	Artificial intelligence
AIS	Air-insulated switchgear
AM	Amplitude modulation
ANN	Artificial neuron network
ART	Adaptive resonance theory
ASP	Advanced signal processing methods
BPNN	Backpropagation neuron network
BW	Bandwidth
CC	Capacitive coupler
CCNN	Cascaded neuron network
CIGRE	International Council on Large Electric Systems
CMAC	Cerebellar model articulation controller
CNN	Convolution neuron network
CPN	Counter propagation network
CRNN	Convolutional recurrent neural network
CT	Current transformer
DAC	Damped alternating current
DC	Direct current
DC	Direct voltage
DCS	Directional coupler sensor
DCS	Directional coupler sensor
DDB	Dodecylbenzene
DGA	Dissolved gas analysis
DKE	Electrotechnical standardization
DL	Deep learning
DOA	Direction of arrival
DSP	Digital signal processor
EE PD	Electrical PD
EM	Electromagnetic
EMC	Electromagnetic compatibility
ES	Expert system
Ex/ATEX	Equipment for potentially explosive atmospheres

EXNN	Extension neuron network
FDA	Fischer discriminant analysis
FIR	Finite impulse response
FPGA	Field programmable gate arrays
GCB	Gas blast circuit breaker
GIB	Gas insulated buses
GIL	Gas-insulated transmission line
GIS	Gas-insulated substation, gas-insulated switchgear
GNSS	Global navigation satellite systems
GPS	Global positioning system
HAIP	Highly available IP
HF	High frequency
HFCT	High-frequency current transformers
HMM	Hidden Markov models
HPFF	High-pressure fluid-filled
HPGF	High-pressure gas–filled
HV	High voltage
HVDC	High-voltage direct current
IEC	International Electrotechnical Commission
IEEE	Institute of Electrical and Electronics Engineers
IGBT	Insulated gate bipolar transistor
IIR	Infinite impulse response
IoT	Internet of Things
ISM	Industrial, scientific, and medical
LI	lightning
K-NN	K-nearest neighbor algorithm
MCC	Modified cross-correlation
ML	Machine learning
MV	Medium voltage
OIP	Oil-impregnated paper
PAT	Phased array theory
PCA	Principle component analysis
PCB	Printed circuit board
PD	Partial discharges
PDEV	Partial discharge extinction voltage
PDIV	Partial discharge inception voltage
PDM	Partial discharge monitoring
PLC	Power line carrier
PMT	Photomultiplier
PNN	Probabilistic neuron network
PRPD	Phase resolved partial discharge
PRPDA	PRPD analysis
PSA	Pulse sequence analyzing
PWM	Pulse width modulation
R&D	Research and Development with parallel to R&D

RBFNN	Radial basis function neuron network
RC	Rogowski coil
RF	Radio frequency
RIV	Radio interference voltage
RNN	Recurrent neural networks
RPDIV	Repetitive partial discharge inception voltage
RST	Rough set theory
RTU	Remote terminal unit
S/N	Signal-to-noise
SCLF	Self-contained liquid-filled
SI	Switching
SNR	Signal-to-noise ratio
SOM	Self-organizing maps
SSC	Stator slot coupler
SVM	Support vector machine
TB	Technical brochures
TE	Transverse electric
TEAM	Temperature stress, electrical stress, mechanical stress, and ambient
TEM	Transverse electromagnetic
TEV	Transient earth voltage
TF	Time-frequency
TM	Transverse magnetic
TOA	Time of arrival
TS	Technical specification
TVA	Corona electromagnetic probe tests
UAV	Unmanned aerial vehicle
UHF	Ultra-high frequency
UTC	Coordinated Universal Time
UV	Ultraviolet
UWC	Ultrasonic wave concentrator
VFF	Very fast front time
VHF	Very high frequency
VLF	Very low frequency
VT	Voltage transformer
W-CNN	Wavelet kernel-based convolution neural network
WLAN	Wireless local area network
WSN	Wireless sensor network
XPLE	Cross-linked polyethylene

1

Introduction

This book is a comprehensive introduction of partial discharges (PDs) – detection, identification, and localization. Dielectric insulations used in high-voltage equipment have to counter the effect of PDs caused by inhomogeneous field configurations or inhomogeneous dielectric material. These continuous stress from PDs can increase, damage the insulation, and lead to power outages within a short time. Therefore, detection, identification, and localization of PDs are required to evaluate the dielectric insulation performance of electrical power equipment. Moreover, PD measurement is used for quality assurance during high-voltage tests in order to confirm that the high-voltage equipment will operate with high reliably and at high efficiency by which the PD level obtained from the factory test must be within the limits of the equipment standards. In addition, PD measurement is efficient and important for diagnosis and research.

PD phenomena have been recognized since the beginning of high-voltage technology in the twentieth century. Building on loss angle measurements in the 1930s, Frederik Hendrik Kreuger introduced modern PD testing in the 1960s through charge-based measurement. This moved quickly to an international IEC standard (IEC 60270). Since then, PD measurement has occupied an important place in research and industrial application as a nondestructive high-voltage test of the insulation systems of power equipment. With the progress in microelectronics and computer development, new methods have become common practice, such as phase-resolved PD measurement. Furthermore, considerable progress has been made in PD detection techniques such as acoustic method and the UHF method to apply in field tests.

In the standard (IEC 60270), the measurement of PDs is a charge-based measurement. The values are expressed in Coulomb (C). PDs can also be detected through electromagnetic (EM) waves, light, heat, pressure, noise, and chemical changes in the dielectric materials. All of these physical and chemical processes can be used to measure the discharge phenomena. Above all, the acoustical and UHF range measurements range (IEC TS 62478) have prevailed in the PD detection in recent years. There are also different recommendations for PD detection by these methods in the equipment standards.

Another important aspect is the behavior of PDs governed by direct current (DC) voltage. DC is coming up more and more, especially for long transmission lines, but there is a lack of knowledge and testing about the behavior of PDs under DC stress. Therefore, there is a

Partial Discharges (PD) - Detection, Identification, and Localization, First Edition. Norasage Pattanadech, Rainer Haller, Stefan Kornhuber, and Michael Muhr.

great deal of catching up to do in research and evaluation in this field. This is reflected in numerous publications, recommendations, regulations, and standards published.

1.1 Overview

PD has been presented in the chapters of many books in recent years. However, the last book that dealt exclusively with PD was published years ago. The authors' intention in writing this book is to provide a far more up-to-date discussion of the topic, beginning with the physical behavior and classification of various PD types in Chapter 2, before proceeding to discuss modeling of PD behavior in Chapter 3. There follows a discussion of the well-known classical (PD) model using capacitors (and resistors), followed by discussion of the dipole model, which may bring new and interesting aspects to this field. Chapter 4 describes the measurement of PDs with decoupling, acquisition, processing, and pattern classification – one of the important chapters in this book. Another important concept is the denoising for PD measurement, which is very significant, especially for onsite PD measurement and the PD monitoring.

Chapter 5 deals with EM methods and Chapter 6 with nonelectrical methods of PD detection. It includes EM and acoustical methods but also optical and chemical methods. Especially acoustical detection and the PD detection with UHF are widely used in on-site PD testing and diagnosis. Besides the detection of PDs, the location of the PDs is very important. On the one hand, the levels or patterns of PD are informative, but on the other hand it is the PD source location that indicates how dangerous PD is. An evaluation of PDs is discussed as well: Are they harmless or harmful? And what are the limiting values given in the standards? The chapter wraps up with an outlook for the industry going forward.

The primary objective of this book is to present the current status of the knowledge in PD detection, identification, and localization in both research and industrial application. Chapter 7 deals with the localization of PD effects, which is important for knowing the location of PDs. Despite all the advances provided by new technologies and new methods of PD detections, phenomena associated with signal propagations and PD activities remain problematical with regard to the various insulation systems in high-voltage equipment (Chapters 9 and 10), as these phenomena cannot be fully explained by traveling wave theory or high-frequency signal analysis. Therefore, difficulties persist in the analysis and evaluation of PD test results. The parameters measured at the terminals do not directly correlate with the phenomena occurring at the PD site. For example, while PD phenomena are measured at the terminals of the test sample as a transient mechanism, the test parameters do not depend on the PD phenomena solely occurring at the breakdown location, but also on the setup of the test sample, and therefore standard PD measurement only quantifies what is called the apparent charge (Chapter 11). Moreover, the measurement of PDs is different, depending on types of power apparatus, because of the inherent capacitive and/or inductive behavior present. Thus, the test sample must be considered electrically as a lump or distributed parameter.

Besides universities and research institutes, there are several national and international societies that deal with PD questions. These include the International Electrotechnical Commission (IEC), Institute of Electrical and Electronics Engineers (IEEE), Conseil International des Grands Réseaux Électriques (in English, International Council for Large Electric Systems, CIGRE), and Deutsche Elektrotechnische Kommission (in English,

German Electrotechnology Commission, DKE). Moreover, the contemporary issues of both PD phenomena and PD measurement are discussed, not only for practical applications but also to provide theoretical background about PD behavior. The reflection on the fundamental knowledge of PD is more and more important to thoroughly understand PD phenomena and their influence on insulation materials and systems. It is especially important to research DC stress and PD behavior.

1.2 Acknowledgments

We would like to thank all our expert colleagues and friends who have brought us more thoroughly into to the field of partial discharges and deepened our knowledge, expertise, and interest of it through conversations, discussions, publications, and development in this field.

Participation in many international (IEC, CIGRE) working groups, committees, meetings, and conferences has also helped us to bring the field of PDs close enough that we could devote ourselves to the work of writing this book. We therefore want to thank everyone from universities, industry, and economy once again, but we would particularly like to give our special thanks to our excellent experts and colleagues Eberhard Lemke and Detlev Gross.

1.3 Users

This book is intended for all interested readers a monography, a reference book, or a state-of-art report. Therefore, this book can be of interest to:

- Electrical engineers engaged in the manufacturing of high-voltage equipment, such as transformers, generators, switchgears, cables, and so on. Engineers working in utilities will also find it useful.
- Students in universities specializing in high-voltage engineering.
- Personnel in research and development employed in design of power apparatus, testing high-voltage apparatus, or working in independent test laboratories with concerning with standardization.

August 2022

Norasage Pattanadech
King Mongkut's Institute of Technology
Ladkrabang – Bangkok/Thailand

Rainer Haller
University of West Bohemia
Pilsen/Czech Republic

Stefan Kornhuber
University of Applied Science
Zittau/Germany

Michael Muhr
Graz University of Technology
Graz/Austria

2

Physical Behavior of Partial Discharges

2.1 Introduction

A *partial discharge* (PD) is defined as

> "... localized electrical discharge that only partially bridges the insulation between the conductors and which can or cannot occur adjacent to a conductor. Partial discharges are in general consequences of local electrical stress concentration in the insulation or on the surface of the insulation. Generally, such discharges appear as pulses having duration of much less than 1 μs" [1].

This statement is focused on pulse-like behavior of partial discharges that are generated by ionization processes within the insulation. However, this definition does not consider that in many cases of electrical insulation, there are already charge carriers (mainly ions) below the ionization level[1] due to natural radioactivity or cosmic radiation. For example, under (normal) atmospheric air conditions, the number of charged ions of ~$(10^2$–$10^3)$ $1/m^3$ may be expected [2].

If within the insulation any electric field is acting, those charge carriers move according to physical law toward their counter-electrodes. If there are any insulating interfaces within or near the field-space, the moving carriers could accumulate on their surface for a certain time.

According to the Shockley theorem, each movement of charge carriers leads to an equivalent current in a connected outer circuit [3]. Below the already mentioned physical defined level of self-sustaining ionization process, the value of such currents is minimal, in the range of (p ... n) A. In the case of alternating current (AC) electric fields, as is common practice for quality test procedures under AC conditions, the just-mentioned effect of accumulated charges does not play any significant role due to the recombination processes at polarity change of outer electric field. But in case of applied unipolar voltage stress as in a DC or impulse field, some charge accumulation could be considered on existing interfaces

1 In that context the ionization level is defined as initiating of self-sustaining discharges, what means, that within the field-space more charged carriers will be generated as it is counted according to the adequate equilibrium of generation and recombination processes.

Partial Discharges (PD) - Detection, Identification, and Localization, First Edition. Norasage Pattanadech, Rainer Haller, Stefan Kornhuber, and Michael Muhr.

within the electrode system. From a physical point of view, such charge is acting as an additional field source and, therefore, could have certain influence on initiating self-sustaining ionization processes by influencing the original (background) field within the electrode system. In addition, there is evidence that pulse-less discharges may occur even at AC electric fields [4]. Nevertheless, the majority of PD phenomena is based on the pulse character, as mentioned, and will therefore be described in later chapters.

For local occurrence of any partial discharge, which is ignited within a certain part of electrical insulation, two basic requirements must be fulfilled:

- Local electrical field strength E_{loc} (given by voltage stress) in that part must exceed the (intrinsic) dielectric strength E_d of the insulation material and, therefore, fulfill the equation:

$$E_{\text{loc}} \geq E_{\text{d}} \tag{2.1}$$

- Free charge carrier(s), mainly electron(s), enable any ionization process to be initiated.

As defined, a partial discharge leads to a limited breakdown within a volume-part of insulation necessarily associated with an ionization process of gas molecules.[2] This event can take place not only in ambient air or other gases but also in gas-filled cavities of solid insulation or in micro bubbles and water vapor of insulation liquids. Likewise, partial discharges may occur on interfaces between different dielectrics as pressboard barrier in liquids, insulator surfaces in gases, or interfacial discharges in slots of solids. Those partial discharges have a typical characteristic related to the connected surface: They "glide" over the participated surface led by electrical field strength, and therefore, they are often called *gliding* (or *creeping*) discharges. A simplified overview about such possible "PD-sources" is shown in Figure 2.1, whereby at "homogeneous" air insulation[3] those parts are mainly located near electrodes,[4] but at other insulation type the condition acc. to (2.1) may be fulfilled also within the insulation by possible imperfections like bubbles, impurities, etc. Note, that the physical nature of partial discharges is mainly based on physics of gas discharges, discussed in this chapter.

Any partial discharge process is dependent on parameters, which are influenced not only by magnitude of electrical field strength $\boldsymbol{E}$ in the insulation caused by electrode configuration (field type, type of imperfect size, and its location) but also by structure and type of insulating material (dielectric parameters, dielectric (intrinsic) field strength), evidence of interfaces, and even by ambient conditions (pressure, humidity, pollution). The requirement of available charged carriers is mainly influenced by type and duration of stressed voltage (alternating, direct, impulse or combined), the type of insulation material (gaseous, liquid, solid, hybrid), and the evidence of any other sources of charged carriers caused by cosmic, natural, or artificial radiation.

---

2 Other physical processes that might lead to electrical breakdown (e.g. vacuum insulation or similar) are not the focus of this book and, therefore, will not be further discussed.

3 "Homogeneous" insulation is characterized by one insulation material only, e.g. atmospheric air.

4 For PD-inception is only the electrical field strength according to (2.1) important, what might be independent on electrode's potential. That's also why "earthed" electrodes PD could occur.

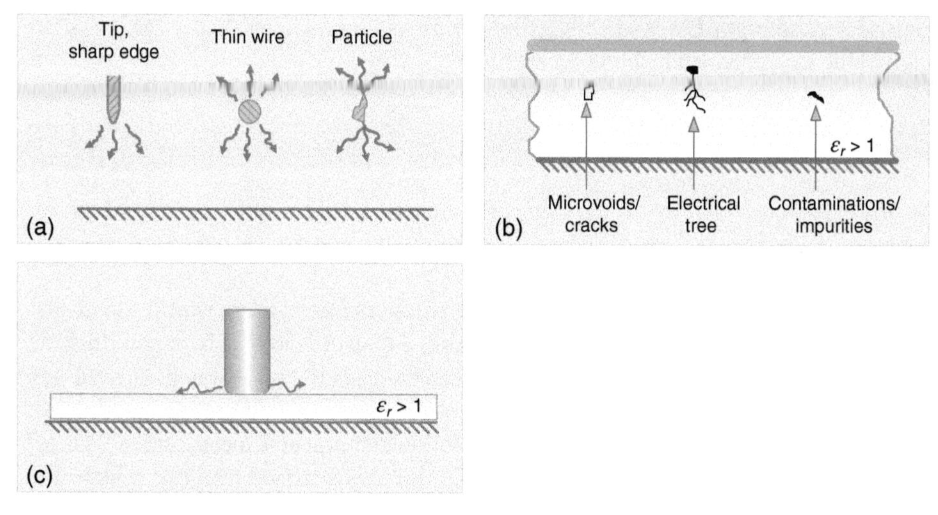

Figure 2.1 Typical PD sources within the insulation (schematically) (a) external, (b) internal, (c) gliding-type.

Partial discharges caused by self-sustaining ionization process are accompanied with a large variety of physical phenomena like light emission, mechanical and chemical reactions, as well as acoustic phenomena. All these phenomena will be applied for recognition, detection, and interpretation of PD measurement. It is obvious, that the PD behavior is several for different types of insulating material, electrical field configuration and imperfections, therefore, for a common description a certain classification of PD phenomena seems to be necessary. There are various criteria for such a classification: the location of PD source (imperfection) relating to the insulation (external, internal, surface); the insulating material (gaseous, liquid, solid, hybrid); the initiating electrical field (AC, DC, impulse); physical phenomena (electrical, optical, chemical, mechanical, acoustical); or even the electrical equipment in which the PD occur (e.g. switchgear, transformer, rotating machines, cables, insulators). For this chapter, the first classification will be preferred.

2.2 External Discharges

As defined, external PD occur "outside" of any insulation equipment preferable on sharp edges or points, but also on long electrodes with small curvature (e.g. ropes on overhead lines) or on surfaces of solid insulation (e.g. insulators). Such PD are typical gas discharges and may occur if electrical field strength $\boldsymbol{E}$ is high enough to initiate a self-sustaining ionization process, that value commonly termed as *inception value.*

The physics of gas discharges were intensively investigated already in last centuries [5–9], covering a large variety of characteristic discharges like *glow-, Townsend-, streamer-, leader discharges,* etc., sometimes also characterized as *Corona discharges.* Such PDs occur, if the electrical field has a certain degree of nonuniformity, which might be characterized, for example, by the field efficiency factor η. The field efficiency factor η is defined by the ratio of electric field strength $\boldsymbol{E}$ within an electrode arrangement (voltage U, gap distance d) as average value $\boldsymbol{E}_{mean}$ divided by maximum value $\boldsymbol{E}_{max}$ [10]:

$$\eta = \frac{\hat{E}_{mean}}{\hat{E}_{max}} = \frac{\hat{U}}{d} \cdot \frac{1}{\hat{E}_{max}} \tag{2.2}$$

For any evidence of (stable) PD the value of η should be, for example, in atmospheric air in the range or less than 0.2, mainly dependent on pressure and may be different for various gas type.

Each gas discharge is characterized by movement of charge carriers within the insulation gap, caused by formation of electron avalanches and drift of ions of both polarities. As already mentioned, each movement of charge carriers leads to an equivalent current in a connected outer circuit. Due to a several mobility of electrons and ions and their equivalent current is different as shown by theoretical calculation and practical measurement [11, 12] (Figure 2.2). A typical PD current is a very fast pulse characterized by a rise time in the nanosecond (ns) range (electrons shown in the red dotted ovals) and with longer duration time of ns to μs range (ions, shown in green dotted ovals). That time behavior corresponds in the frequency domain with equivalent spectra, which reach up to a few GHz, depending on discharge type (Figure 2.3) [14].

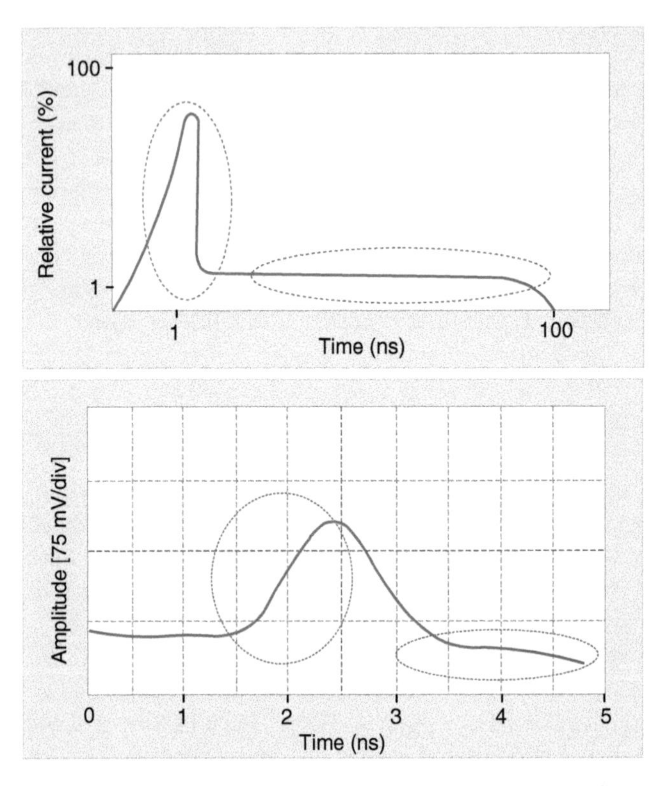

Figure 2.2 Time behavior of PD current pulse (acc. to [11, 12]).

For better explanation of PD behavior, a simple tip-to-plane arrangement with $\eta < 0.2$ in air is considered (Figure 2.4). The expected discharge current will be detected by measuring resistance (e.g. using an oscilloscope).

If the voltage $\boldsymbol{U}$ and, therefore, the electrical field strength $\boldsymbol{E}$, reaches its *inception value*, a self-standing ionization process is initiating. That may be described with the generation of electron avalanches caused by collisions of accelerated electrons with neutral gas molecules [7]. Likewise, ions of both polarities will be generated. Note that electrons will be delivered not only by collision actions but also by *secondary effects* such as impact of ions with the electrode surface or by high-energized photons [9]. The generated space charge carriers – electrons and ions – have different drift velocity (e.g. the mobility of electrons is much higher than for ions). Therefore, the further development of discharge depends on

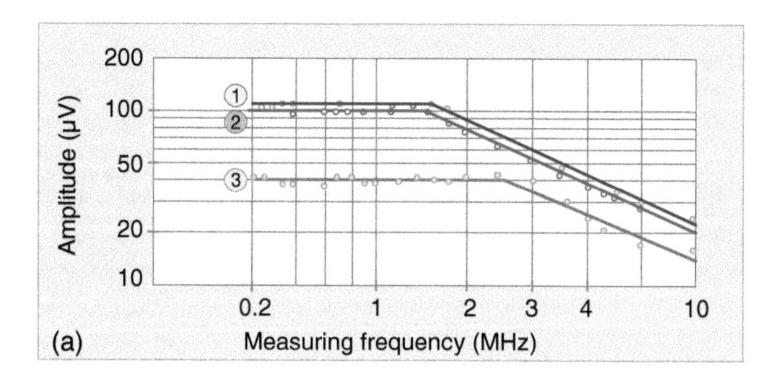

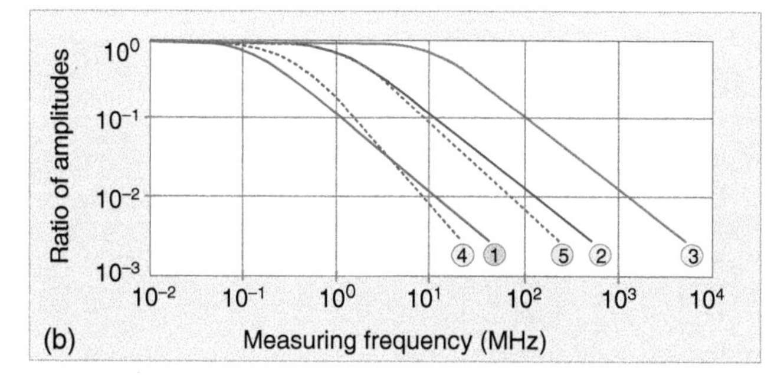

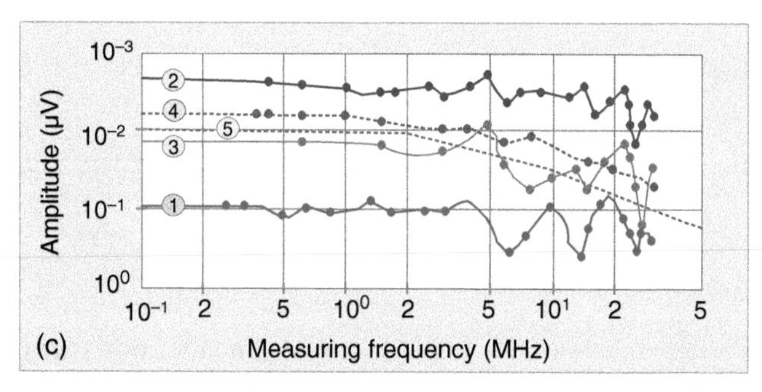

Figure 2.3 Frequency spectra of PD measured and calculated by several authors, after [14] (a, b, c) (1,2 – PE-cable core, 3,4 – rod-plane-arrangement, 5 – calculated for cable core). Behavior of simulated PD current pulses based on (d) time and (e) frequency.

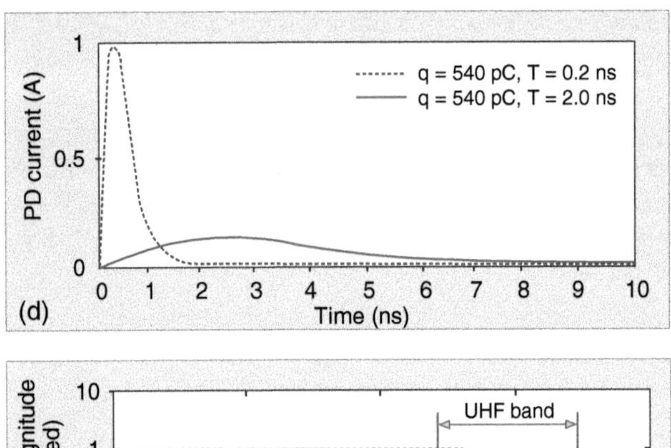

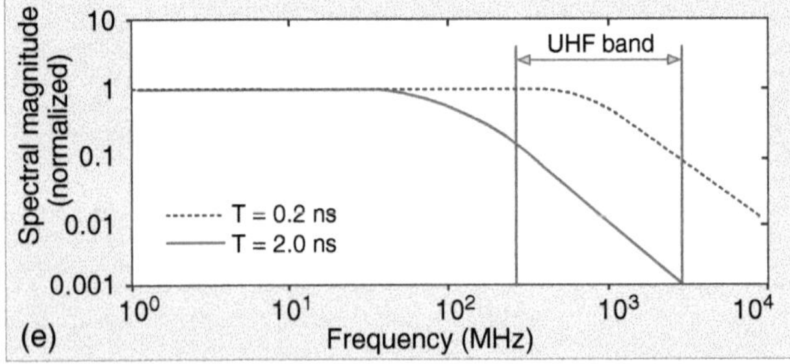

Figure 2.3 (Continued)

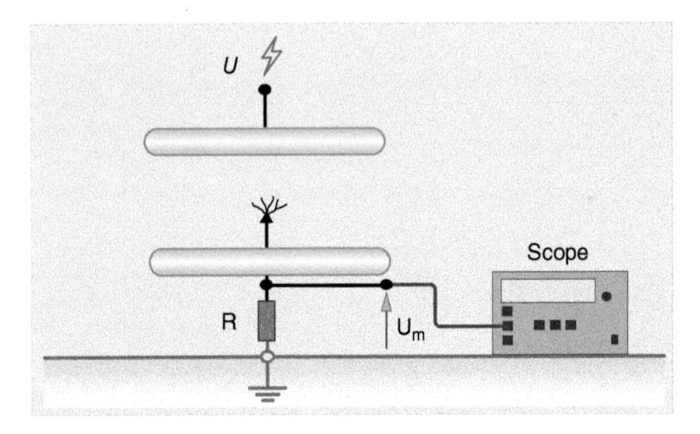

Figure 2.4 Tip-to-plane arrangement in air.

the polarity of the tip, which influences the movement of the generated charge carriers and should be discussed separately.

2.2.1 Tip with Negative Polarity

If the tip polarity is negative, the generated positive ions are collected in a (critical) area near the tip, but the negative ions, mostly produced by attachment of electrons, are in certain distance from the positive ones, forming a kind of *dipole field* (Figure 2.5a, b). The dipole field is superimposed with the background (*Laplacian-*) field $E_g(x)$ and might be termed as *Poisson-field* $E(x)$ as schematically shown in Figure 2.5c. The area for ionization activities is given by the field-dependent ionization coefficient α_e and the exceeding of E

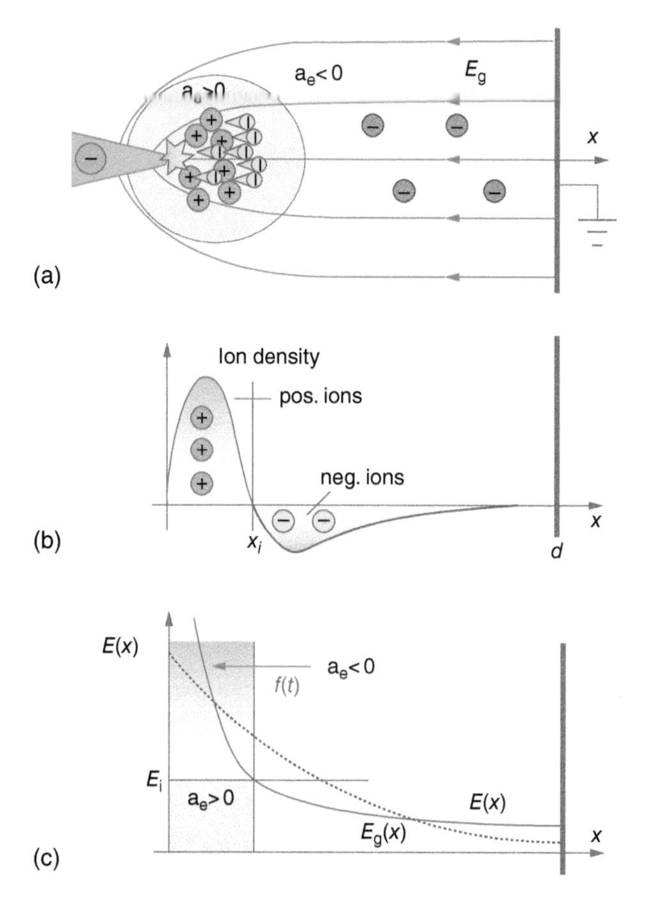

Figure 2.5 Charge behavior and electric field distribution after ionization process (negative tip polarity).

over E_i, the minimal necessary ionization field strength [5]. On one hand, the field strength is increased within a nearest distance to the tip forcing the ionization activity in that area ($\alpha_e > 0$), but likewise, at a certain distance near the tip, the formed negative ions reduce the field strength magnitude ($\alpha_e < 0$). That leads to an extinction of the ionization activity, and the discharge is quenched. After a certain time, given by the drift velocity of the ions, the field strength again increases and the described process may be repeated. If the gap voltage is kept unchanged, a train of small, current pulses might be measured (Figure 2.6). That discharge is often termed as Trichel pulses [8].

Trichel pulses are small discharges with relative low magnitude of current pulses and charge respectively, time duration in ns-range and relatively high repetition rate dependent on voltage magnitude. The measurable charge of such PDs do not exceed some 100 pC.

2.2.2 Tip with Positive Polarity

If the tip polarity is positive the generated electron avalanches move toward the tip (anode) in an increasing field (Figure 2.7a) and will be, finally, absorbed by the electrode. The (slow) positive ions are drifting toward the counter electrode (Figure 2.7b), which leads in first order to a decreasing field strength $E(x)$ near the tip (Figure 2.7c).

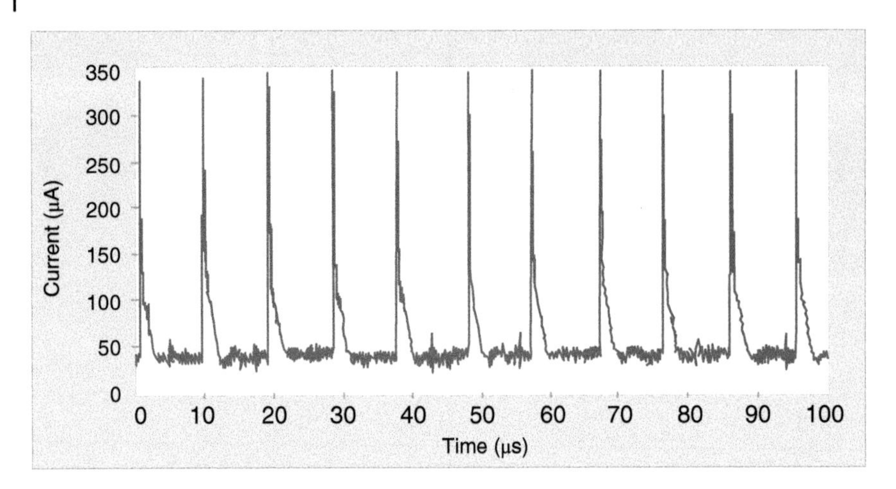

Figure 2.6 Trichel pulses (negative tip polarity).

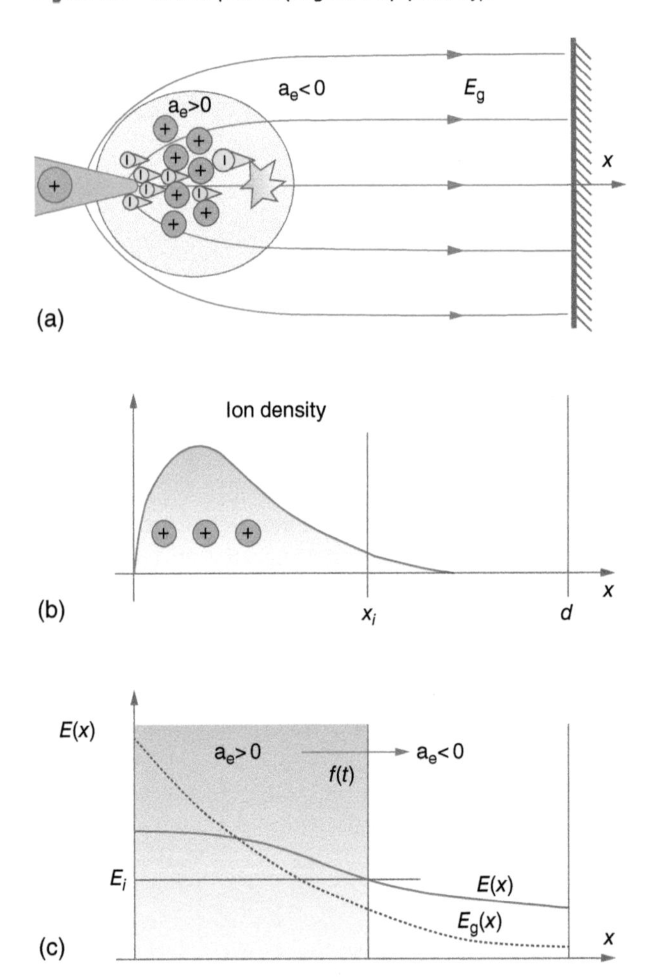

Figure 2.7 Space charge and electric field distribution after ionization process (positive tip polarity).

This means analogous to negative tip polarity, that the discharge process stops until the positive ions move far enough from the tip and the field strength may be increased again. The difference in discharge behavior compared with the negative tip polarity is above all the drift direction of ion space charge, which influences the discharge process ("moving" ionization border, Figure 2.7c). In case of positive tip, the ion movement is directed toward a field space with a more divergent electric field. This also leads to discharges of more "disruptive" character compared with negative tip polarity.

In both cases of tip polarity, the ion space charge changes the resulting electric field near the tip. At negative tip the field concentration and, at the same time, also the *ionization probability* is higher than at positive tip. This is additionally supported by space charge movement below the ionization level. In conclusion, it is understandable that in case of negative tip, the ionization process is initiated at lower voltage values compared with positive tip. That means at negative tip polarity, lower partial discharge inception voltage (PDIV) values will be obtained compared with the opposite one. This behavior could also be observed in case of altered voltage conditions as, e.g., at AC. If the gap voltage is increased continuously, the inception level (PDIV) is reached at first in negative half cycle[5] and, only later, at higher voltage values, might it also discharge in positive half cycle. This behavior is more significant if the field nonuniformity is large ($\eta << 0.2$).

The described phenomenon is characterized as *low-intensity discharge* according to its small charge magnitude and often termed as *Avalanche or Townsend*-discharges. Besides that mechanism, the discharge field in the critical area near the tip has no significant impact on the background field generated by gap voltage. This is true if the gap distance is in the range of some mm to cm. At larger distances as well as higher magnitude of electric field, the intensity of electron avalanches will be increased, accompanied by forced emission of photons, which leads to an additional ionization activity (mainly caused by forced photo-ionization). Simplified, one may imagine that from original (primary) electron avalanches, a higher number of photons were emitted, which contributes to more electrons than by described collision process and weak photo-ionization only. Therefore, much more avalanches are generated, creating a "filament" of discharges – termed as *streamer discharge* (Figure 2.8), [13, 15]. Quantitatively, that threshold is reached if the approximate number of electrons in generated avalanches exceeds a value of ~(10^6–10^8) [6].

The discharge development is faster than under Townsend conditions; the moving avalanches are also generated locally parallel. An impression of that discharge activity is given by a streamer-photogram, made in atmospheric air at nonuniform field conditions (Figure 2.9).

At streamer discharges higher values of discharge current and charge will be obtained compared with Townsend discharges, therefore, they might be characterized as *strong-intensity* discharge with typical charge magnitudes up to few nC.

The described behavior is strongly dependent on value of electric field and might be easily studied at AC-field with different instantaneous values during voltage half-cycle, but also under relevant values at DC condition (Figure 2.10).

5 Related to the tip polarity.

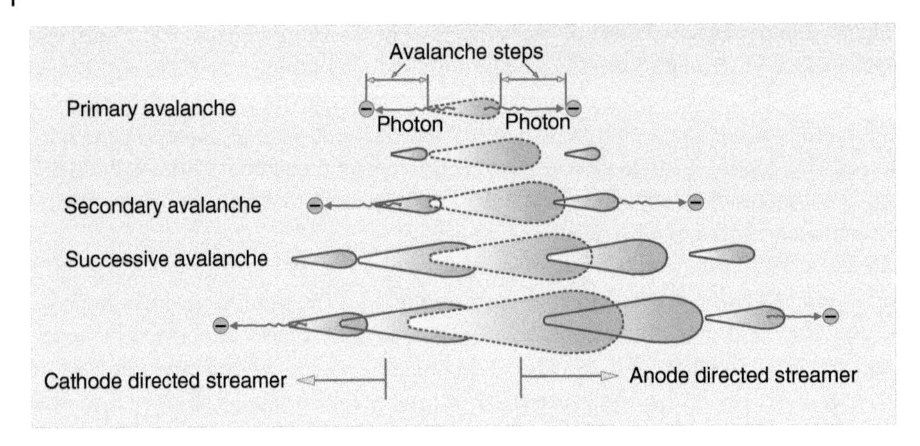

Figure 2.8 Development of streamer discharge (schematically).

Figure 2.9 Streamer in atmospheric air (photography [16]_LEMKE_1967).

If the gap voltage will be increased over the PDIV value, the first discharges (Trichel pulses) are starting in the negative half-cycle. They have high pulse rate and low magnitude of current pulses. Above a certain voltage value, the pulse character of that discharge partly disappears and a "pulseless" or *glow discharge* may be observed. If the gap voltage is further increased, the first streamers occur even in positive half-cycle, characterized by their unstable behavior in-time. They are often termed as *pre-onset streamers*, mainly accompanied by strong audible noise. At smaller gap distances in cm-range, a small increase in voltage may lead to the final breakdown.

At longer gap distances (>1 m), an additional type of discharges may be observed. If the voltage is increased and, therefore, also the electrical field strength, a further physical process of charge carrier generation must be considered – thermal ionization. That is caused by highly energized charge carriers moving within a channel of small radius, which lead to a high temperature in the relevant gas volume and, therefore, to additional generated electrons. As a result of that process, a discharge channel with high current density is generated – mainly termed as *leader* discharge. Due to the highly ionized discharge channel accompanied with strong light emission, the discharge current and, therefore, the charge values are large in the range of nC to μC. It should be noticed that *leader* discharges might be generated even at higher[6] frequency of applied field. The leader discharge

6 For common AC case a frequency of (50/60) Hz is assumed.

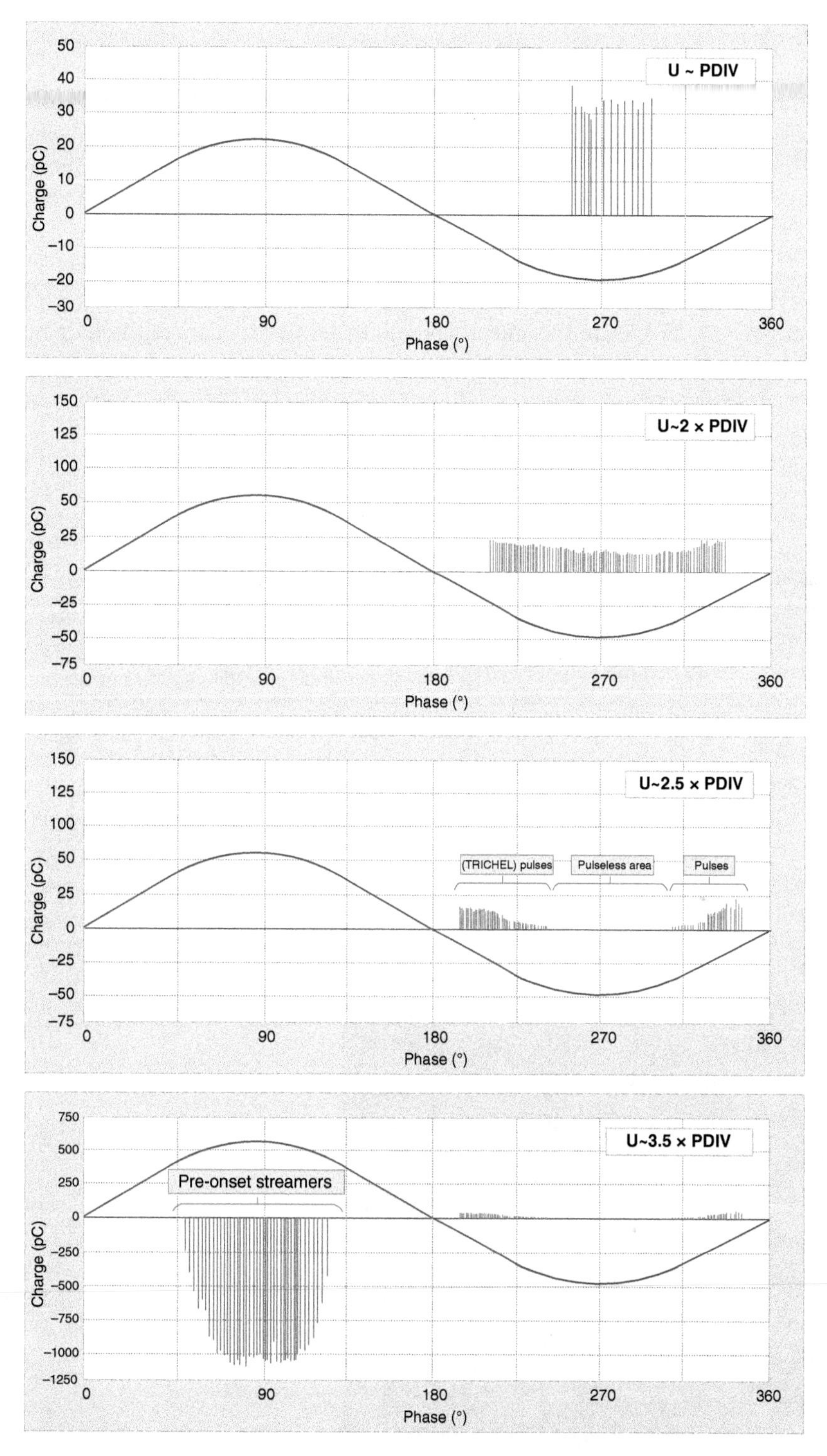

Figure 2.10 PD pulses at various magnitudes of gap voltage (U ≥ PDIV).

is the last mode of self-sustaining discharges before breakdown of entire insulation takes place. In atmospheric air, stable leader discharges may be observed only under AC condition at strong nonuniform electrical field condition and long gap distances (>1 m) [16] or, as mentioned above, at smaller gap distances but at much higher value of frequency than used for AC power transmission.

As an overview the discussed types of discharge in atmospheric air are summarized in Figure 2.11.

It should be noticed that the described ionization behavior may be observed also at other type of electrode configuration e.g. on thin wires (ropes) of overhead lines at higher operational voltage (>110 kV). Even this effect (commonly termed as *Corona* discharge) was extensively investigated in the past [18, 19]. Corona discharge is accompanied by light emission in ultraviolet (UV) range, which might be usable also for diagnostic purposes [17, 20], (Figure 2.12).

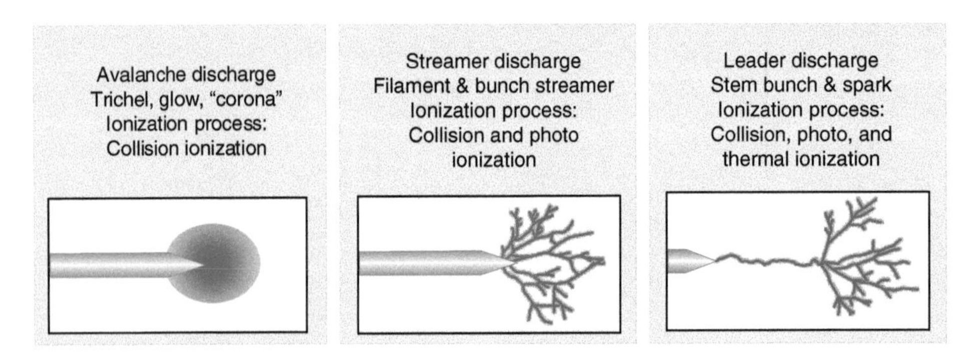

Figure 2.11 Typical discharges in atmospheric air, according to [17].

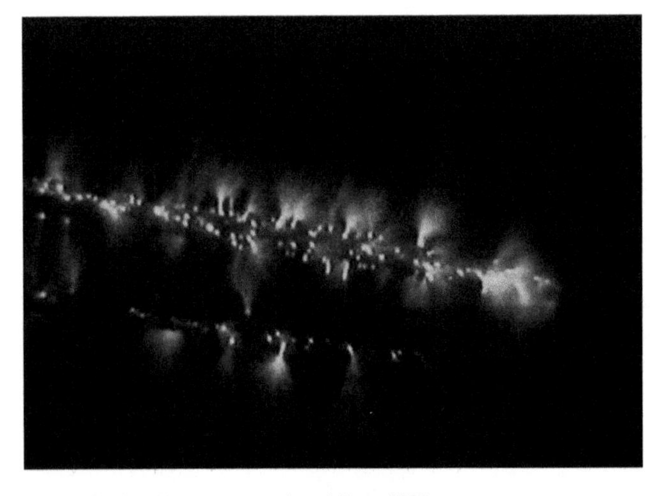

Figure 2.12 Corona at overhead lines [20].

The discussed types of gas discharges were described at air conditions, but research has shown that equivalent or similar discharges also in other type of gaseous insulation should be considered – e.g. in SF_6, N_2 or in several mixtures of them [21–24].

2.3 Internal Discharges

Partial discharges that are partly or complete embedded in liquid or solid insulations are termed *internal discharges*. As already mentioned, such discharges might be initiated also within the insulation at any imperfections, not necessarily adjacent to electrodes. Besides the required electrical conditions (field strength, free charge carrier), the initiation of any first discharges is strongly dependent on material properties as molecular structure, ionization ability as well as moisture, temperature etc. Even the evidence of impurities or cavities may have a significant impact on discharge behavior.

Note that the mentioned condition for ionization – the evidence of free charge carrier – may be different compared with the situation at external discharges. Especially within new manufactured materials, free electrons are available with low probability only; therefore, free charge carriers are probably mainly provided by natural radioactivity or cosmic radiation. Besides the expected ions in atmospheric air (see Section 2.1), the number of free electrons available for internal discharges may be approximated at $2*10^6\,m^{-3}s^{-1}$ up to "some hundreds or thousands per second," dependent on location with respect to imperfect size or dimension, and material [25, 26]. It should be mentioned that this availability of free electrons has a direct influence on the time behavior of partial discharges and may lead to certain delay of ionization process despite exceeding the threshold value of local electric field strength.

2.3.1 Discharges in Liquid Insulation

A once-initiated discharge in liquids commonly leads to a gas discharge occurring in micro bubbles or gas-filled cavities mainly caused by vaporization and local overheating [27]. If the electrical field within such a gas-filled space is high enough for further ionization, a streamer discharge is formed, accompanied by some transient currents (bursts) and emitted light signals as well. At the same time, some shock waves are generated. The streamer is moving to its counter electrode and stops when the local electric field becomes insufficient for further movement. Even a string of micro bubbles is generated, which may be dissolved in the liquid.

Streamer characteristics of insulating liquids depend on rise time and polarity of the applied voltage, as well on pressure, temperature, chemical structure, and physical properties of the liquid [28, 29]. In many investigations, of mineral oil in particular, shape and drift velocity of streamers were shown to be different, dependent on polarity of applied voltage [30, 31]. That was also demonstrated at strong nonuniform electric field ($\eta << 0.2$) under AC conditions, at which different streamer types were obtained depending on the polarity (Figure 2.13) [30].

Typical for both types of streamers are their current shape, which can be measured in an outer circuit. After initial pulses, a certain number of bursts occurs with mainly ascending

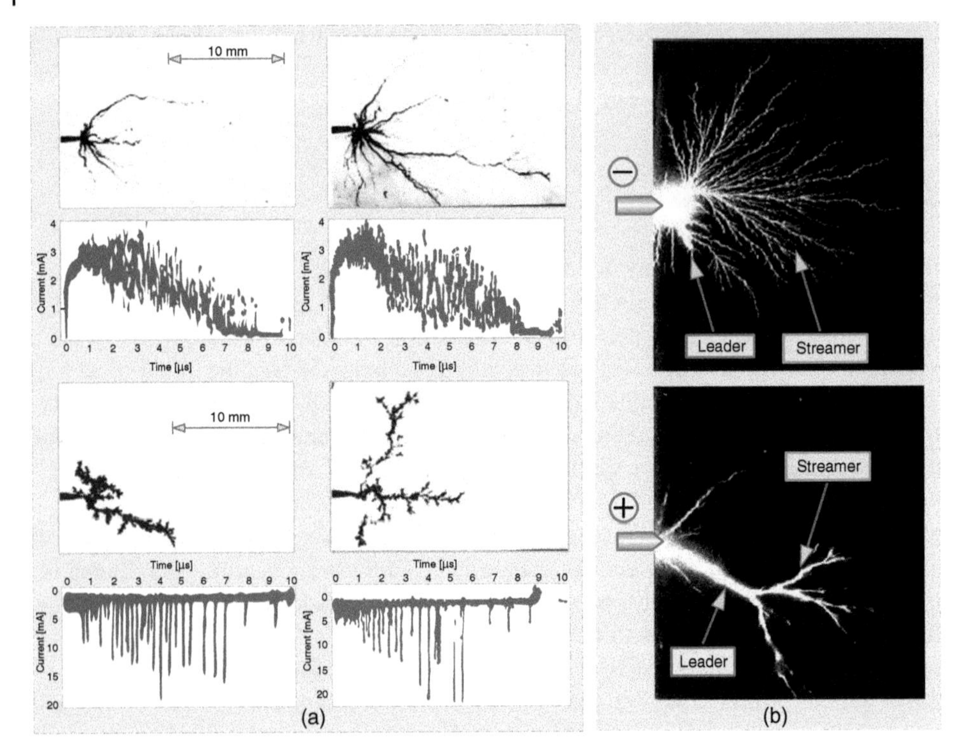

Figure 2.13 Discharges in mineral oil at different tip polarity shown as (a) graphs of bursts with ascending magnitude and (b) photograms with leaders and streamers identified [29, 30].

magnitude forming a kind of "group pulse" (Figure 2.13a). The duration of that pulse is in μs-range with measurable charge values up to 300–400 nC. If the number of generated streamers and charge carriers, respectively, are high enough, a leader discharge may be observed, which is stable only in strong nonuniform fields.

Such a "complete" discharge is visible also on photograms (Figure 2.13b), whereby the first part of discharge is a leader (connected to electrode) and from its top additional streamer discharges are visible. The length of that streamer discharge is dependent on the field distribution within the entire electric field space. If the electric field strength is high enough and the streamer discharges may reach the counter-electrode, a main backward discharge initiates the breakdown of the entire insulation.

It should be noticed that in insulating liquids as well, a certain part of discharge activity without impulse character should be considered. Analogous to pulseless discharges at gaseous insulation, that fact could have a certain influence on partial discharge behavior and finally on breakdown characteristic of entire insulation. That space charge evidence should be especially considered at unipolar field stress condition.

2.3.2 Discharges in Solid Insulation

Partial discharges in solid dielectrics may occur in voids or other impurities at which the intrinsic field strength may be exceeded already at lower voltage values compared with the

Figure 2.14 Tree structure in PMMA at strong nonuniform electric field [33].

whole insulation gap voltage. Commonly higher permittivity of the adjacent insulating material is what leads to higher electrical strength in the void. Typical imperfections in solid insulation are cracks or cavities filled with lower-molecular components produced during the manufacture of the insulation as well as air-filled cavities caused by chemical reactions. In the same way, such voids may be generated by embrittlement or degradation of molecule chain by highly energized space charges caused by mechanical or thermal shock, which could take place in high-polymeric insulation material at electric field stress [32]. Two main types of defects for generating partial discharges are usually considered – voids and cracks or small slots having a typical tree-structure (Figure 2.14). The latter ones are usually called *electric trees*.[7] In both cases, gas discharges could be generated if the electric field strength in those imperfections reaches or exceeds its threshold values.

But it should be noted that, especially in solid insulations, a key requirement for generation of discharges must be fulfilled – the evidence of free charge carriers (electrons) (see Section 2.1). This fact may necessarily lead to a time delay in initiating PDs (PD inception), although the electric field strength is high enough and exceeds the intrinsic field strength of imperfection. That behavior of time lag was often observed, for example, at new manufactured insulation [25]. Especially in such cases, the type of applied voltage also has a main influence on the occurrence of PDs. One may expect severe PD evidence even at the same electric field strength given by maximal value of gap voltage, when the insulation will be stressed by AC-, DC- or impulse voltage. This behavior will be discussed in more detail in later chapters.

Investigations related to voids and trees, respectively, were intensively forced by worldwide applied polymeric insulation, especially for cables and switchgears [35, 36]. The principal PD behavior in such cases should be described exemplarily for a small void at AC stress (Figure 2.15).

7 Note, that such electric trees could be developed even from water-filled cavities with tree structure, which was a huge problem in earlier introduced PE cables [34].

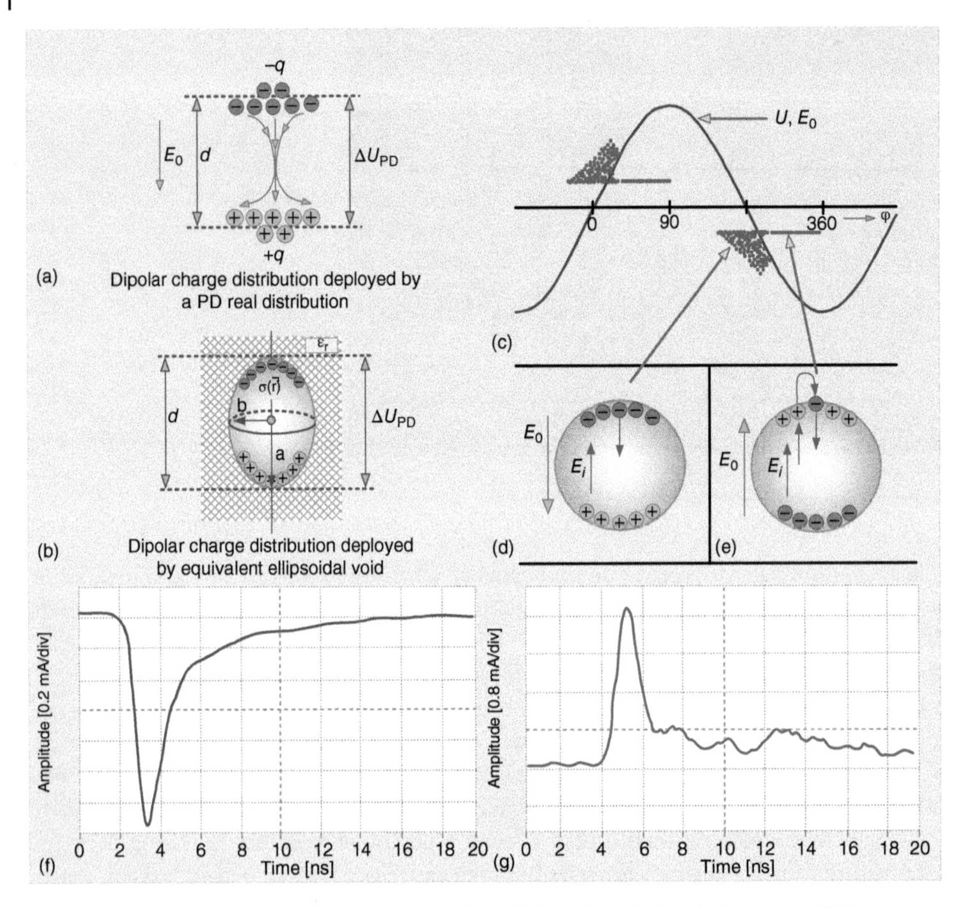

Figure 2.15 Typical PD behavior within a small void in polymeric insulation, acc.to [25]. Bipolar charge distribution deployed by (a) PD real distribution and (b) equivalent ellipsoidal void. Schematically illustrated by charge deployment on the inner void surface for different polarities and resulting field strength values (d, e) as well as discharge pulse distribution over AC voltage (c) with respect to related single PD pulses (f, g), acc. to [44].

If the applied electric field $\boldsymbol{E_0}$ exceeds the intrinsic field strength of void insulation and a free electron is available, a gas discharge (avalanche) could be generated, caused by collision and photon ionization processes. The free electrons could be generated by photons that are hitting the gas molecules (volume ionization) and/or by detrapping processes from evident surface charge inside the void (surface ionization) [25]. For new (fresh) manufactured insulation, no surface charge is still available, so the volume ionization must be considered. In that case, any PD activity could start with a significant delay in second–minute range, despite the electric field already exceeding the intrinsic value of the void.

The moving avalanche discharge is detectable in an outer circuit by current pulse, as shown in Figure 2.15f, g. Those pulses have a very fast rise time (<1 ns) and duration of widely scattered values in ns-range [38]. The significant time parameters of such PD pulses depend on the type of void (gas pressure, location within the insulation, size dimension) as well as magnitude and type of applied voltage and stressing time [39]. Analogously to external gas discharges, those internal discharges are often termed *streamer-like* or *Townsend-like* discharges [40].

Following the physical behavior of the described gas discharge, bipolar charge sheets will be deployed on the void walls (Figure 2.15a). According to Poisson's law those charge sheets generate their own electric field E_q, which is directed against the applied (background) field E_0. If the value of the Poisson-field E_q is comparable with that of applied field E_0, the discharge will be quenched and no further movement of charged carriers inside the void takes place. Some of the charged carriers will recombine, while others will make it to the adjacent surface, forming a surface charge. This surface charge causes an internal electrical field that is superimposed to the external field. It might be assumed, then, that the field strength inside the void is reduced on a residual value E_{res}. Further discharge activity depends on the behavior of applied electric field as well as on the given availability of free electrons and the evidence of surface charge. In case of an AC field and high number of available free electrons, a possible sequence of that process is depicted in Figure 2.16.

The first initiated discharge causes a displacement of the internal "void" field related to the applied (background) field, which changes with every discharge event. Upon reversal of the external applied field, the described discharge activity occurs in the opposite polarity, and so on. However, this process is dependent on many factors, such as material related availability of free electrons, residual surface charge, as well as external applied field, and it was observed with significant results [41, 42].

If the applied field stress is permanent at AC conditions, the discharge activity in such voids may be changed over the stressing time. It was observed that, at first, streamer-like pulses in ns-range with small rise time (<1 ns) occur (Figure 2.17a) [43]. After a certain stressing time, the character of pulses was changed to longer time duration but with less amplitude, similar to Townsend-like pulses (Figure 2.17b) and after continuing field stress changed again to smaller discharge pulses compared with the previous one but with higher pulse rate (Figure 2.17c). It was assumed that the latter type of those pulses was responsible for a certain deterioration of the void surface.

Due to the wide application of polymeric materials in cable insulation with weakly uniform electric field, such small voids lead to significant quality issues, because once

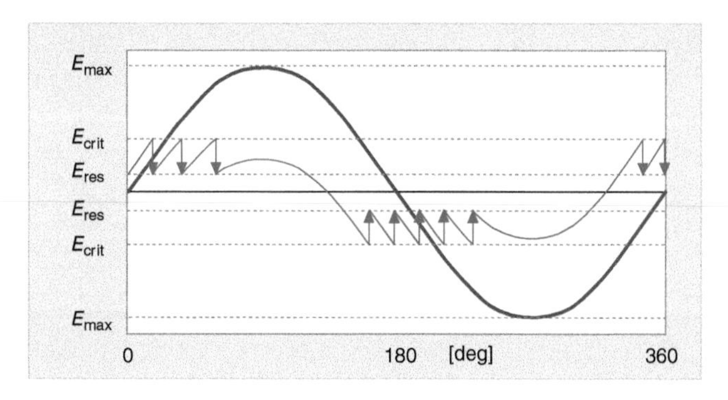

Figure 2.16 Typical characteristic of the electric void field at AC condition.

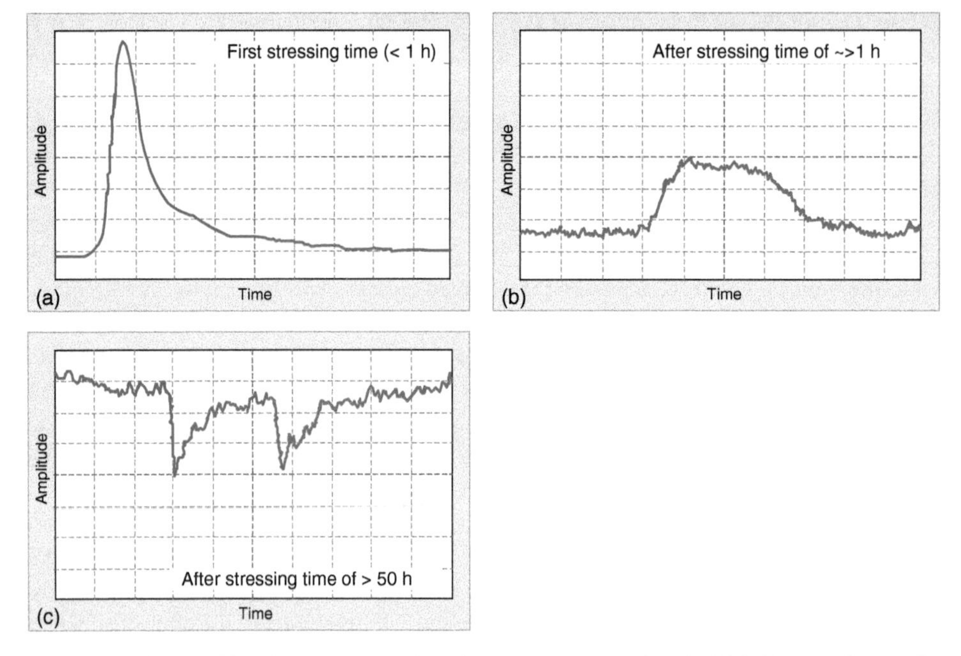

Figure 2.17 Typical PD pulses over stressing time at permanent electric AC field stress for small void in polymeric insulation. (a) Streamer-like pulses in ns-range with small rise time (<1 ns); (b) pulses with longer time duration but with less amplitude, similar to Townsend-like pulses; (c) smaller discharge pulses but higher pulse rate after continuing field stress [43].

initiated, partial discharges might lead to final breakdown in unexpected short operation time. However, huge progress has been reached in estimating of PD activities in such type of voids[8] [44, 45].

Another typical example for imperfections in polymeric materials like such as polyethylene, polypropylene, epoxy resin, and many more is a tree-like structure caused by, for example, water inclusions, embrittlement or degradation of molecule chains, and cracks. Under electrical stress condition on those inhomogeneities, a local increasing value of electric field occurs, which could lead to a *treeing discharge* (see Figure 2.14). Depending on the applied (background) field, such trees could growth in relatively short time to the counter electrode, leading to a complete breakdown of the insulation. The PD behavior is different during the treeing growth, which very often is detectable in three stages (Figure 2.18), [46].

In the first stage, treeing growth is initiating. The PD activity takes place mostly within small, gas-filled branches of tree structure, and the measurable pulses are of very small magnitude (few pC). The second stage might be characterized by mainly "volume-enhancement" of tree structure with slow development toward the counter electrode. In that stage, relatively low PD activity may be observed, mainly characterized by

8 It should be stated, that by the same reason also a huge progress in developing of adequate measuring systems was reached [37].

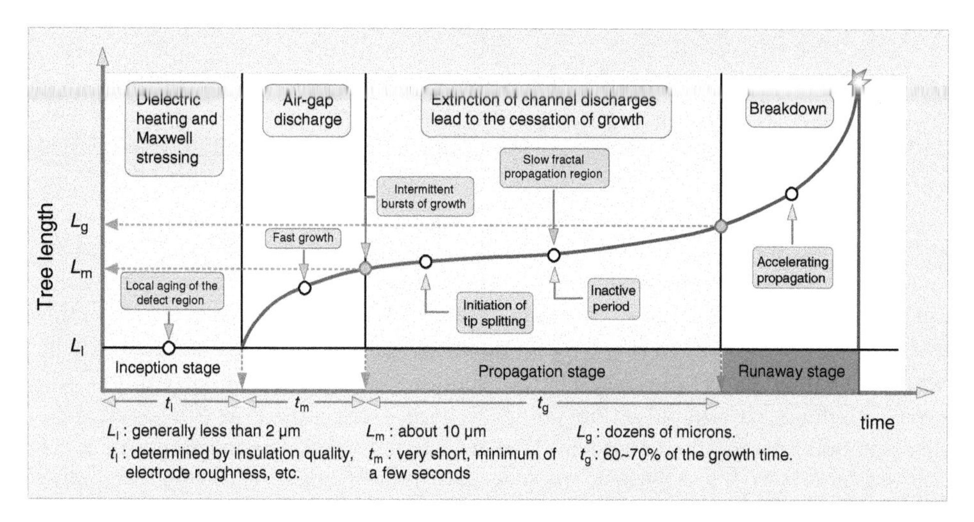

Figure 2.18 Characteristic treeing growth over stressing time at permanent electric AC field stress in polymeric insulation *acc. to* [46].

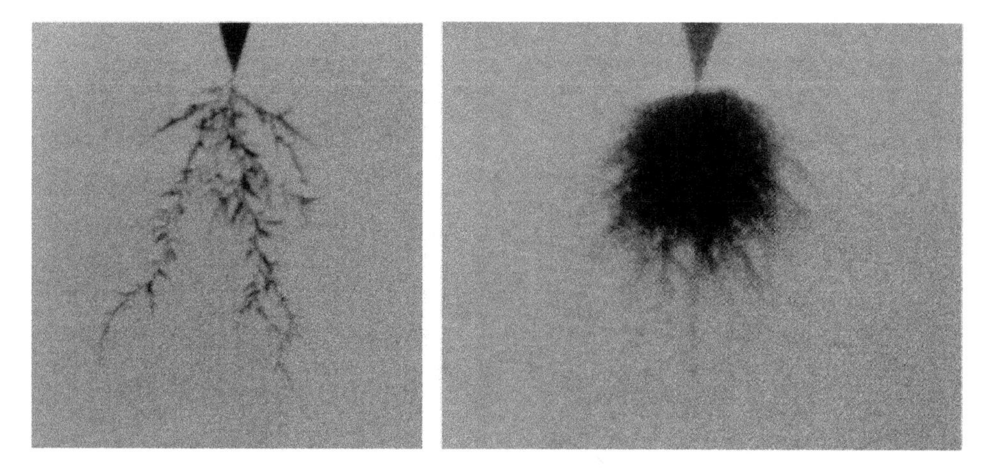

Figure 2.19 Typical types of treeing structure in polymeric insulation acc. to [48].

"impulse-less" discharges [47, 48]. The last stage is accompanied by an accelerated growth directed to the counter electrode, characterized by pulses again with small magnitude but relatively high repetition rate. PD behavior might be influenced mainly by the structure of the tree, which could be typical classified by "branch"-like or "bush"-like trees (Figure 2.19), [49].

In conclusion it may be stated that for detection of complete PD behavior of insulation with tree-defects, it is necessary to measure pulse and pulseless PD-activity.

It should also be mentioned that moving particles or swarming impurities within a solid or liquid insulation might be classified as internal discharges, but this type of PD is not in the scope of this book.

2.4 Gliding Discharges

Partial discharges on a surface may occur in any gaseous medium at their interfaces with solid dielectrics as well as at barriers in liquids. Characteristically for such so-called gliding (or creeping) discharges is the presence of high tangential component of electric field along the interface, which can be typically found in bushings, cable terminations, and sites in transformers, but also at outdoor and indoor insulators as principally shown in Figure 2.20.

The discharges may be initiated in case of a sharp edge (Figure 2.20 left) frequently with small discharges (glow-discharge), whereas in case of a more "rounded" electrode initiated at a so-called-triple point (Figure 2.20 right) immediate streamer discharge may be observed. The interface is acting like a barrier conducting the discharge current.

Typical discharge current pulses for gliding discharges have a rise time of a few ns and a longer duration time of about 100 ns to a few μs (Figure 2.21a). The magnitude of those pulses is relatively high and depends (besides many other factors) on the type of applied voltage ($\Delta U/\Delta t$). At AC-conditions, a changing of time parameters over a longer stressing time was observed (Figure 2.21b).

2.5 PD Quantities

To evaluate PD processes in technical insulations for purposes of detection, identification and location of weak points the estimation of characteristic PD quantities is necessary. In most cases[9] the current pulses caused by severe PD sources will be measured via lead terminals of the entire insulation, which means that only a certain fraction of the original PD pulse may be obtained. Additionally, the measured current pulse propagating through the insulation may be distorted in its amplitude and duration – in certain cases, also accompanied by oscillations. That's why a quantitative description of PD behavior based on pulse current shape parameter as rise- or fall time as well as magnitude seems to be problematically due to the strongly dependence on parameter of insulation design and measuring circuit. On the other hand, the current–time integral of PD

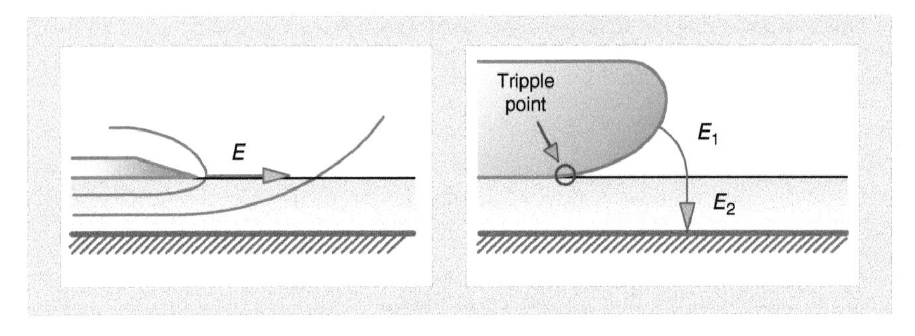

Figure 2.20 Typical field configuration for occurrence of gliding discharges.

9 Other optional measurement is for example the arrangements depicted on Figures 2.4 and 2.21.

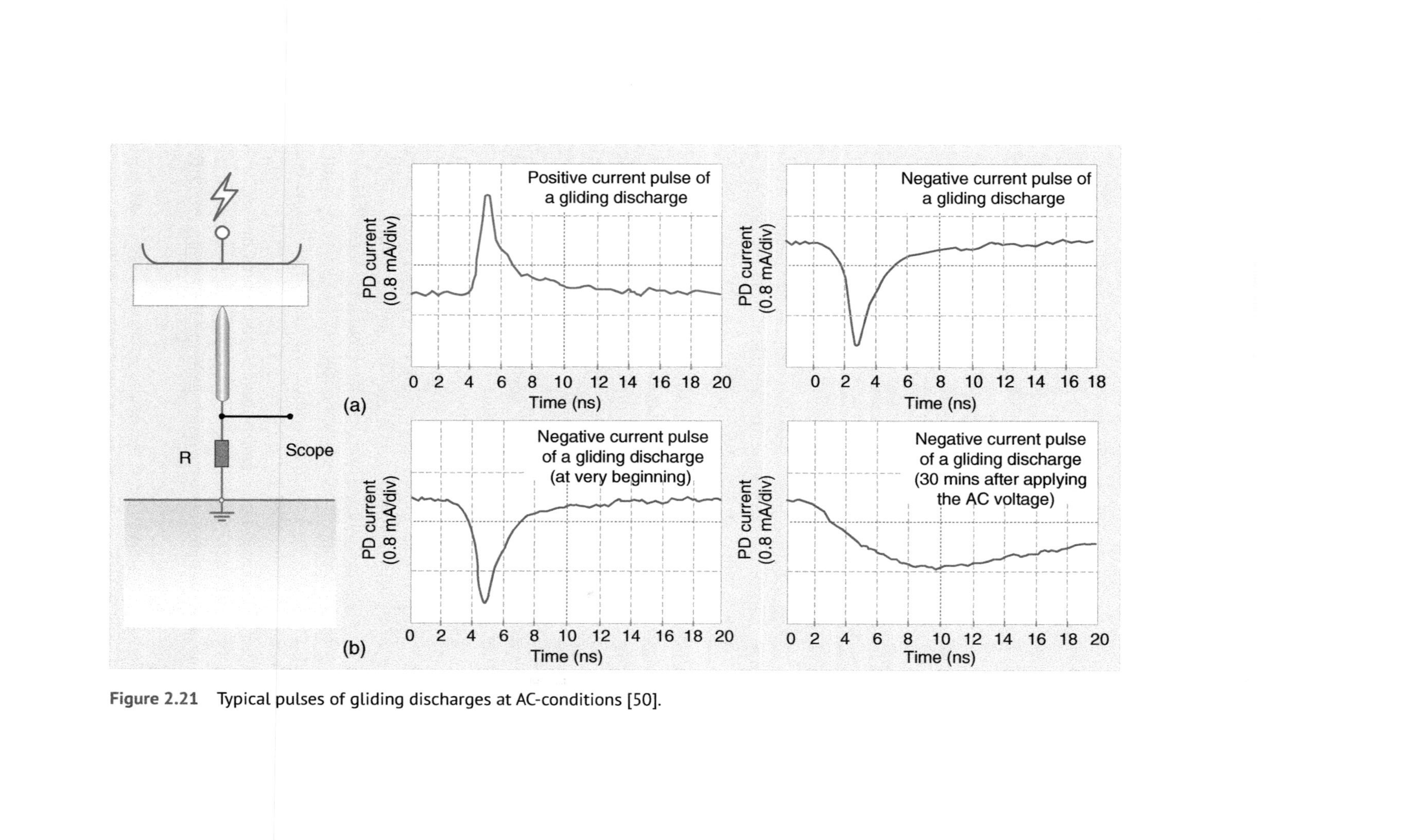

Figure 2.21 Typical pulses of gliding discharges at AC-conditions [50].

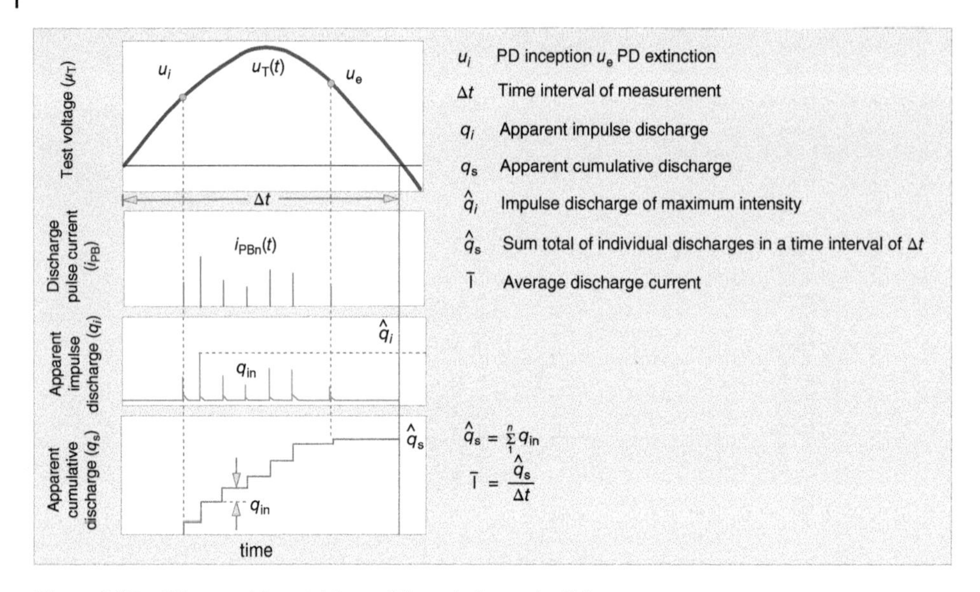

Figure 2.22 PD quantities at AC conditions (schematically).

pulse i.e. the pulse charge is in many cases independent from the mentioned influence and, therefore, recommended as the most suitable PD parameter for evaluation of PD processes [35, 51]. Nevertheless, the problem of measuring a certain fraction of PD pulse remains, so the measured charge is called *apparent charge* q_{app}. This parameter is focused on a single PD pulse event. It is obvious that for estimation of the entire PD activity, the whole process within a certain time interval should be considered. In that case, one could distinguish between parameters derived from the PD process and those related to the applied voltage. For better illustration, an overview on PD quantities at AC conditions is shown in Figure 2.22.

It may be assumed that a current pulse $\boldsymbol{i(t)}$ caused by any PD activity is detected by a suitable measuring unit as a part of the whole PD testing circuit. Using appropriate procedures and measuring equipment, the integral of the detected current pulse $\int i(t)dt$ is formed as q_i – like the *apparent charge* of a single PD pulse (i). It could be termed as a charge pulse $q_i(t)$. Further is assumed, that a train of m pulses occurs within a certain time interval Δt, so a *pulse repetition rate n* may be defined as the ratio between the number of PD pulses and the value of this time interval to $n = m/\Delta t$.[10] For estimation of whole PD activity the recording of the so-called *cumulative charge* q_Σ is useful [52, 53]. The value *cumulative charge* q_Σ requires the measurement of all single pulse q_i occurring during a certain time interval Δt to $q_\Sigma = \Sigma\, q_i|_{\Delta t}$. Note, that in some cases, the recording of entire PD activity-pulse and pulseless discharges during Δt seems to be useful, whereas the parameter q_Σ should be termed as q_{sum}. However, the record of q_{sum} requires a special measurement circuit, which is in quality testing practice difficult to

10 In some cases, the pulse repetition frequency N is defined as number of equidistant pulses per second. N is mostly associated with the situation in calibration of PD measuring circuit [1].

realize. If considering the chosen time interval Δt as value related to the q_Σ or q_{sum}, one gets the PD current I_{PD} as the ratio between the measured charge and the related time interval to $I_{PD} = q_\Sigma/\Delta t$ or $I_{PDs} = q_{sum}/\Delta t$, respectively.

From a physical point of view also the parameter PD energy w is of certain interest. For a single pulse (i) the parameter w_i means the energy, which is dissipated by that pulse and may be defined as $w_i = q_i \cdot u_i$.[11] It seems to be that w_i is responsible for deterioration or damage, especially at internal discharges. Analogously to *cumulative charge* the *cumulative PD energy,* w_Σ may be defined related to a certain time interval Δt. Likewise, as for *PD current* the *PD power* P_{PD} may be obtained. It should be noticed that the recording of *PD energy* is also accompanied by certain measuring efforts and, therefore, not commonly used in PD quality testing practice.

At AC conditions a further parameter for PD occurrence is of certain interest – the instantaneous time t_i at which the PD pulse occurs. Related to the time period of the AC voltage T, the parameter t_i may be expressed by the *phase angle* φ_i according to $\varphi_i = 360\cdot(t_i/T)$. This information might be very useful for estimation of type of PD source as it depicted schematically for external and/or internal discharges (Figure 2.23). Typical for those PDs is the range of phase angle at which they occur, because external PDs are commonly dependent on the magnitude of gap voltage and, therefore, may be detected in area around the peak value (Figure 2.23a; see also Figure 2.15). Likewise, that behavior

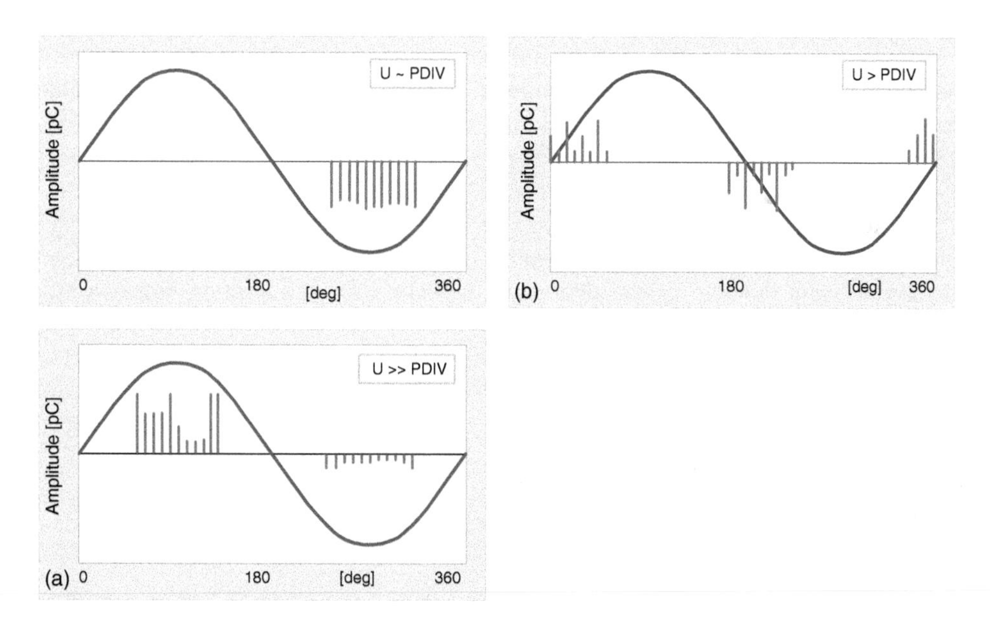

Figure 2.23 Charge pulse oscillograms for simple PD arrangements at AC conditions (schematically) (a) external discharge, (b) internal or surface discharge.

11 u_i is the instantaneous value of voltage at which the PD pulse occurs.

may be obtained at sharp edges in liquid insulation. In opposition to that, internal or surface discharges are dependent on the voltage gradient dU/dt and, therefore, may occur in area of $(dU/dt)_{max}$ (Figure 2.23b). In practical cases, PD may occur in various ways, as for example in values of intensity and pulse polarity, but that *phase dependency* remains for the principle PD sources.

Therefore, it is obvious that the estimation of φ_i was approved as being very useful in many cases for identification and evaluation of PD activity, especially depicted as so-called PRPD pattern [54, 55]. Unfortunately, this kind of information is limited to AC conditions, because for any other type of applied voltage like DC or impulse voltage, the parameter φ_i is not available due to the missing periodic alternating values. Nevertheless, the time of PD occurrence related to the instantaneous voltage value is in many cases very important for understanding and interpreting PD processes and might be obtained by special measuring procedures appropriate to the applied voltage conditions [56, 57]. These issues concerning PD behavior, mostly obtained by appropriate computer-based measuring systems, will be detailed in the following chapters.

Beside the described parameter, there are other important PD quantities related to the applied voltage – the aforementioned inception (PDIV) and partial discharge extinction voltage (PDEV). If the applied voltage is gradually increased from low values up to the value at which any PD intensity is observed, that value is termed as *PDIV*. Likewise, if the applied voltage is gradually decreased from a value at which PD was observed to a value without any PD, that value is termed as *PDEV*.

For electric insulation with no "charge memory," the values of PDIV and PDEV are commonly equal, as for example at insulation in ambient air. But for liquid and solid insulation, especially with insulating interfaces, the initiating of a PD causes space charge that could remain for a certain time within the insulation, having some influence of the extinction of PD. In such cases, the value of PDEV could be lower than the PDIV one showing a kind of *PD hysteresis*, as it is principally shown in Figure 2.24 (a, b).

That behavior also depends on type of voltage stress conditions and might be useful for evaluation of PD activity. It should be noted that the pulse polarity of PD activity in some cases also delivers valuable information about the PD source and location, respectively, which is partly discussed in the following chapters.

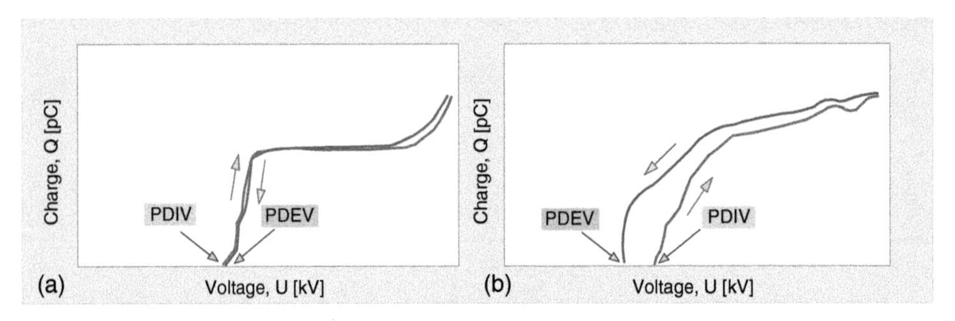

Figure 2.24 Charge pulse oscillograms for simple PD arrangements at AC conditions and PDIV/PDEV values (schematically) in which insulation is (a) ambient air and (b) liquid or solid.

References

1. International Electrotechnical Commission (2015). *IEC 60270: High-Voltage Test Techniques – Partial Discharge Measurements.* Switzerland: International Electrotechnical Commission.
2. Saltzer, M., Gaäfvert, U.,.K.,.B. et al. (2011). Observation of space charge dynamics in air under DC electric fields. In: *2011 Annual Report Conference on Electrical Insulation and Dielectric Phenomena (CEIDP)*, 141–144. Cancun, Mexico: IEEE.
3. Shockley, W. (1939). Current to conductors induced by a moving point charge. *Journal of Applied Physics* 9: 635.
4. Lemke, E. (1976). Contribution to the measurement of pulse- less partial discharge at AC voltage (in German). *ELEKTRIE* 30 (9): 479.
5. Meek, J.M. and Craggs, J.D. (1978). *Electrical Breakdown of Gases.* New York: Wiley.
6. Raether, H. (1941). On the formation of gas discharges (in German). *Zeitschrift für Physik* 117: 375.
7. Townsend, J.S. (1915). *Electricity in Gases.* Oxford University Press.
8. Trichel, G.W. (1938). *Mechanism of the negative point-to-plane corona near onset. Physical Review* 54: 1078.
9. Loeb, L.B. (1939). *Fundamental Processes of Electrical Discharges in Gases.* New York: Wiley.
10. Schwaiger, A. (1923). *The Theory of Electrical Strength* (in German). Berlin: Springer.
11. Degn, P. (1971). Partial discharges in solid dielectrics. PhD thesis. Technical University of Denmark, Lyngby.
12. Boggs, S.A. and Stone, G.C. (1982). *Fundamental limitations in the measurement of corona and partial discharges. IEEE Transaction on Electrical Insulation* 17 (2): 143.
13. Dawson, G.A. and Winn, W.P. (1965). A model for streamer propagation. *Zeitschrift für Physik* 183: 159–171.
14. König, D. and Rao, Y.N. (1993). *Partial Discharges in Electrical Power Apparatus.* Berlin, Offenbach: VDE-Verlag.
15. Gänger, B. (1953). *The Electrical Breakdown of Gases* (in German). Berlin: Springer.
16. Lemke, E. (1967). Breakdown mechanism and characteristics of non-uniform field electrode configuration in air for switching surge (in German). PhD thesis. Technische Universität Dresden.
17. Schwarz, R. (2009). Measurement Technique and Diagnostic for Components of Electrical Power System (in German). Habilitation thesis. Technische Universität Graz.
18. Kuffel, E. and Zaengl, W.S. (1984). *High Voltage Engineering Fundamentals.* Pergamon Press.
19. Rizk, F.A.M. and Trinh, G.N. (2014). *High Voltage Engineering.* CRC Press.
20. Beck, H.P. *Electrical Power Engineering* (in German). Technische Universität Clausthal.
21. Mosch, W. and Hauschild, W. (1979). *High Voltage Insulation System with SF6.* Berlin: Hüthig-Verlag.
22. Pelletier, T.M., Gerrais, Y., and Mukhedkar, D. (1979). *Dielectric Strength of N2-He Mixtures and Comparison with N2-SF6 and CO2-SF6 Mixtures*, 3e. Milano: ISH, paper 31.16.
23. Arora, R. and Mosch, W. (2011). *High Voltage and Electrical Insulation Engineering.* New Jersey: Wiley.

24 Hama, H., Okabe, S., Bühler, R. et al. (2018). Dry air, N_2, CO_2, and N_2/SF_6 mixtures for gas-insulated systems. *CIGRE TB* 730.

25 Niemeyer, L. (1995). *A generalized approach to partial discharge modeling. IEEE Transaction on DEI* 2 (4): 510.

26 Adili, S. (2013).Pulsed X-ray induced partial discharge measurements. PhD thesis. ETH Zurich, Switzerland.

27 Küchler, A. (2018). *High Voltage Engineering*. Berlin: Springer.

28 Liao, T.W. and Anderson, J.G. (1953). *Propagation mechanism of impulse corona and breakdown in oil. AIEE Transaction* 72 (I): 641.

29 Lewis, T.J. (1994). *Basic electric processes in dielectric liquids. IEEE Transaction on DEI* 4: 630.

30 Fiebig, R. (1967). Behavior and mechanism of discharges at AC voltage in insulating fluids at non-uniform fields (in German). PhD thesis. Technische Universität Dresden.

31 Hauschild, W. (1970). About the breakdown of insulating oil in a non-uniform field at SI voltages (in German). PhD thesis. Technische Universität Dresden.

32 Whitehead, S. (1951). *Dielectric Breakdown of Solids*. Clarendon Press.

33 Pilling, J. (1976). Contribution to the interpretation of life-time characteristics of solid insulations (in German). Habilitation thesis. Technische Universität Dresden.

34 Ryan, H.M. (2013). *High-Voltage Engineering and Testing*, 3e. IET.

35 Kreuger, F.H. (1989). *Partial Discharge Detection in High-Voltage Equipment*. London: Butterworths.

36 Dissado, L.A. and Fothergill, J.C. (1992). *Electrical Degradation and Breakdown in Polymers*. P. Peregrinus Ltd., IET.

37 Lemke, E. (1979). A new method for PD measurement on polyethylene insulated power cables. In: *3rd International Symposium on High Voltage Engineering*, vol. I. Milano: IEEE paper 43.13.

38 Morshuis, P.H.F. (1995). *Partial discharge mechanisms in voids related to dielectric degradation. IEE Proceedings – Science, Measurement and Technology* 142 (1): 62–68.

39 Forssén, C. and Edin, H. (2008). *Partial discharges in a cavity at variable applied frequency Part 1: Measurements. IEEE Transaction on DEI* 15 (6): 1601–1609.

40 Devins, J.C. (1984). *The physics of partial discharges in solid dielectrics. IEEE Transaction on EI* 19: 475.

41 Tanaka, T. (1986). *Internal partial discharge and material degradation. IEEE Transaction on EI* 21: 899.

42 Pepper, D. and Kalkner, W. (1997). PD Measurement on Typical Defects on XLPE Insulated Cables at Variable Frequencies. In: *10th International Symposium on High Voltage Engineering*, vol. 4, 389. Montreal, Canada: Institut de Recherche d'Hydro-Quebec.

43 Morshuis, P.H.F. (1993). Partial discharge mechanisms. PhD thesis. Delft University of Technology.

44 Gutfleisch, F. and Niemeyer, L. (1995). *Measurement and simulation of PD in epoxy voids. IEEE Transaction on DEI* 2 (5): 729.

45 Gross, D. (1999). On site partial discharge diagnosis and monitoring on HV power cables, Conf. Proceedings, Versailles, France: Jicable, pp. 509.

46 Yang, Y., Zhi-Min, D., Qi, L., and Jinliang, H. (2020). Self-healing of electrical damage in polymers. *Advanced Science* 7 (21): 2002131.

47 Beyer, M. and Löffelmacher, G. (1975). Investigation of chemical-physical processes at partial discharge channels in polyethylene. In: *2nd International Symposium on High Voltage Engineering*, vol. 1), (in German), 633. Zürich: ISH.

48 Löffelmacher, G. (1976). About physical-chemical processes at discharge channels in Polyethylene and Epoxy resin at non-uniform electrical AC field (in German). PhD thesis. Technische Universität Hannover.

49 Vogelsang, R., Fruth, B., Farr, T., and Fröhlich, K. (2005). Detection of electrical tree propagation by partial discharge measurements. *European Transactions on Electrical Power* 15 (3): 271–284.

50 Lemke, E., Berlijn, S., Gulski, E. et al. (2008). *Guide for Partial Discharge Measurements in Compliance to IEC 20670*. e- CIGRE.

51 Pedersen, A. (1987). Partial discharges in voids in solid dielectrics; an alternative approach. In: *Conference on Electrical Insulation & Dielectric Phenomena — Annual Report*, 58–64. Gaithersburg, MD: IEEE.

52 Lemke, E. (1969). "*A new method for wideband measurement of partial discharges*" (in German),. *ELEKTRIE* 23 (11): 468.

53 International Electrotechnical Commission (2010). *IEC 60270: High-Voltage Test Techniques – Partial Discharge Measurements*. Zurich, Switzerland: IEC.

54 Kranz, H.G. (1982). Partial discharge evaluation of polyethylene cable-material by phase angle and pulse shape analysis. *IEEE Transaction on EI* 17: 151.

55 Fruth, B. and Gross, D. (1994). Combination of frequency spectrum analysis and partial discharge pattern realizing. In: *IEEE Int. Symposium on Electrical Insulation*, 296. Pittsburgh: IEEE.

56 Hoof, M. and Patsch, R. (1994). Analyzing partial discharge pulse sequences – a new approach to investigate degradation phenomena. In: *IEEE Int. Symposium on Electrical Insulation*, 327. Pittsburgh: IEEE.

57 Hoof, M. (1997). Pulse sequence analysing: A new approach for PD diagnosis (in German). PhD thesis. University Siegen (D).

3

Modeling of PD Behavior

3.1 Introduction

For deeper understanding and interpretation of PD measurement results as well as verification of recognized PD phenomena, various models of PD behavior mostly based on equivalent circuits have been developed. One of main topic of interest was focused on modeling PDs within a solid insulation with gas-filled cavities because it was found that the latter one had a harmful impact on insulation's lifetime as well as on power losses. Therefore, developed PD models commonly refer to discharges in such cavities, but in some cases, models were modified and enhanced for other PD sources according to the aforementioned PD classification (see Chapter 2). This chapter describes PD models for internal discharges as well as gliding and external discharges. The current applied models can be mainly classified into two groups – the modeling based on equivalent circuit, or so-called network-based model, and the modeling based on physical relations of electric field within the insulation, or so-called field-based model.

3.2 Network-Based Model

The first PD modeling was described in 1932 by Gemant and Philippoff when power losses in mass-impregnated power cables due to heavy cavity discharges were investigated [1]. The experimental setup comprised a sphere-to-sphere gap F connected in series with a capacitance C_1 and in parallel with a capacitance C_2[1] a protection resistance R and fed by AC test transformer, as depicted in Figure 3.1.

If the gap voltage reaches the ignition value, the capacitance C_2 will be discharged via F and, after extinction of spark, again recharged, and so on. That behavior leads to transients in the connected circuit. The original oscilloscopic records gained for two test-voltage levels are shown in Figure 3.2.

To simulate the situation in real solid insulation containing gaseous cavities and evaluate the PD charge transfer, the equivalent circuit of Figure 3.1 was enhanced and modified as depicted in Figure 3.3 [3, 4].

1 effective capacitance of the used scope and an additional divider capacitance.

Partial Discharges (PD) - Detection, Identification, and Localization, First Edition. Norasage Pattanadech, Rainer Haller, Stefan Kornhuber, and Michael Muhr.

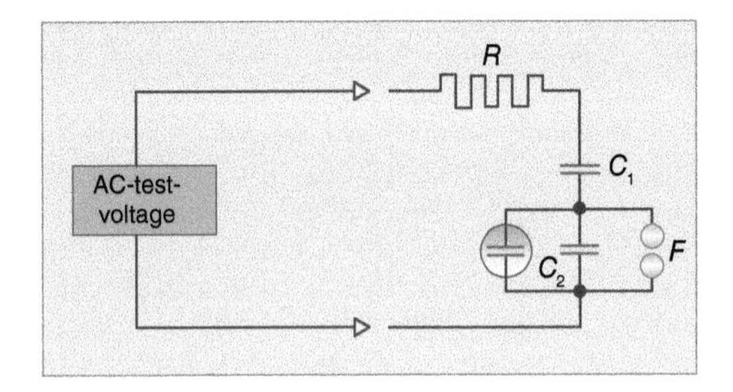

Figure 3.1 Experimental setup according to [1] for studying cavity discharges in mass-impregnated power cables.

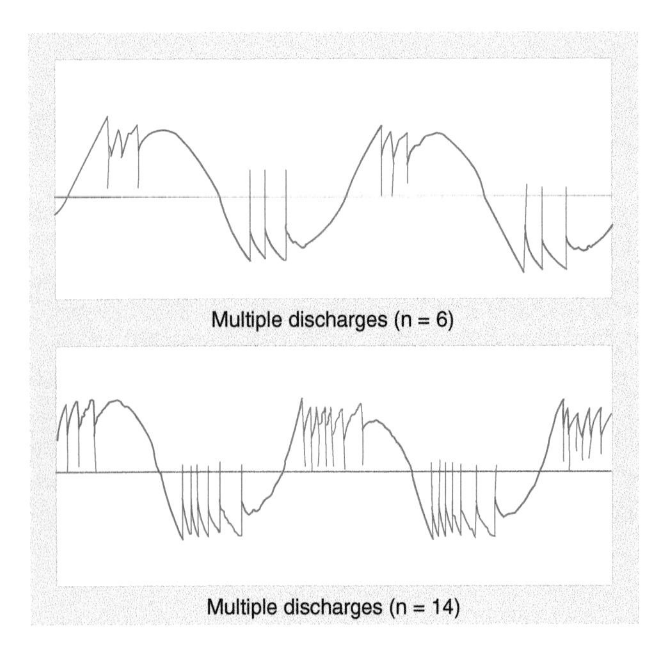

Figure 3.2 Typical oscilloscope screenshots of the voltage across the spark gap *F* at different voltage magnitude, acc. to [1].

Here C_a represents the equivalent capacitance of the test object, while C_{b1} and C_{b2} simulate the equivalent capacitances of the healthy solid dielectric between the gaseous cavity and the electrodes of the test object (fig. 3.3a). According to Figure 3.3b, the resulting capacitance is given by

$$C_b = \frac{C_{b1} . C_{b2}}{C_{b1} + C_{b2}} \tag{3.1}$$

The gaseous cavity itself is also represented by the equivalent capacitance C_c (see Figure 3.3b, c) and the spark-gap was replaced by an electronic switching device. Despite (later) modifications based on the conductance of the healthy insulation and the void by shunts, this network-based PD model serves as the main components with three characteristic capacitances and was commonly referred to as abc-model [5].

Regarding the ignition of a PD event, it is postulated that an assumed cavity capacitance C_c is entirely discharged via the parallel connected spark gap (Figure 3.3b, c, see also Figure 3.1). As a result, the transient voltage across C_c collapses down from the

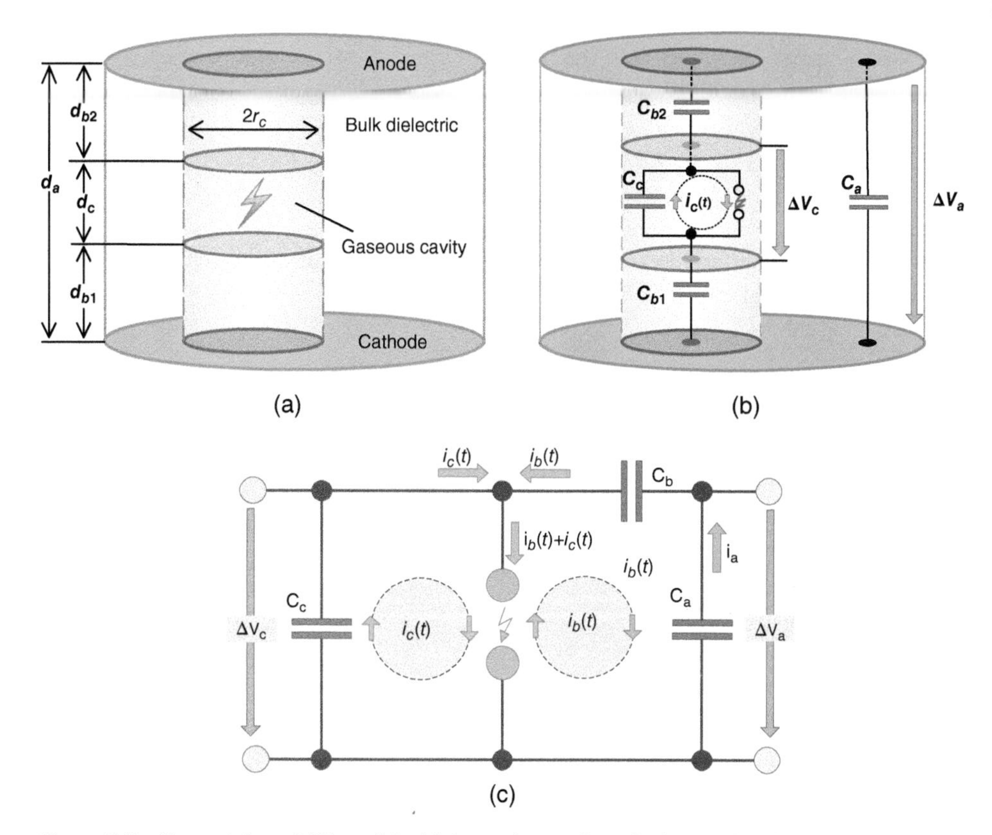

Figure 3.3 Network-based PD model widely used to analyze the internal and external current transients on account of a cavity discharge under AC voltage, acc. to [2]. (a) Gas-filled cavity embedded in the solid dielectric, (b) equivalent capacitances where $C_a >> C_c >> C_b$, (c) Equivalent network.

inception voltage V_i to a residual (extinction) voltage V_e and leads to a voltage drop ΔV_c across C_c. If the residual voltage is assumed to be zero, the magnitude of transient voltage drop ΔV_c is equal to the inception voltage V_i of the spark gap. Based on this, the charge q_i transferred inside the gaseous cavity and often termed as internal or true charge is commonly expressed by

$$q_i = \Delta V_C \cdot \left(C_C + \frac{C_a \cdot C_b}{C_a + C_b} \right) \approx \Delta V_C \cdot (C_C + C_b) \tag{3.2}$$

As can readily be deduced from Figure 3.3c, only a small fraction of the voltage step ΔV_c is transferred via the equivalent capacitance C_b to the object capacitance C_a, which follows from the capacitive divider ratio as

$$\Delta V_a = \Delta V_C \cdot \frac{C_b}{C_a + C_b} \tag{3.3}$$

Multiplying this by the equivalent test object capacitance C_a, the external charge q_a measurable at the electrodes of the test object, and hence detectable in the connection leads, becomes

$$q_a = \Delta V_a . C_a = \Delta V_C . \left(\frac{C_a . C_b}{C_a + C_b} \right) \tag{3.4}$$

Dividing this by the internal charge q_i given by Eq. (3.2), the ratio between external and internal charge becomes

$$\frac{q_a}{q_i} = \frac{C_a . C_b}{(C_a + C_b).(C_C + C_b)} \tag{3.5a}$$

As for gaseous cavities embedded in the bulk dielectric of high-voltage (HV) apparatus and their components, the inequalities $C_b << C_a$ and $C_b << C_c$ are generally satisfied. Eq. (3.5a) can also be expressed by

$$\frac{q_a}{q_i} \approx \frac{C_a . C_b}{C_a . C_C} \approx \frac{C_b}{C_C} << 1 \tag{3.5b}$$

Based on this, it is stated in the relevant literature as well as in standard IEC 60270 [6] that the external charge q_a amounts to only a small fraction of the internal charge q_i transferred inside the gaseous cavity. Thus, the term *apparent charge* has been introduced for the external charge q_a detectable at the terminal of the test object. Nevertheless, from a physical point of view, the immeasurable internal or true charge is responsible for affecting the insulation by any evidence of PD – i.e. degradation or suffering effects. Therefore, the so-defined parameter apparent charge does not represent the complete harmful impact on insulation by its value. Limiting values for q_a given in some standards for technical insulations are therefore not physically derived and based on life-experience data obtained at relevant equipment.

The described model is usable even for assessing other PD quantities – as, for example, the *pulse repetition rate n*. As defined, n is the ratio between the total number of PD pulses recorded in a selected time interval and the duration of this time interval [6]. Analyzing the breakdown sequence of the spark gap bridging the equivalent cavity capacitance C_c in Figure 3.3 the following theoretical relations can be found [1]:

$$\begin{aligned} n_C &= 4 \quad \text{for} \quad 1 \le \frac{V_p}{V_i} < 2, \\ n_C &= 8 \quad \text{for} \quad 2 \le \frac{V_p}{V_i} < 3, \\ n_C &= 12 \quad \text{for} \quad 3 \le \frac{V_p}{V_i} < 4, \end{aligned} \tag{3.6}$$

where n_c is the breakdown sequence per cycle, V_p the peak value of the AC test voltage applied, and V_i the inception voltage, which equals the breakdown voltage of the spark gap. However, these relations (3.6) are valid only under the assumption of entire discharge of the equivalent cavity capacitance C_c, which means the residual voltage is zero. If it is assumed that the discharge of C_c is not complete, after a breakdown the value of residual voltage $V_e > 0$ over C_c must be considered. Defining the ratio as

$$p = \frac{V_e}{V_i} \tag{3.7}$$

it was shown that the ratio V_e/V_i influences the repetition rate significantly [7]. That impact has led to a modified equivalent circuit, the so-called *five-capacitance model* proposed by Böning [7], Figure 3.4. The evidence of V_e is realized by the network elements R_4 and C_4, assuming some practical relations. Without going in details, the value for p can then be approximated as

$$p \sim \frac{1}{1 + \frac{C_1}{C_4}} \tag{3.8}$$

with C_4 as additional cavity capacitance and R_4 as resistive connection between C_1 and C_4 inside the cavity.

As an example, the repetition rate for different values of p is depicted in Figure 3.5 (fig.3.5 a,b,c), where it can be seen, that e.g. for $p \sim 0.7$ the repetition rate is twice as high as for $p \sim 0.5$. That behavior could be explained by evidence of a too-large cavity, when only a fraction is discharged or any space charge from previous PD event remained in the cavity.

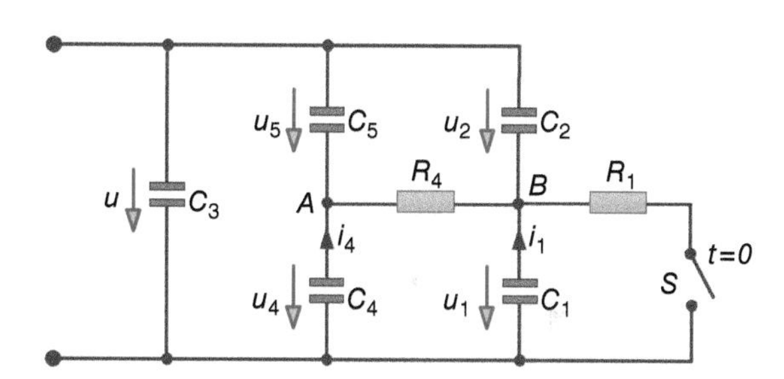

Figure 3.4 Modified five-capacitance network model acc. to [7].

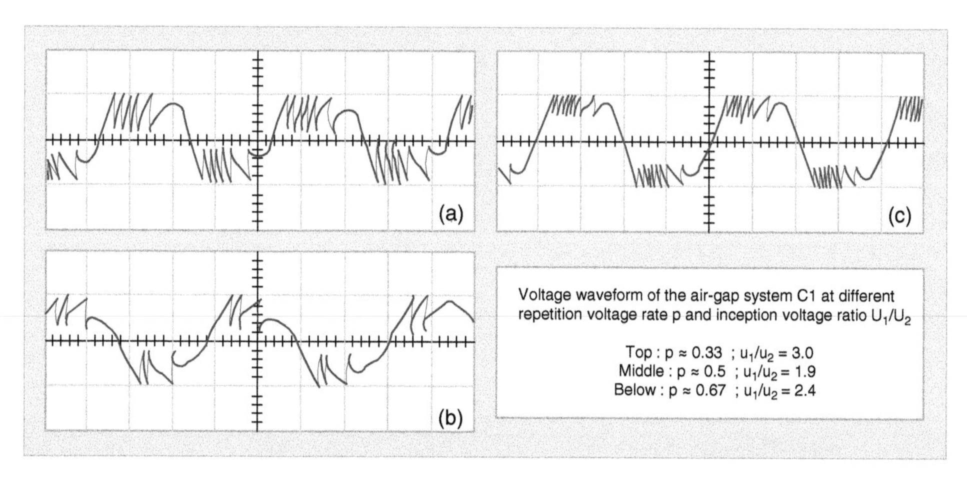

Figure 3.5 Impact of ratio p on repetition rate of PD transients for equivalent network. Voltage waveform over the air-gap capacitance C1 at various p and several inception voltage ratio u_1/u_2 (a) p = 0.33, u_1/u_2 = 3.0; (b) p = 0.5, u_1/u_2 = 1.9; (c) p = 0.67, u_1/u_2 = 2.4 [7].

Despite the better-modeled repetition rate at large cavities, the five-capacitance network model was not widely applied, probably because of too many difficulties in estimation of additional separate network elements. Further applications of PD modeling based on equivalent circuit were mainly provided by the *three-capacitance*, or abc-model.

Although the described abc-model is simple, it may simulate the measurable transients in an outer measuring circuit related to a discharge event, such as the apparent charge magnitude and PD current. Therefore, it was applied even for further development of relevant measuring technique and determining of derived requirements for measuring circuits. Specifically, the modeled recurrence of discharges related to alternating voltage was used for recognition of PD creating charts with pulse counts over phase angle of the AC test voltage. These phase-resolved PD patterns (PRPD) are in many cases typical for type and source of PD events and, therefore, usable as valuable diagnostic tool under practical test relations [4, 8, 9]. So, for example, the recurrence of discharges as shown in Figure 3.6 is typical for cavity discharges under AC conditions.

The described abc-model is also usable for external discharges if marginal modifications in the equivalent circuit are made [10] (Figure 3.7). The resistance R_2 should represent the discharge channel moving toward the counter electrode having a certain conductivity, C_1 and C_3 equals to the abc-model (C_c, C_a). In opposition to the "cavity"-modeling the resistance R_2 is also dependent on the discharge type e.g. Trichel, streamer, and so on, which was a further obstacle for wider using of that model. Nevertheless, the recurrence of PD events

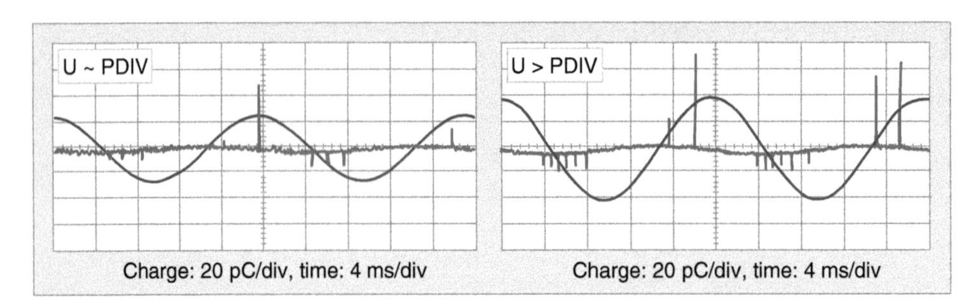

Figure 3.6 Recurrence of PD at cavity discharges and different test voltage magnitude.

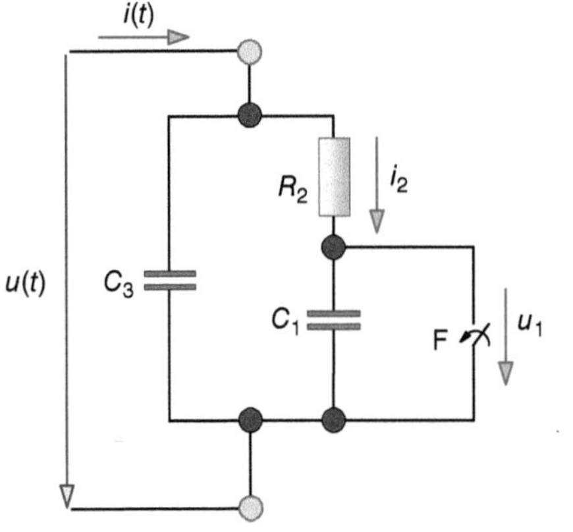

Figure 3.7 Modified abc-model for external discharges.

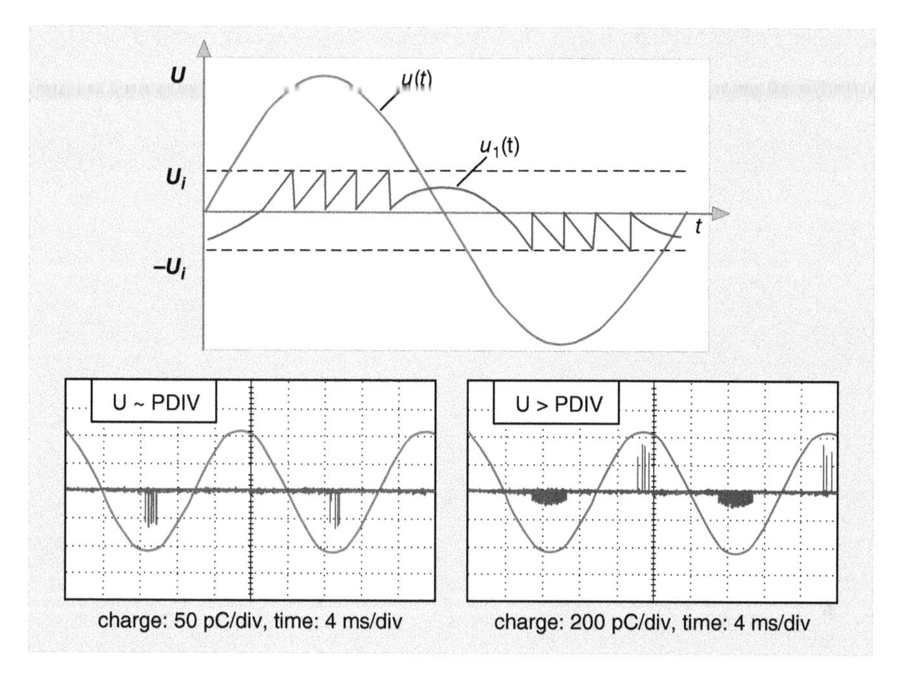

Figure 3.8 Recurrence of PD at external discharges and different voltage magnitude (point-to-plane in air).

in relation to the AC test voltage is modeled in a correct manner (Figure 3.8), what might be used for diagnostic purposes.

Due to the growing importance of quality testing for HVDC equipment, the developed modeling was proved for application even under direct voltage conditions. Generally, at DC conditions it must be considered that the potential distribution between the electrodes is no longer controlled by the permittivity but rather, by the volume and surface conductance of the insulating material. Consequently, modeling of cavity discharges under direct voltage, the network-based model according to Figure 3.3 was modified accordingly, as proposed already by Kreuger [4], Figure 3.9.

Here the equivalent capacitances C_a, C_b, and C_c are bridged by the resistances R_a, R_b, and R_c to simulate the resistive potential distribution, which dominates not only the inception voltage but also the repetition rate of the PD pulses. However, to analyze the very fast PD transients, it seems to be adequate to consider only the capacitive elements. That means the previously introduced Eqs. (3.1–3.5a, b) are also applicable without restriction. Considering cavity discharges under direct voltage, in addition to considering internal and external charge, the *recovery time* T_c is of interest. This represents the time span elapsing between two consecutive PD pulses, as has also been treated by Kreuger [4] and, in more detail, in [11, 12], and illustrated in Figure 3.10.

Based on the classical network theory, it can readily be shown that recovery time is affected not only by the volume and surface conductivity of the solid dielectric but also by the inception voltage V_i and even by the magnitude V_a of the DC test voltage applied. With reference to the equivalent circuit shown in Figure 3.9, the time-dependent voltage $V_c(t)$

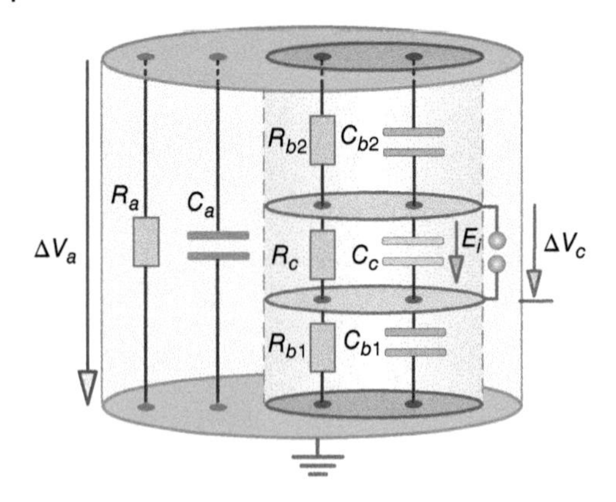

Figure 3.9 Cavity network-model at DC conditions, acc. to [4].

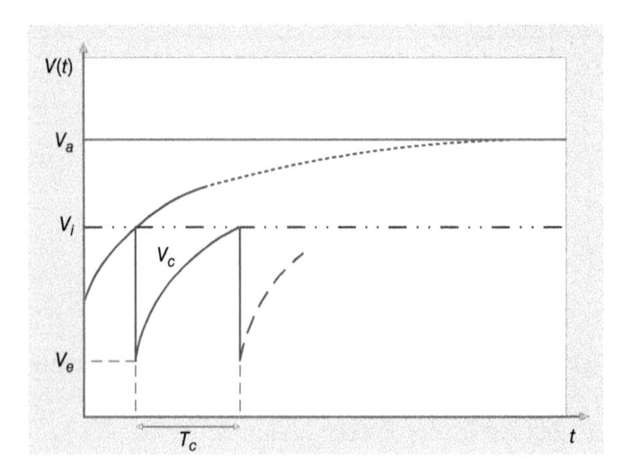

Figure 3.10 Voltage across cavity capacitance and recovery time between two consecutive discharges at DC conditions (parameter acc. to Figure 3.9).

rising across the cavity capacitance C_c after each spark breakdown can simply be predicted based on the classical circuit theory, which leads to the following expression:

$$V_C(t) = V_a \frac{R_C}{R_b + R_C}\left(1 - e^{-\frac{t}{\tau_C}}\right) \tag{3.9}$$

where τ_c represents a characteristic time constant given by

$$\tau_C = \frac{R_b R_C}{R_b + R_C}(C_b + C_C) \tag{3.10}$$

Based on this, the minimum time span T_c elapsing between two consecutive spark breakdowns can roughly be predicted using the following approach:

$$T_C \approx 3\tau_C = \frac{3R_b R_C}{R_b + R_C}(C_b + C_C) \tag{3.11}$$

To accomplish the initialization of a spark breakdown at inception voltage V_i, the DC voltage to be applied to the network according to Figure 3.9 must approach the following magnitude:

$$V_a \geq V_i \cdot \frac{R_b + R_C}{R_C} = V_i \cdot \left(1 + \frac{R_b}{R_C}\right) \tag{3.12}$$

It should be noted that this simple approach of repeating charge of cavity capacitance and igniting the discharge at constant value of inception voltage V_i does not consider the random nature of the discharge process and the time delay before the necessary first free electron is available (see Chapter 2). In measuring that behavior, it was shown that both the inception voltage V_i as well as the extinction voltage V_e are of stochastic character, which leads to certain time delay for pulse inception [13, 14] (Figure 3.11). Additionally, considering the free electron availability, it is understandable that the PD behavior at DC voltage might be more complicated, as in the case of AC stress voltage. That issue will be highlighted later in the chapter, as well as in Chapter 8.

The modeling of PD events by described equivalent circuits was also made for gliding discharges. Under different test conditions (e.g. AC, DC), the distribution of electrical field along the insulating surface was modeled. The generation of discharges is explained by Figure 3.12.

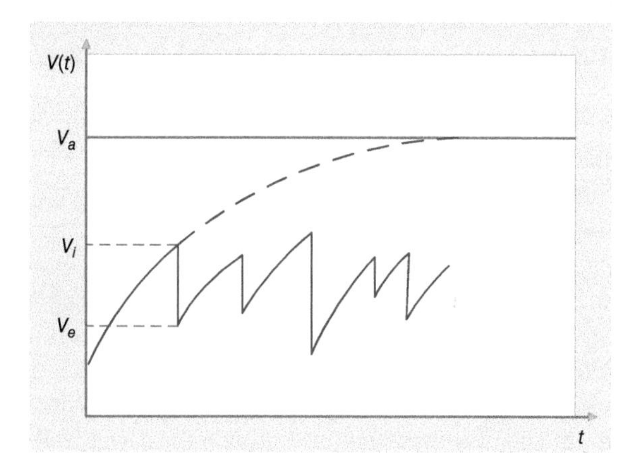

Figure 3.11 Voltage across cavity capacitance at various magnitudes V_i and V_e with consecutive discharges at DC conditions, acc. to [13].

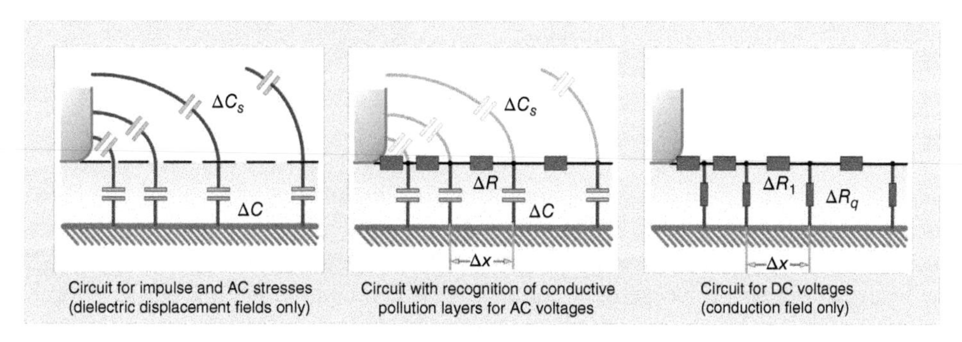

Figure 3.12 Equivalent circuit for gliding discharges at AC- and DC-conditions acc. to [15].

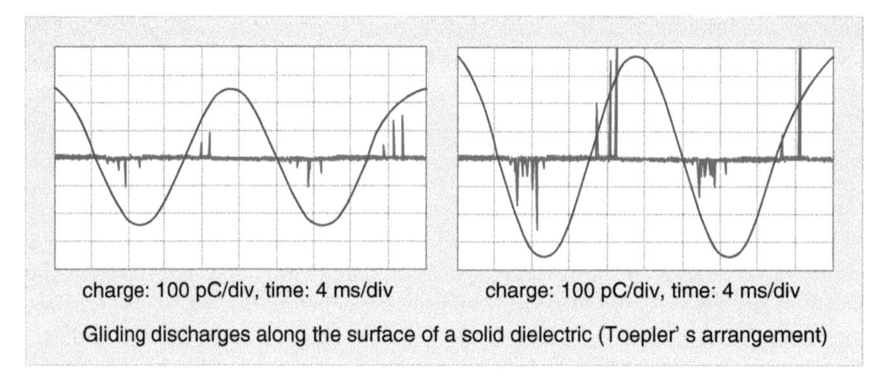

Figure 3.13 Recurrence of PD at gliding discharges.

At AC-conditions the arrangement might be modeled by an equivalent circuit according to Figure 3.12, at which the capacitances of the insulator (ΔC) have higher values compared with the surrounding medium (ΔC_S). Therefore, the displacement current part of *streamer* discharge is relatively high, which enables a transition to *leader* discharge already at short distances of few cm [16]. In case of DC conditions the discharge current will be determined by the surface resistances resp. their conductivity.

Also, as already at external and internal discharges, the recurrence of PD events is one of the most usable results with that kind of modeling for diagnostic purposes (Figure 3.13). The main disadvantage of that approach is the missing agreement with physical phenomena within the insulation – as, for example, the generation rate of real discharge in a cavity, including the stochastic behavior of PD initiation or the surface charge accumulation along the void during a sequence of PD events.

Note, that the basic idea of abc-model is the changing of systems charge balance, including capacities due to PD event. Thus, one of the major arguments against the equivalent circuit model was that the void surface is treated as an equipotential surface, which is not the case in reality. Pedersen first argued that the concept of void capacitance and its use by an equivalent circuit is addressed to a subject that is in essence a field problem [17]. As an alternative, a proposed field theoretical approach treats the PD transients as consequence of a dipole moment established by the bipolar charge carriers created in the case of a PD event and deposited just thereafter on the cavity walls. The main benefit of this concept is that it can also be adopted for quantitative estimations of physical parameters [18, 19]. That approach is related to field-based modeling.

3.3 Field-Based Model

The basic aim of that field theoretical approach is the quantitative estimation of physical parameters of partial discharges by the measurable quantity induced charge, which occurs on electrodes of the insulation system caused by a discharge within the insulation [20, 21]. The measurable *induced charge* or, according to the commonly used diagnostic practice termed as *apparent charge q*, is caused by the discharge due to the insulation defect (e.g. cavity). The discharge event within the defect is accompanied by movement of bipolar charge carriers (electrons, ions), which will deposit on defect walls leading to a certain surface charge distribution on them. This surface charge is considered one of the most responsible parameters for degradation and aging of surrounding insulation and thus dangerousness of PD. It may be termed as true charge, q_{true}. One of main tasks for interpretation of measured PD quantities is to estimate the relation of measurable induced or apparent charge q to the immeasurable true charge q_{true}. As discussed below, the field-based model fulfills that task under certain simplifications and assumptions.

Looking from any electrode of the insulation system, the surface charge caused by a discharge may be considered under certain conditions as an electric dipole having a certain dipole moment p. This dipole moment p is established by drifting of bipolar charge carriers and their deposition on defect walls. It is directed from negative to positive charges and may be calculated by

$$p = \int_S r\sigma(r)\,dS \tag{3.13}$$

where r and dS are related to the defect element, $\sigma(r)$ equals the accumulated surface charge on the walls. The surface charge distribution $\sigma(r)$ is heteropolar located, which means that, for example, positive ions will be deposited on the cathode side of the wall. The induced charge can be estimated by

$$q = -\boldsymbol{p} \cdot \boldsymbol{grad}\lambda \tag{3.14}$$

where λ is a normalized potential function obtained by solving the applied (Laplacian-) field equation, depending on insulation design as well as defect size and location. To calculate the dipole moment p, according to Eq. (3.13) the surface charge magnitude as well as its spatial distribution related to the defect size must be estimated, for that the Poisson equation should be solved. According to various relevant studies, it can be stated that hitherto only solutions for simple defect configurations were known [22, 23]. Therefore, the approach should be explained for a simple arrangement – a small void embedded in the bulk dielectric insulation between plane parallel electrodes, as illustrated in Figure 3.14. The main characteristics of that void should be its maximal extension d in the direction of

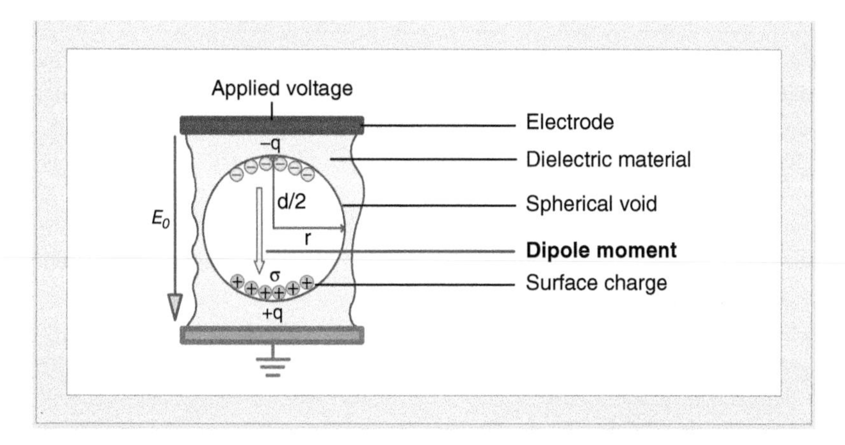

Figure 3.14 Arrangement for explanation of field-based modeling.

the applied background field $E_0(t)$, its average diameter $2r$ perpendicular to E_0, the ionization properties of the gas within the void, represented by inception field strength E_{inc} and the local field strength inside the void $E_{void}(t)$ as determined by the voltage U_0 applied to the insulation system, its geometry, and the location and shape of the void.

As already mentioned, for any occurrence of a PD, two conditions are required: a free electron must be presented to start an ionization process and the field E_{void} in the void must exceed the discharge inception strength E_{inc}. Regarding the first condition, it should be assumed that (at least) one free electron necessary for any ionization process is available without any restrictions. To estimate the surface charge magnitude, it is necessary to assume a certain discharge mechanism that could be characteristic for the void discharges.

Various studies indicate that one of the major discharges in voids is of streamer type [13, 24, 25]. A streamer discharge occurs if the streamer inception field strength E_{inc} is exceeded. The latter might be approximated by

$$E_{inc} = E\left[1 + \frac{B}{\sqrt{pd}}\right] \tag{3.15}$$

where E is the limiting value i.e. the field below where ionization grow is impossible, B is a gas property, d is the void maximum extension diameter, and p the gas pressure. If the streamer length is comparable with void extension the field in the void is reduced to a residual value E_{res} what means that the electric field in void collapses. That collapse $\Delta E = E - E_{res}$ is associated with a space charge transport through the void according to

$$q \sim \varepsilon\varepsilon_0 r^2 \left(E - E_{res}\right) \tag{3.16}$$

This charge, q, was already introduced as *true charge* and is deposited in a thin surface layer on the void walls. Associated with that surface charge is an electric field Eq, which will oppose the applied field within the cavity $E_a(t)$ (Figure 3.14). The electric field in the void E_{void} is superimposed on time-dependent field $E_a(t)$ caused by external applied voltage with increasing charge field $Eq(t)$ and may be expressed by

$$E = E_a(t) + E_q \tag{3.17}$$

The void field $E_a(t)$ is related to the background field $E_0(t)$ at the void location (but in absence of the void) and the impact of void shape as well as the permittivity of surrounding dielectric. The latter is expressed by a geometry factor f depending on the void shape; thus, the field $E_a(t)$ yields $f\,E_0$. The principal situation before, at, and after one discharge event is schematically depicted in Figure 3.15.

After estimation of true charge q and with respect to the void extension also of dipole moment the induced charge on the electrodes or the insulation system may be calculated according to Eq. (3.15) by

$$q' = q\left[\frac{E}{U}\right]dg \tag{3.18}$$

where E/U is the normalized background field at void location, d is the void extension, and g the geometry factor determined by the void shape and the relative permittivity of the surrounding dielectric.

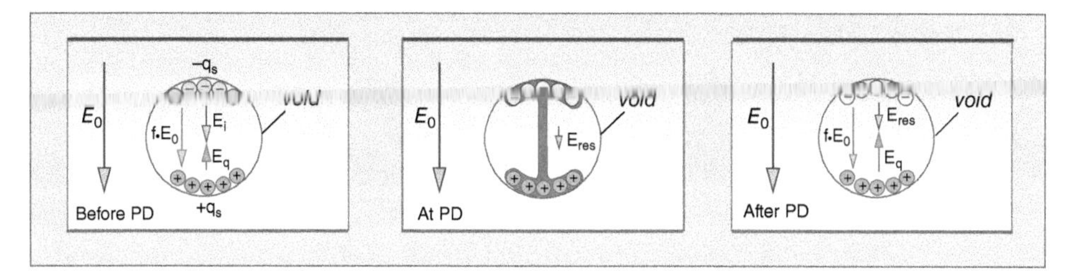

Figure 3.15 Surface charge deposition and associated electric fields, acc. to [26].

This example of applied electrostatic field equations illustrates the possibility to calculate the values of true charge and induced charge caused by a PD within a void. For calculation, the void parameter-like shape and extension as well as parameter of included gas like ionization coefficient and pressure etc. must be known. Under those conditions it is possible to classify measured (induced) charge values at practical insulation i.e. with unknown void parameters by comparison with results of simulated well-known void configurations.

It should be noticed that the calculated results differ in absolute values of true and induced charge, in particular, the inequality $q_{\text{true}} > q'$ has to be considered. The reason for that might be caused by using of electrostatic field equations – (Laplace, Poisson) – because the latter consider not the moving space charges within the void, but implicit also the polarization change of surrounding dielectric expressed by bound charges which may contribute also to the induced charge on the electrodes [17].

Another approach for estimation of charge values caused by PD within a void was described by Lemke [2, 19, 27]. However, that approach is based on the induced charge concept instead of just the electrostatic charge, and thus the transient discharge currents are analyzed. It was shown that under certain conditions the true charge could be estimated without knowledge of exact void geometry.

As mentioned above, the main difference with the electrostatic approach is the detailed analyzing of moving charges at a PD event and, therefore, convective and displacement currents have to be considered. It is assumed that the partial discharge within the defect is of streamer type occurring in a spheroidal void embedded in the bulk dielectric between plane parallel electrodes, as depicted in Figure 3.16.

Provided that a streamer-like discharge is launched adjacent to the anode-side dielectric boundary, the propagation will finally be hampered at the instant when the streamer head approaches the cathode-side dielectric wall. Hence, the maximum streamer length becomes equal to the cavity length d_c aligned with the basic field E_b, as indicated in Figure 3.16 (left). Due to the Coulomb forces, the electrons and positive ions, initially distributed more or less uniformly inside the streamer filament [28, 29], are attracted by the anode and cathode. As a result, the bipolar charge carriers will be deposited on the anode-side and cathode-side dielectric walls, as illustrated in Figure 3.16 (right). That means that after a certain transition time, which is in the order of nanoseconds, an electrical dipole will be established. As the dipole field opposes the basic field E_b caused by the applied test voltage, the resulting field strength inside the gaseous cavity is reduced accordingly. Hence further ionization processes will be quenched for a certain time interval until the field caused by the rising AC test voltage again approaches the inception field strength. As known from the physics of gas discharges, electrons and positive ions, which are moving along the streamer filament

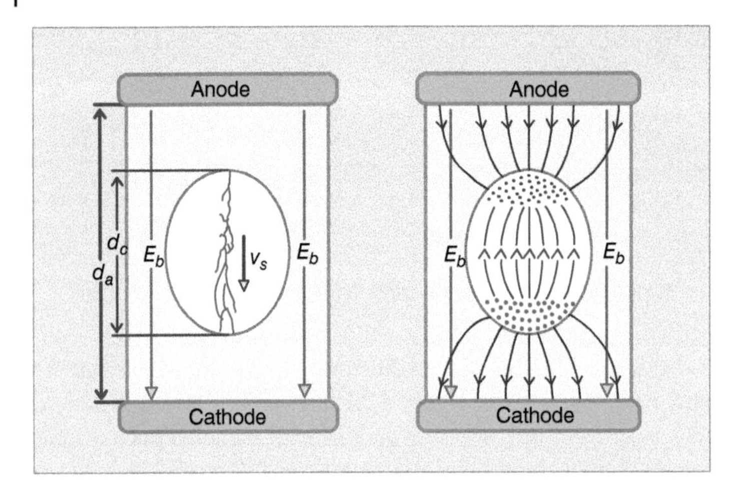

Figure 3.16 Streamer-like discharge bridging a spherical void (left) and deposited space charge on the void walls (right) acc. to [27].

on account of the Coulomb force, cause a convective current, in the following denoted as $i_c(t)$. This continues like a displacement current $i_b(t)$ through the residual dielectric column between cavity and the electrodes of the test object, and hence also as a displacement current through the equivalent test object capacitance C_a, in the following denoted as $i_a(t)$, so that the *continuity equation* applies:

$$i_c(t) = i_b(t) = i_a(t) \tag{3.19}$$

Upon integration over the transition time t_p elapsing until the dipole moment is established, the entire charge flown through the gaseous cavity and hence through the test object capacitance can be expressed by:

$$\int_0^{t_p} i_c(t) \cdot dt = \int_0^{t_p} i_b(t) \cdot dt = \int_0^{t_p} i_a(t) \cdot dt \tag{3.20a}$$

which means

$$q_c = q_b = q_a \tag{3.20b}$$

This underlines that for the test specimen illustrated in Figure 3.17 the internal (true) charge q_c equals the external charge q_a. If it is assumed that the "internal transfer" of charge to the electrodes is carried out without any losses, then the *measurable charge* is equal the *true charge*. Obviously, this contradicts the widely accepted statement that the apparent charge measurable in the connection leads of the test object attains only a small fraction of the internal charge transferred in the gaseous cavity, which is also the main outcome of the traditional abc-model.

Considering now a single PD event leading to the ionization of n_i gas molecules, it must be realized that the net charge created inside the gaseous cavity remains zero. In fact, each electron released from a neutral gas molecule creates a positive ion. That means the number of positive charge carriers created in case of a PD event is always equal to the number

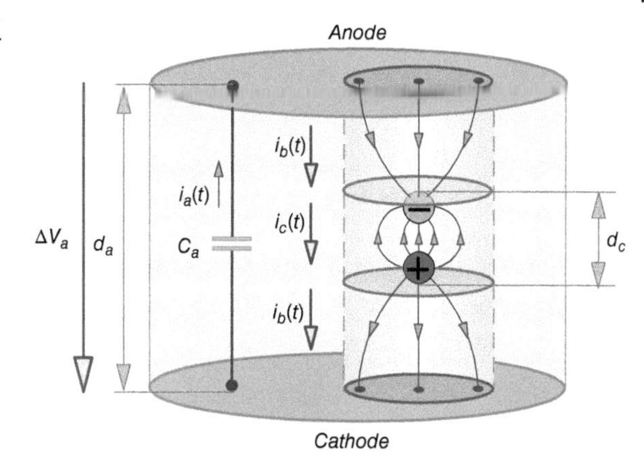

Figure 3.17 Dipole-based PD model for void discharges under AC condition.

of negative charge carriers and given by the number of ionized gas molecules n_i. As a result, n_i positive ions and n_i electrons (including those attached to neutral gas molecules and hence forming negative ions) are drifting from the site of origin into cathode and anode direction on account of the Coulomb force. Finally, the bipolar charge carriers will be deposited on the anode-side and cathode-side dielectric boundaries. As a first approximation, the field energy W_i transferred to the negative and positive charge carriers under this condition can reasonably be approximated by

$$W_i = e \cdot n_i \cdot E_i \cdot d_c \tag{3.21}$$

where E_i is the inception field strength and d_c the cavity length aligned with the field lines (Figure 3.17). Furthermore, the relevant literature indicates that the field energy transferred to the bipolar charge carriers is about three orders of magnitude greater than that energy required to ionize these gas molecules [30]. That means the energy dissipating inside the gaseous cavity on account of collision and photo ionization processes must not be taken into consideration. To ensure that the measurable charge magnitude q_a according to Eq. (3.20b) is not significantly reduced by the AC test facility energizing the test object, the relevant standard IEC 60270 recommends the usage of a HV series impedance to be connected between test object and test facility [6]. Under this condition, the test object can be considered as disconnected from HV test facility, which applies for the transition time required to accomplish the establishment of the dipole moment, which occurs at a nanosecond time scale. That means the entire field energy transferred to the bipolar charge carriers is exclusively provided by the test object capacitance C_a. Hence, Eq. (3.21) can also be expressed by

$$W_i = V_i \int_0^{t_p} i_c(t)\,dt = V_i \cdot q_c = e \cdot n_i \cdot E_i \cdot d_c \tag{3.22}$$

Separating the charge q_c, one yields

$$q_c = e \cdot n_i \cdot d_c \cdot \frac{E_i}{V_i} \tag{3.23}$$

With reference to Eq. (3.20a,b), it can also be written:

$$q_c = q_a = C_a \cdot \Delta V_a = e \cdot n_i \cdot d_c \cdot \frac{E_i}{V_i} \tag{3.24}$$

where ΔV_a is the magnitude of the transient voltage collapsing across the equivalent test object capacitance C_a as consequence of a PD event, which is the origin PD signal captured from the terminals of the test object to measure the pulse charge q_a.

In this context, it is worth mentioning that Eq. (3.24) is in accordance with the Shockley-Ramo theorem [31] and would also be obtained based on a field-theoretical (electrostatic) treatment of the problem, as carried out by Pedersen et al. [20]. As will be shown, Eq. (3.24) provides a means to assess the link between the measurable pulse charge q_a, and the number n_i of gas molecules ionized in the gaseous cavity in case of a PD event. The prediction of n_i becomes simple for field configurations that can be treated analytically, such as the uniform field between plane-parallel electrodes, as well as slightly nonuniform fields between coaxial cylinder electrodes and concentric sphere electrodes, as discussed below. Thus, it should be emphasized that the pulse charge q_a measured for such simple electrode configurations represents the true charge flown through the gaseous cavity, and not just an apparent charge, which amounts only a small fraction of the true charge, as discussed previously. However, performing PD tests of large-scale high-voltage equipment containing inductive elements, such as power transformers and rotating machines, the PD signal transferred from the site of origin to the terminals of the test object will considerably be distorted. This is not only due to the exciting of oscillations and the occurrence of traveling waves, but also to the attenuation and dispersion phenomena. In such cases, the measurable charge decoupled in an outer measuring circuit is indeed not equal to the true charge flowing through the gaseous cavity and has a reduced value. However, this is not related to the apparent charge deduced from the traditional abc-model or mentioned in the relevant standard.

3.3.1 Stages of PD Behavior Modeling for DC Conditions

Because of increasing importance for PD measuring under DC conditions, the described approach should be discussed more in details even for DC conditions. At DC conditions, the PD behavior can be distinguished between the following major stages.

3.3.1.1 Stage 1: Inception of Ionization Processes

Subjecting HV equipment and their components to a constant direct voltage, the potential distribution inside the bulk dielectric is no longer controlled by the dielectric permittivity, as under alternating voltage, but rather by the volume and surface resistance. Consequently, the inception of ionization processes is dominated by the resistive potential distribution. This is commonly simulated by a resistive network, as sketched in Figure 3.18 (left). However, an analytical prediction of the inception field strength seems to be impossible, because this is strongly affected not only by the local field strength but also by the temperature distribution in the bulk dielectric and environmental conditions, as well.

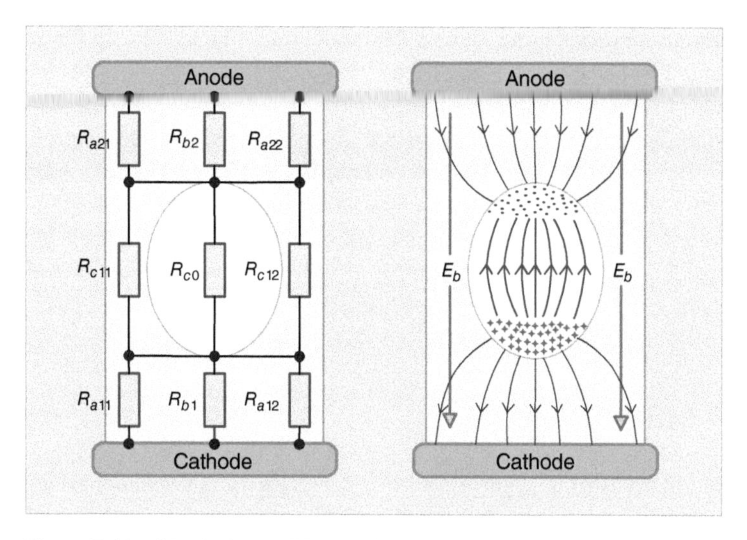

Figure 3.18 Dipole-based PD model for void discharges under DC condition acc. to [27].

3.3.1.2 Stage 2: Establishment of an Electrical Dipole

Provided that the field strength inside the gaseous inclusion exceeds the inception field strength E_i, the phenomena of void discharges under DC voltage are well comparable with those under AC voltage. That means an electrical dipole will also be established as consequence of a single PD event, as illustrated in Figure 3.18 (right). Hence, the field-based PD model according to Figure 3.17 as well as the associated Eqs. (3.20a,b–3.24) are applicable without restriction to analyze the PD transients under direct voltage.

3.3.1.3 Stage 3: Dissipation of the Electrical Dipole

Even if the conductivity of insulating materials and hence the electron mobility is extremely low, it seems to be likely that the positive ions deposited on the cathode-side dielectric boundary will finally be neutralized by electrons de-trapped from the solid dielectric adjacent to the positive space charge cloud, as will be described in more detail below. Moreover, it is supposed that the electrons deposited on the anode-side dielectric wall are entering the solid dielectric wall and drifting further into anode direction. Practical experiences reveal that the decomposition of the electrical dipole and hence the re-establishment of the initial field conditions required to initialize the succeeding PD event may last up to some hours.

3.3.2 Extended Modeling Parameters

As mentioned above, the discussed field-based PD modeling provides quantitative estimations of physical parameters of an insulation with cavity defects. For example, the number of ionized charge carriers – ions and electrons – might be predicted based on measured pulse charge q_a in case of a PD event. Additionally, under AC voltage conditions it could be shown that this space charge intensifies the field in the bulk dielectric, which may become extremely high in proximity to each cavity tip. That means the solid dielectric suffers not only from the permanent electron/ion bombardment but may additionally be deteriorated on account of the field intensification.

But even under DC voltage conditions, the applied PD model might be led to important physical information about PD behavior, so for example the PD pulse repetition rate. As already mentioned, at cavity discharges under direct voltage the recovery time between subsequent PD pulses is of significant interest. To analyze that, first, the density of the electron current in the solid dielectric adjacent to each positive and negative charged cavity tip shall be considered. For simplicity, the positive and negative charge carriers shall again be considered as concentrated in spherical space charge clouds of radius r_c as illustrated in Figure 3.19.

Provided that r_c is considerably lower than the cavity length d_c in field direction, the density $J_e(t)$ of the electron current following from Maxwell's equations can be approximated by

$$\boldsymbol{J}_{\mathbf{e}}(t) = \kappa_v \cdot E_p(t) = \kappa_v q_p(t) / 4\pi \cdot \varepsilon_r \cdot \varepsilon_0 \cdot r_c^2 \tag{3.25}$$

where κ_v is the volume conductivity of the solid dielectric, $E_p(t)$ is the time-dependent field strength adjacent to the dielectric boundary and ε_r is the relative dielectric permittivity of the solid dielectric. In the following, it shall be assumed that most of the electrons are entering the dielectric boundary via an area A_e, which is comparable to that of a semi-sphere of radius of r_c and can hence be approximated by $A_e = 2\pi r_c^2$, see Figure 3.19. Electrons attracted on account of the Coulomb force by the positive ion cloud deposited on the cathode-side cavity wall. Under this condition the electron current $i_e(t)$ can be accounted for, if the current density $J_e(t)$ according to Eq. (3.25) is multiplied by this area:

$$i_e(t) = \left[\boldsymbol{J}_{\mathbf{e}}(t)\right] 2\pi r_c^2 = \kappa_v q_p(t) / 2\varepsilon_r \varepsilon_0 \tag{3.26}$$

Taking into account that the number of positive ions in the following denoted as $n_p(t)$ is reduced continuously, the charge $e\, n_p(t)$ carried by the positive ions will also be reduced accordingly. Hence, the electron current can also be expressed as follows:

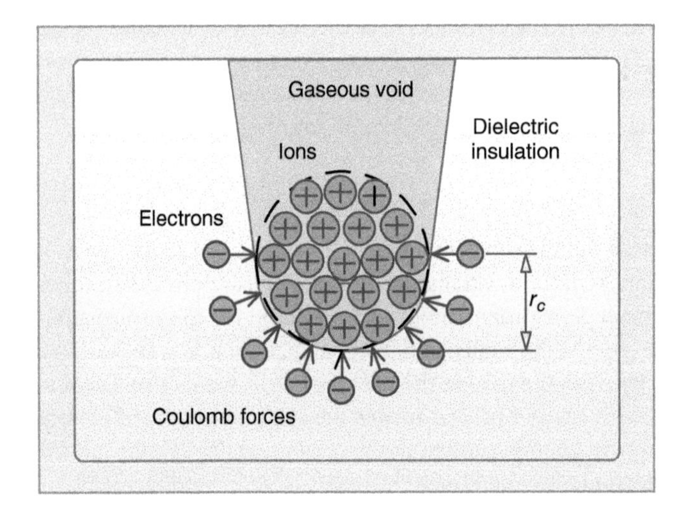

Figure 3.19 Electrons attracted by Coulomb forces to positive ions deposited on the cathode-side of void wall.

$$i_e(t) = -\frac{dq_P(t)}{dt} = -e\frac{dn_P(t)}{dt} \tag{3.27}$$

Combining this with Eq. (3.26), it can also be written

$$\frac{dn_P(t)}{dt} = \kappa_V \frac{n_P(t)}{2\varepsilon_0\varepsilon_r} \tag{3.28}$$

Upon separating the variables, one gets

$$-\frac{dn_P(t)}{n_P(t)} = \kappa_V \frac{dt}{2\varepsilon_0\varepsilon_r} \tag{3.29}$$

The resolution of this differential equation reads

$$n_P(t) = n_i e^{-\frac{t}{\tau_e}} \tag{3.30}$$

where n_i is the total number of ionized gas molecules, which equals the number of positive ions deposited initially on the cathode-side dielectric boundary, and τ_e is a characteristic time constant given by the dielectric properties:

$$\tau_e = \frac{2 \cdot \varepsilon_r \cdot \varepsilon_0}{\kappa_V} \tag{3.31}$$

The above discussed approaches have one common characteristic – they are deterministically in case of fulfilling the necessary conditions for PD, namely the evidence of an initiatory free electron. That "first" electron determines the statistical characteristics of PD events such as time of occurrence, inception delay, and thus the distribution of charge magnitudes – for example, in the PRPD-pattern at AC voltage. The introduction of such a statistical approach into the modeling praxis was first done by Niemeyer and co-workers [18, 26]. In general, for providing an initiating electron two main processes can be distinguished, namely volume and surface electron generation. The volume generation includes radiative gas ionization by energetic photons and field detachment of electrons from negative ions. In both cases, the electron generation rate might be approximated by

$$\dot{N}_e = \eta_i(\text{gas}, E, \ldots) p V_{\text{eff}} \left(1 - \frac{\eta}{\alpha}\right) \tag{3.32}$$

where p is the gas pressure and V_{eff} is the effective gas volume exposed to the radiation and to the electric field. The function η_i describes the particular ionization mechanism and generally depends on the type of gas, the electric field E and further parameters [18], η is gas attachment coefficient and α is gas ionization coefficient.

Notice that the approach was used in some modeling simulations when the randomness of discharges are considered [23, 26]. The generation of free electrons by volume ionization might be problematic in cavities embedded in solid insulation. Considering a virgin cavity, the time delay for that initiating electron, often termed as *waiting time* or time lag, might be extremely long, so the PD inception measured by the value of PDIV can lead to erroneous results if the time delay is much higher than the testing time of the applied test voltage. As an example, the time lag in correlation with the number of free electrons vs. the shape of spherical voids indicates the reciprocal dependence of initiating electrons with the void size, i.e. that for small defects with diameter less than 1 mm and a gas pressure of ~75 kPa the inception time can be delayed much more than 10^3 second (Figure 3.20). This fact may lead to an important issue at PD quality testing with limited testing time.

Once the first discharge has occurred, the second option for producing of free electrons should be considered, namely the emission of de-trappable electrons from the accumulated surface charge. Without going in detail, it was observed that this process may have an important influence on the statistics of PD events depending on the "life-time" of the surface charge on the void walls [14]. Especially under DC conditions, that effect plays a significant role at PD quality testing and should be considered.

3.3.3 Summary

The detailed discussion of various PD models shows a different explanation of the PD phenomena, especially for cavity discharges under various voltage stress conditions. The traditional network-based PD model describes the measurable PD transients on the

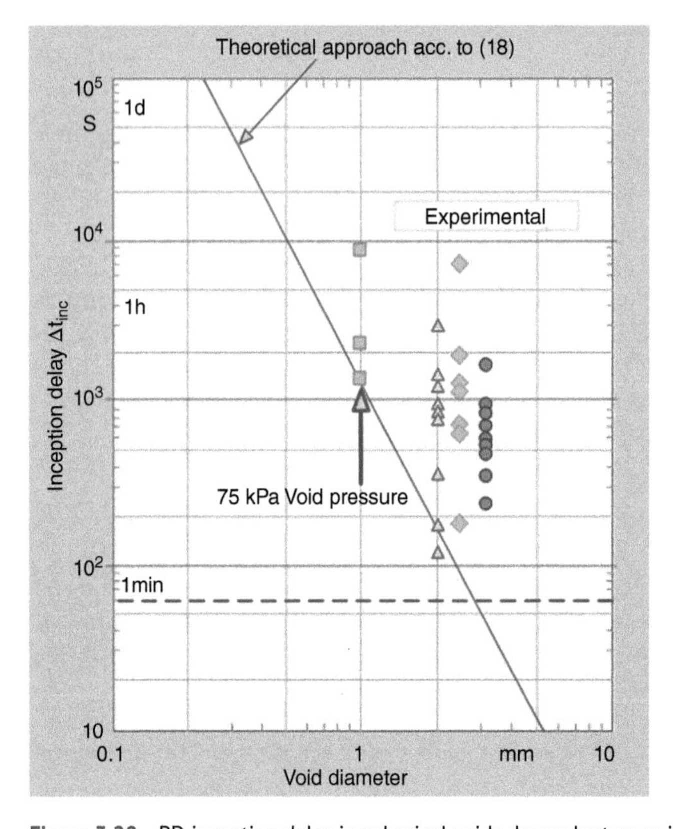

Figure 3.20 PD inception delay in spherical voids dependent on void diameter acc. to [23].

electrodes of the insulation system with a discharge of an internal virtual capacity representing the defect. That internal "short-circuit" leads to a change of entire charge balance and to a very fast-time response on the electrodes of the insulation system. If the external voltage source is practically disconnected from the test object due to a large impedance of connection, the voltage on the electrodes will drop. This voltage drop leads in an outer measuring circuit to a transient current pulse, which might be decoupled by an appropriate PD measuring device. After adequate processing of that signal, the PD quantities (e.g. the parameter apparent charge) is available. However, this model is not related to the physical phenomenon of internal discharge event and does not consider any statistics of PD events.

The field-based PD model explains the discharge phenomena based on physics of gas discharge and enables an analytical treatment of the PD transients, for instance, to estimate the link between the measured pulse charge and the number of gas molecules ionized in case of a PD event, as well as to predict the field enhancement in the bulk dielectric adjacent to the cavity walls, which could induce electrical trees in front of each cavity tip.

References

1 Gemant, A. and v. Philippov, W. (1932). *Die Funkenstrecke mit Vorkondensator* (in German). *Zeitschrift für Technische Physik* 13 (9): 425–430.

2 Lemke, E. (2012). *A critical review of partial-discharge models. IEEE-DEIS* 28 (6): 11.

3 Heller, B. (1961). *Die Ionisierungsvorgänge in der geschichteten Anordnung festes Dielektrikum- Luft* (in German). *Acta Technika CSAV* 6: 203.

4 Kreuger, F.H. (1964). *Partial Discharge Detection in High-Voltage Equipment*. London: Temple Press Books.

5 Whitehead, S. (1951). *Dielectric Breakdown of Solids*. Oxford: Clarendon Press.

6 International Electrotechnical Commission (2000). *IEC 60270: High-Voltage Test Techniques – Partial Discharge Measurements*. Geneva, Switzerland: IEC.

7 Böning, W. (1963). *Luftgehalt und Luftspaltverteilung geschichteter Dielektrika* (in German). *Archiv für Elektrotechnik Bd.* XLVIII (H.1): 7.

8 Lemke, E., Berlijn, S. Gulski, E. et al. (2008). Guide for partial discharge measurements in compliance to IEC 60270. *Technical Brochure No. 366*.

9 Fruth, B. and Fuhr, J. (1990). Partial discharge pattern recognition – a tool for diagnosis and monitoring of aging. *CIGRE Conference Proceedings*. Paris (Auguest 1990).

10 König, D. and Rao, Y.N. (1993). *Partial Discharges in Electrical Power Apparatus*. Berlin, Offenbach: VDE-Verlag.

11 Fromm, U. (1995). Partial discharge and breakdown testing at high DC voltage. PhD thesis. Technische Universität Delft.

12 Fromm, U. and Morshuis, P.H.F. (1995). The discharge mechanism in gaseous voids at DC voltage. In: *9thInt. Symp. on HV Engineering*, 4154/1. Graz: ISH.

13 Devins, J.C. (1984). The physics of partial discharges in solid dielectrics. *IEEE TDEI* 19: 475.

14 Van Brunt, R.J. (1991). Stochastic properties of partial discharge phenomena. *IEEE TDEI* 26 (5): 902–948.

15 Küchler, A. (2018). *High Voltage Engineering*. Berlin: Springer.

16 Alston, L.L. (1968). *High Voltage Technology*. Oxford University Press.
17 Pedersen, A. (1987). *Partial Discharges in Voids in Solid Dielectrics: An Alternative Approach*, 58. CDEIP. Annual Report.
18 Niemeyer, L. (1995). *A generalized approach to partial discharge modeling. IEEE TDEI* 2 (4): 510–528.
19 Lemke, E. *A critical review on partial discharge models. IEEE Electrical Insulation Magazine* 28 (6): 11–16.
20 Pedersen, A., Crichton, G.C., and Mc. Allister, I.W. (1991). *The theory and measurement of partial discharge transients. IEEE TDEI* 26: 487–497.
21 Pedersen, A., Crichton, G.C., and Mc. Allister, I.W. (1995). *The functional relation between partial discharges and induced charge. IEEE TDEI* 2 (4): 535–543.
22 Crichton, G.C., Karlsson, P.W., and Pedersen, A. (1989). *Partial discharges in ellipsoidal and spheroidal voids. IEEE TDEI* 24: 335–342.
23 Gutfleisch, F. and Niemeyer, L. (1995). *Measurement and simulation of PD in epoxy voids. IEEE TDEI* 2 (5): 729–743.
24 Tanaka, T. (1986). *Internal partial discharge and material degradation. IEEE TDEI* 21: 899–905.
25 Luczynski, B. (1979). Partial discharges in artifical gas-filled cavities in high-voltage insulation. PhD thesis. Techn. University of Denmark, Lyngby.
26 Fruth, B. and Niemeyer, L. (1992). *The importance of statistical characteristics of partial discharge data. IEEE TDEI* 27 (1): 60–69.
27 Lemke, E., Muhr, M., and Hauschild, W. (2022). *Modeling of cavity discharges under AC and DC voltage – Part II: Opportunities of the dipole-based PD model. IEEE TDEI.*
28 Dawson, G.A. and Winn, W.P. (1965). A model of streamer propagation. *Zeitschrift f. Physik* 183: 159–171.
29 Lemke, E. (1967). Breakdown mechanism and breakdown vs. gap- distance characteristics of non- uniform air gaps at SI voltages (in German). PhD thesis. Technische Universität Dresden.
30 Raether, H. (1964). *Electron Avalanches and Breakdown in Gases*. London: Butterworths.
31 Shockley, W. (1939). Current to conductors induced by a moving point charge. *Journal of Applied Physics* 9: 635.

4

Measurement of Partial Discharges

4.1 Introduction

As discussed in Chapter 2 partial discharges cover several physical effects based on the discharge process. However, those physical effects have different intensity depending on the insulation material and the place of occurrence. An overview about these effects is shown in Figure 4.1:

- *Conducted currents.* After an internal discharge a rebalancing of charges between the capacitances is needed. This process leads to a high frequency current, which can be measured in the effected branches of the network. The measurement of the conducted currents is usually done by a well-defined measurement circuit [1] discussed later in this chapter.
- *High-frequency EM waves.* The high-frequency content of discharge current leads to high-frequency EM waves emitted by different frequency parts, which can be measured. The high-frequency EM waves are usually measured by appropriate inductive, capacitive sensors or UHF-antenna. The application of such measurement techniques is discussed deeper in Chapter 5.
- *Dielectric losses.* The partial discharges provide power losses within the fault area, which lead to additional heat power and locally increased temperature. The additional power losses can be measured by using loss factor measurement systems. During the measurement of characteristic parameters permittivity ε_r and loss factor tan δ by special measurement bridges (e.g. Schering Bridge) or digital capacitance and loss factor measurement systems the increasing power losses will be evaluated (Figure 4.2), [3]. The partial discharge inception (PDIV) can be examined by the changing gradient of the loss factor over the voltage (so called tan-δ – increment). It should be noticed that the parameter loss factor shows such a behavior on partial discharges only in case of large or multiple discharges as typical for slot discharges within insulation of rotating machines.
- *Pressure wave.* The discharges cause micropressure waves that vibrate the material and lead to further erosion or undercutting of the material and might be detected by acoustic sensors. The measurement of such vibrations can be done by recording of vibrations via

Partial Discharges (PD) - Detection, Identification, and Localization, First Edition. Norasage Pattanadech, Rainer Haller, Stefan Kornhuber, and Michael Muhr.

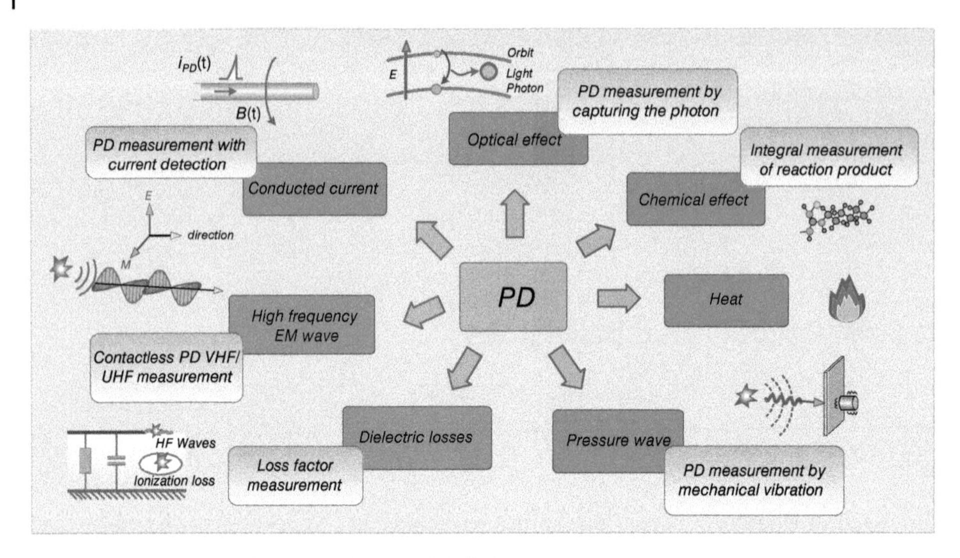

Figure 4.1 Physical effects based on partial discharges.

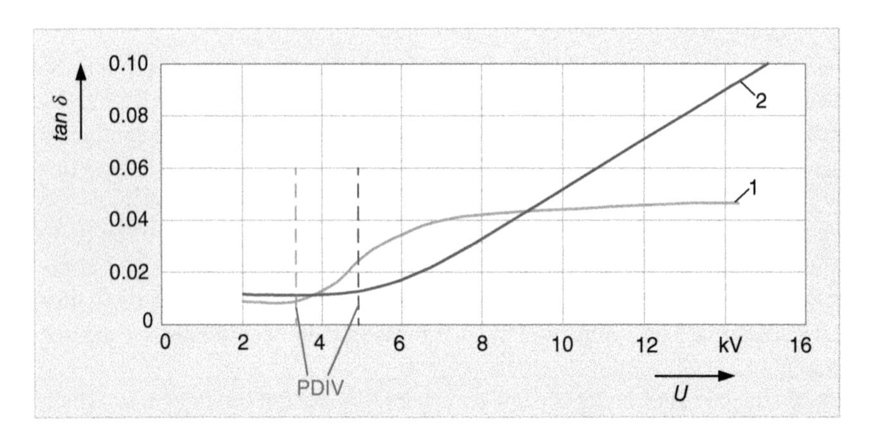

Figure 4.2 Dependence of loss factor of different slot insulation (1, 2) on test voltage, acc. to [2].

ultrasonic structure-borne sound sensors or microphones. The usage of these measurement techniques will be discussed closer in Chapter 6.

- *Heat losses*. Caused by increasing of temperature, which were already discussed above (dielectric losses). The increasing of temperature only locally (~hot spot), which could lead to changing of material behavior as well as further partial discharge resistivity. However, a measurement of smaller partial discharge (PD) events (e.g. smaller than 1 nC) by using temperature measurement systems (e.g. infrared cameras) is quite unusual.
- *Chemical effects*. The introduction of energy (heat, radiation, accelerated particles) can enable chemical reactions (e.g. formation of gases from insulation oil, carbonization of defects). The measurement of the partial discharge activity is done by the determination of the chemical reaction products (e.g. O_3 content measurement, fault gas measurement (essentially H_2, C_2H_2) in mineral insulating oils) [4, 5].

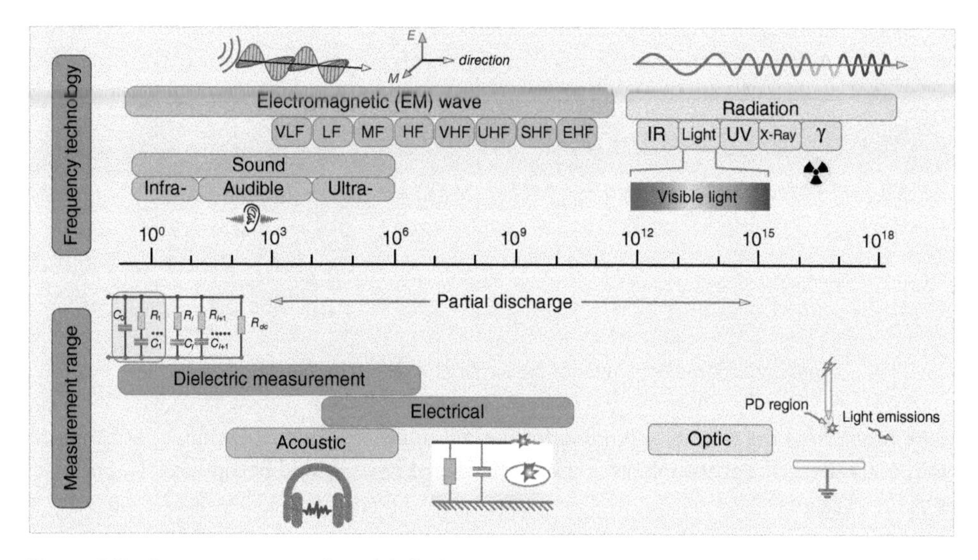

Figure 4.3 Frequency range of partial discharges and detection possibilities.

- *Optical effects*. During the discharge process photons are emitted. These photons can be measured using adequate photon multiplier and measuring systems [6, 7].

As already recognized from the description of the physical effects a very different frequency spectra for the evaluation can be examined. In Figure 4.3, a principal overview of the different areas of the affected frequency spectra and the usability for the different measurement principles is shown.

The frequency range is mainly affected by the process, the participating material itself, and the transmission path.

4.2 Signal Properties

The partial discharge signal is caused by a discharge current within the insulation as discussed in previous chapters. For measurement of that signal by appropriate measurement systems it must be considered that the signal described by its properties like shape, magnitude, frequency, and so on will be influenced on the "path" to the evaluation of final PD parameters like *apparent charge* by many effects (Figure 4.4).

4.2.1 Device Under Test

Important for the influence on signal properties is the internal signal path to the "output" of the tested object, which might be very different depending on type of device. In principle, there are two significant types:

- Concentrated (compact) volume/size (e.g. switch gear, transformers, machines)
- Extended (elongated) volume/size (cables, GIL, overhead lines)

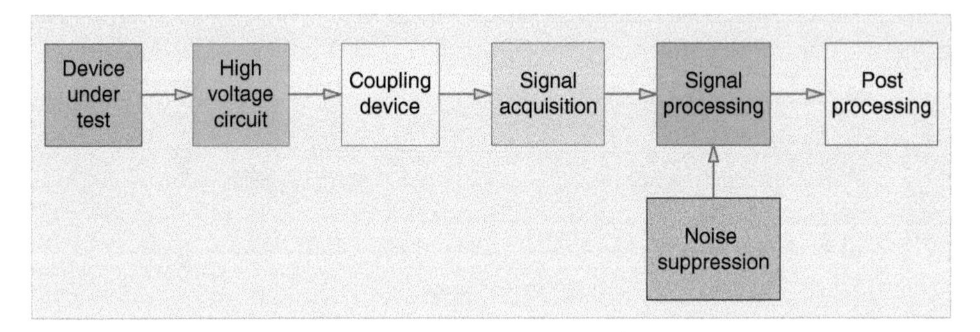

Figure 4.4 Signal path for measurement of partial discharges (schematically).

The problem is the possible signal deformation along the (internal) signal path and/or additional reflections or resonances. Therefore, each device should be analyzed by possible paths for PD signals.

4.2.2 High Voltage Circuit

The measurable PD signals also might be influenced by type of high-voltage supply and the size of complete high-voltage test circuit.

The type of high-voltage supply may be characterized by alternating voltage (AC), direct voltage (DC), impulse voltage (standardized like LI or SI and repetition pulses), and/or combination:

- AC. Introduced in PD measurement techniques and implemented in different standards is the (clean) sinusoidal wave shape. However, in some cases the supply has significant distortions caused by harmonics, which might have an influence on the PD behavior of the tested object [8, 9].
- DC. Hence under investigation worldwide (see Chapter 8), it should be considered that in some cases the DC test voltage is superimposed by a small AC component (ripple). This also might have an influence on the PD activity [10, 11].
- Impulse. The influence by the test voltage on PD behavior is discussed more in details in Chapter 8.
- Size of test circuit. Its influence on the signal properties might be described by two characteristic parameters – *inductance* of connection leads from the high-voltage source to the test object and (parasitic) *capacitance* to Earth or other potentials.

For better explanation of that influence, Figure 4.5 shows how an injected current pulse with constant charge will be deformed by high-voltage testing circuit of different size (set-ups).

It could be demonstrated that the parameter of the test circuit size, in particular inductance and the parasitic capacitances, have a significant influence on the time and frequency behavior of the PD current signal. This situation must be recognized and well understood especially for the right choose of the decoupling unit and of the setup of the frequency bandwidths of the measuring device. Additionally, the example is showing the necessity of the calibration of the measurement circuit when the test object is changed, or the test setup is modified.

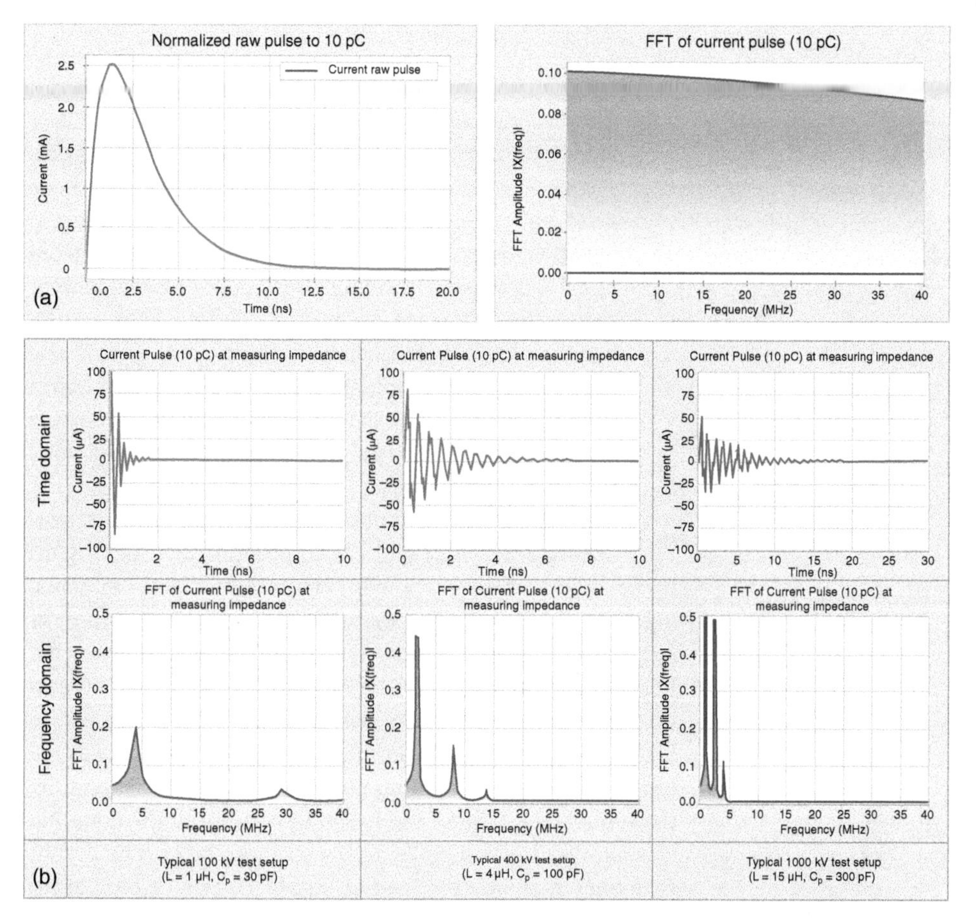

Figure 4.5 Time domain pulse behavior and frequency spectra of injected current pulse at different high-voltage test setups. (a) Injected current pulse as reference value (left: time domain signal, right: frequency spectrum). (b) Time and frequency response at different high-voltage PD test circuits.

The influence on PD signal by the other units of the "measuring path" depicted in Figure 4.4 will be described below in more detail.

4.3 Coupling Methods

For measuring of partial discharges, it is necessary to decouple the PD signal from the high voltage testing circuit. According to the overview about the frequency characteristic of PDs (Figure 4.3) the methods are different dependent on the applied frequency range. For (VHF) range up to ~ (10–30) MHz the commonly applied decoupling methods for partial discharge signals are capacitive coupling in connection with a measuring impedance and inductive coupling. Both principles are discussed here.

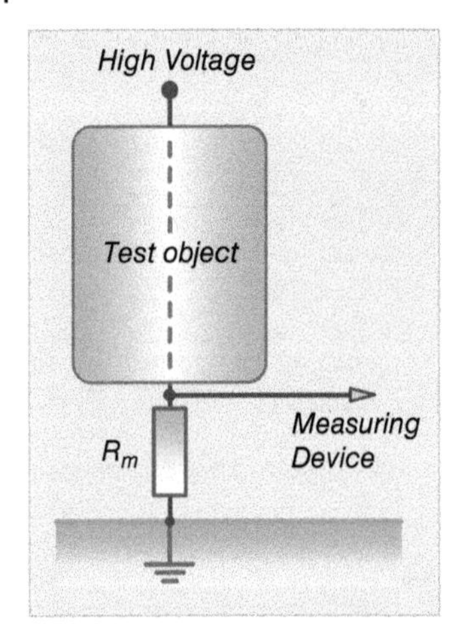

Figure 4.6 Direct decoupling method by using a resistance R_m.

4.3.1 Capacitive Coupling with Measuring Impedance

The decoupling by using a measuring impedance provides a transformation of the conducted current to a corresponding voltage signal. As one of the easiest methods a resistance R_m is used connected with the measuring device by measuring cable (Figure 4.6), [12]. To avoid possible issues of traveling waves at connected components, the resistance needs to be designed for a 50 Ω-compatible high-frequency network. It should be noticed that this option needs a no-grounded test object (device under test) together with safety requirements and, therefore, is used mainly for basic PD investigations in research laboratories.

Besides that, a commonly designed measuring impedance Z_m for AC- and DC-application consists of an inductance L and a resistor R (Figure 4.7). These elements, together with coupling capacitor, provide a high-pass filter of second order whereby its corner frequency must be below the lower cut-off frequency of the "measuring path" of the measuring device. At the same manner, that frequency must be high enough to avoid any influence of test voltage, such as AC (power frequency including possible harmonics). Therefore, the inductance L must be chosen in relation to the coupling capacitor C_k in such a way to match those tasks. The corner frequency may be expressed by Eq. (4.1), as an example for AC testing; with a typical capacitor of 1 nF, the inductance value should be about 15 mH, which meets a corner frequency of ~40 kHz.

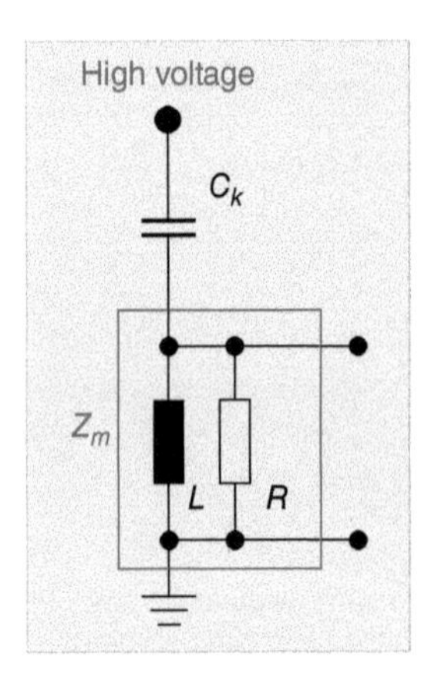

Figure 4.7 Basic design of measuring impedance Z_m.

$$f_{\substack{-3db \\ lower}} = \frac{1}{2\pi}\sqrt{\frac{1}{LC}} \tag{4.1}$$

The elements are commonly arranged in a shielded box with four clamps – often termed a *quadrupole*. Also, overvoltage protection by appropriate arrestors and/or spark gaps should be included. In some cases of AC testing, the design of the quadrupole is extended with a low voltage capacitance. In connection with an appropriate high-voltage capacitor, this capacitance is acting as a voltage divider and enables the measuring of test voltage at the same time as performing the PD measurement. It should be noticed that in some cases the measuring impedance is equipped with a pre-amplifier, which might increase the measuring sensitivity dependent on EM environmental conditions and further processing principles.

The measuring impedance Z_m or quadrupole within the high-voltage test circuit[1] can be positioned on three principal options (Figure 4.8) [13]:

1) The coupling capacitor branch (Figure 4.8a)
2) The test object branch (on the "ground side" of test object, Figure 4.8b)
3) Both branches using a bridge impedance and appropriate bridge measuring methods (Figure 4.8c)

These options are explained in more detail as follows:

- *Measuring impedance* Z_m *in the coupling capacitor branch.* If Z_m is in series to the coupling capacitor the high frequency PD current signal is transferred over this branch to the measuring device. The advantage of this setup is that if a test object fails (e.g. by breakdown), the measuring impedance will not be affected or damaged. At the same manner this setup should be used for test objects, which cannot be disconnected from the ground like rotating machines. The disadvantage is that the measuring sensitivity is less compared with the setup, where the measuring impedance is implemented in the test object branch.
- *Measuring impedance* Z_m *in the test object branch.* If Z_m is in series to the test object, the high-frequency PD current signal is directly transferred over this branch to the measuring device. The advantage is that the measuring sensitivity is higher compared to the previous setup, where Z_m is implemented in the coupling capacitor branch. The disadvantage of this setup is that if the test object fails, the measuring impedance and eventually the connected measuring device could be affected or damaged. It is also possible to place Z_m on the high-voltage side, but in that case the transfer of measured data must be realized from high-voltage potential to low-voltage potential of measuring systems (commonly provided by optical transfer systems).
- *Measuring impedance* Z_m *in both branches (test object and coupling capacitor).* In this case, two measuring impedances (or a special bridge impedance) are needed. This arrangement has the advantage that the measuring sensitivity is higher due to the differential probe gain and the common mode rejection. Additionally, it can be used for polarity discrimination of the measured PD pulses. The disadvantage of this setup is that if the test object fails, the measuring impedances might be affected or damaged. In addition, this arrangement is much more complicated and expensive than the other two options.

Figure 4.9 summarizes advantages and disadvantages of different setups of measuring impedances. It might be concluded that in most cases, the position of measuring impedance Z_m in coupling capacitor branch is applied; however, the other setups also provide their benefit.

The basic requirement for design of measuring impedance is the appropriate pulse transfer characteristic for PD signals. This property may be described by the bandwidth

1 High-voltage test circuit for PD measurement according to [13] consists of characteristic elements as high-voltage supply/blocking impedance/coupling capacitor/coupling device/PD measuring instrument, including measuring cable.

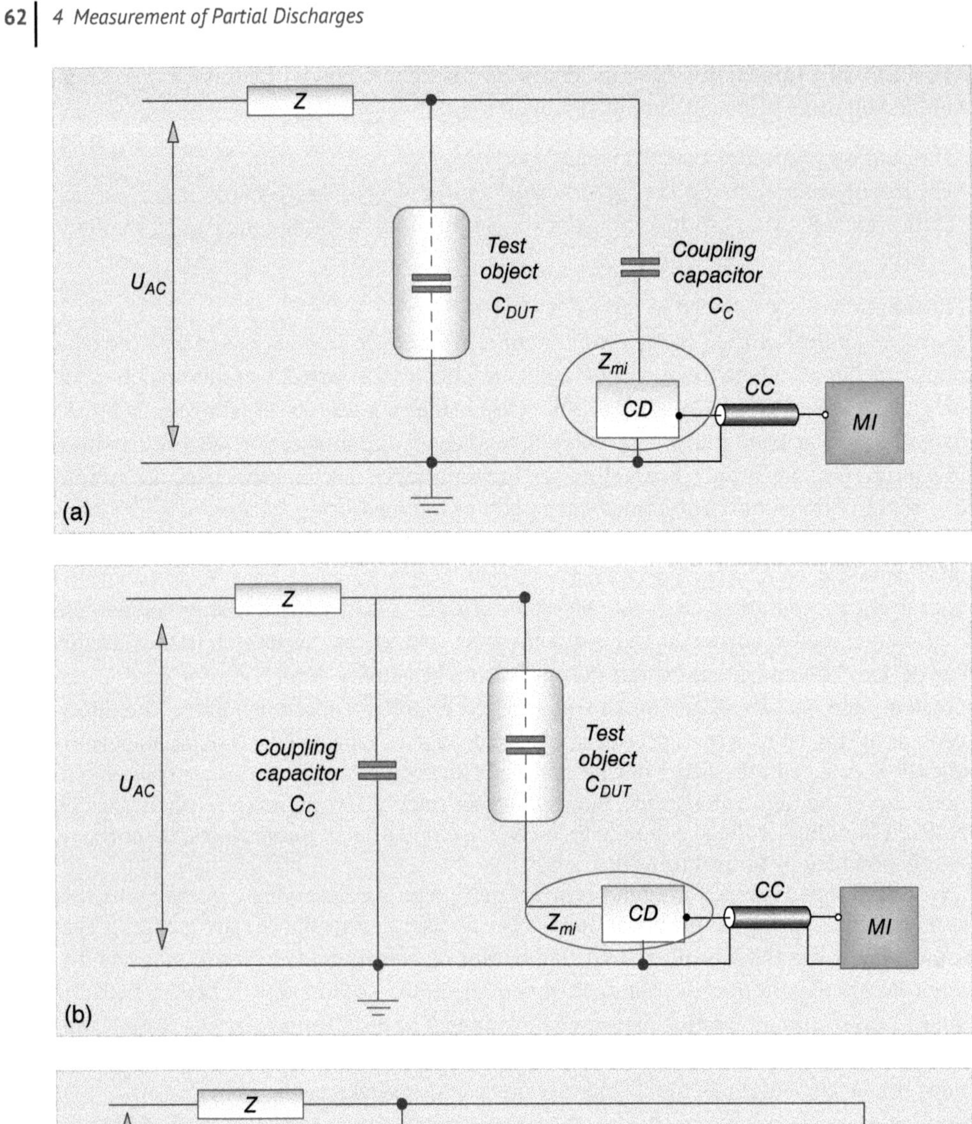

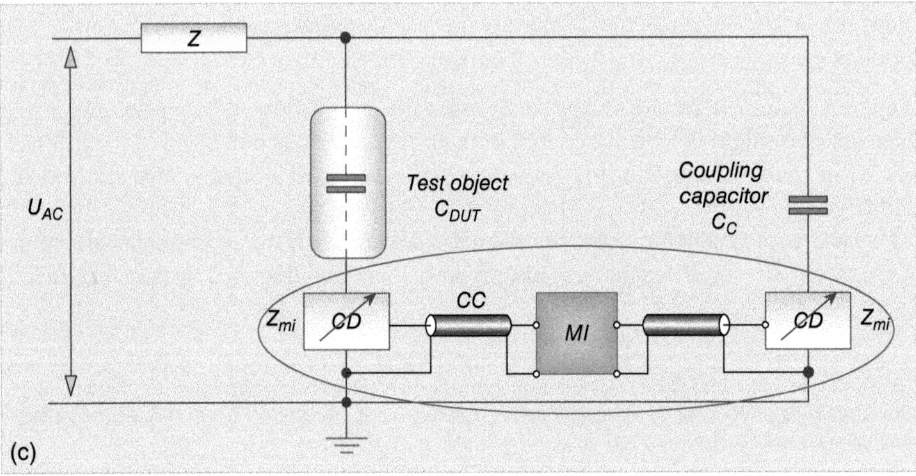

Figure 4.8 Possible positions of measuring impedance in the high-voltage test circuit: (a) in the coupling capacitor branch; (b) in series with the test object; (c) in both branches using a bridge impedance and appropriate bridge measuring methods.

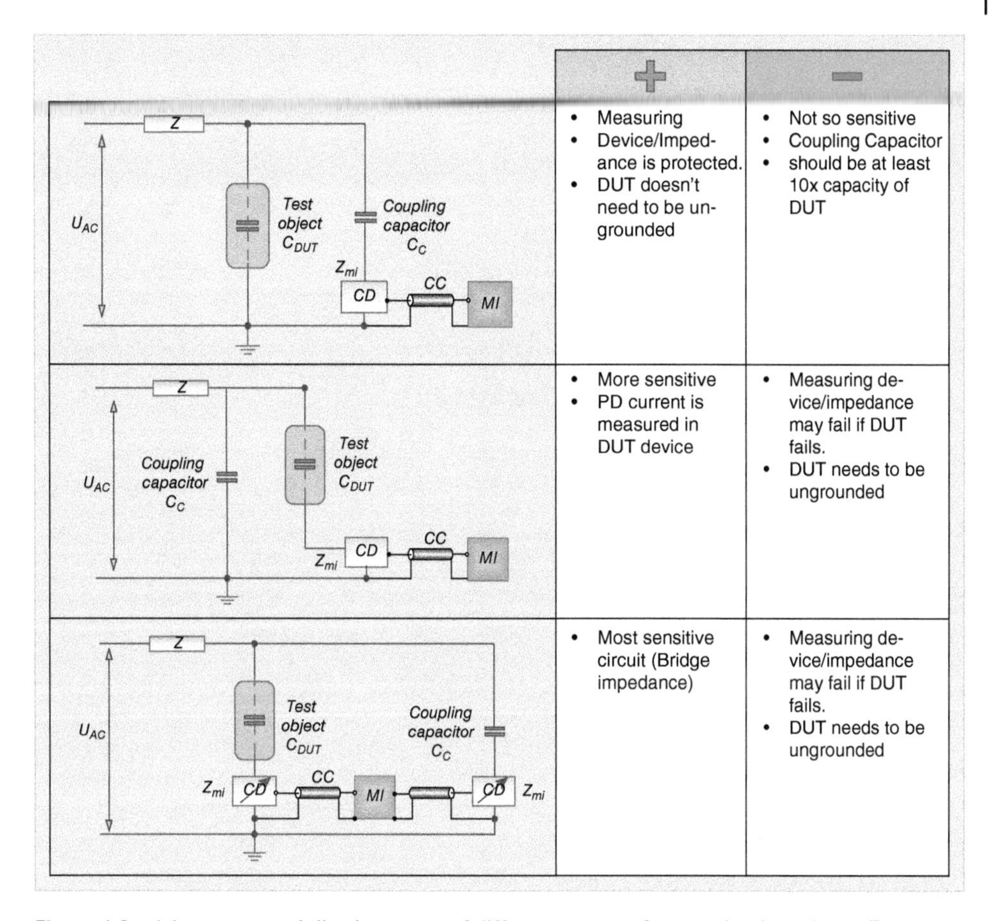

Figure 4.9 Advantages and disadvantages of different setups of measuring impedance Z_m.

characteristic of Z_m (Figure 4.10). It can be seen that in comparison with the behavior of the "resistive shunt method" mentioned above the bandwidth of measuring impedance is characterized by the "transfer window" from ~100 kHz to ~ 30 MHz, which fully covers the requirements of the appropriate standard [13]. The transfer characteristic of measuring impedance depend on the "inner" elements like parasitic inductance and capacitance. These elements should be matched to the transfer characteristic of the connected measuring instrument characterized by the wave impedance of the measuring cable, which is commonly 50 Ω.

If it is impossible to match the design of Z_m to the connected measuring instruments (e.g. at its high resistive input), a compromise must be found between

- Sensitivity
- High pass characteristic
- Traveling wave reflection minimization

As an example, the pulse transfer behavior in case of mismatching of measuring impedance with a connected measuring device of high resistive input is depicted in Figure 4.11. Reflected pulses occur that might lead to wrong interpretation of PD measurement.

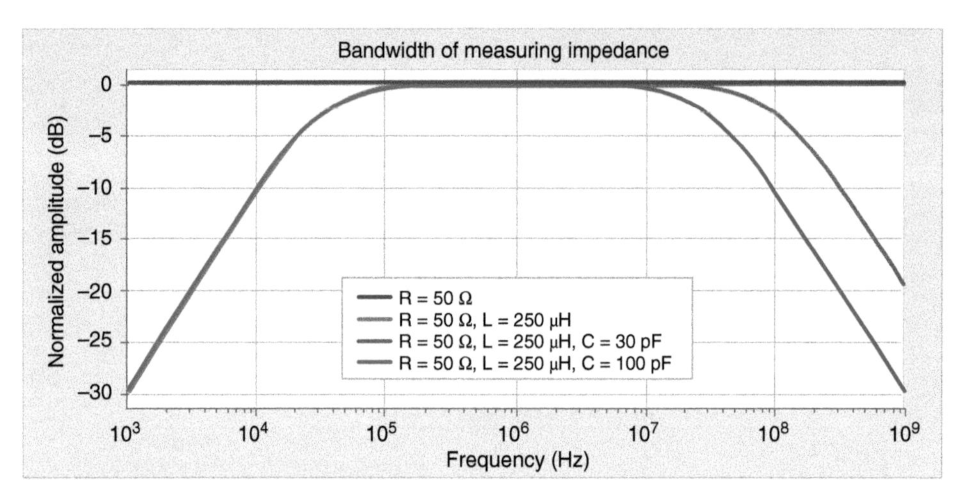

Figure 4.10 Impact on frequency behavior of measuring impedance by its "inner" elements.

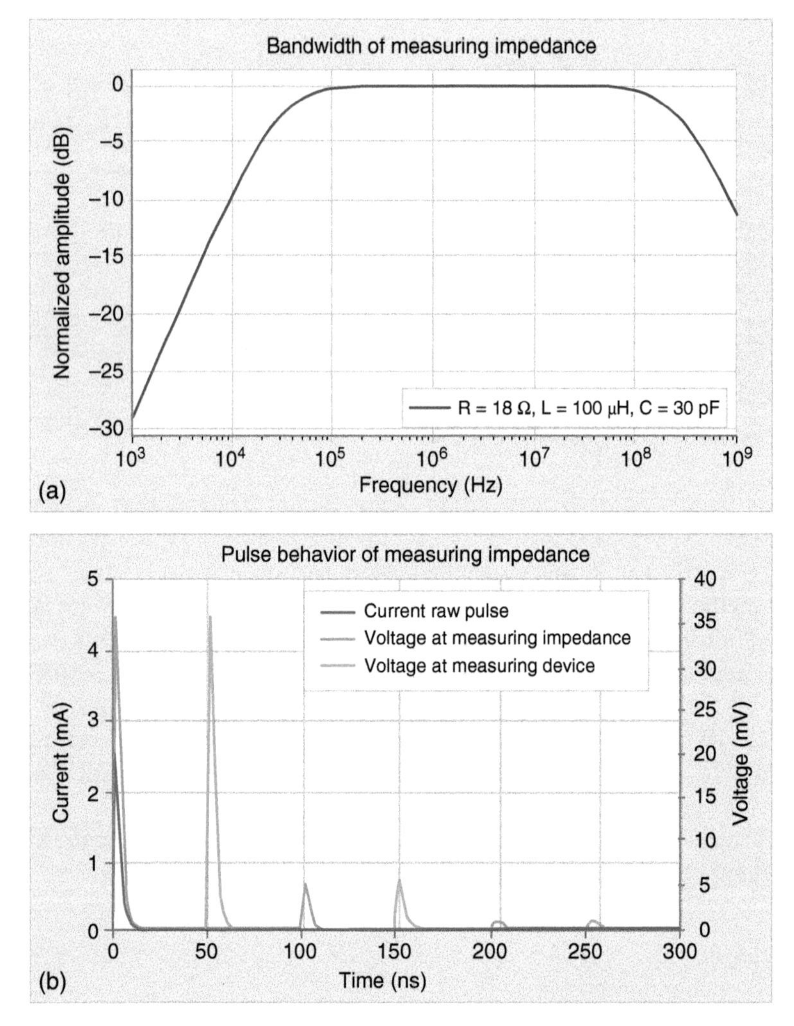

Figure 4.11 Effect of pulse reflections in case of connected measuring device with high resistive (1 MΩ) input: (a) bandwidth and (b) pulse behavior.

4.3.2 Inductive Coupling with High-Frequency Current Transformer

In contrast to the previous method, the decoupling of PD signals may be realised by a current transformer, which means by an inductive coupling method. According to the high-frequency character of PD signals, the current transformer must have an appropriate frequency characteristic (bandwidth), which enables the transfer of detected PD signals to the measuring device. Such transformers are typically equipped by a ferromagnetic core and screened windings, commonly termed as high-frequency current transformers (HFCT) (Figure 4.12).[2] They are adapted to the input impedance of connected PD measuring device, which is commonly 50 ohms.

The principle of an HFCT, shown in Figure 4.12, is to act as an inductive sensor. The sensor is installed around the conductor where high-frequency PD current is passing through. The PD current is inducing a magnetic field within the ferromagnetic core of the sensor. The measuring principle is shown at a simple equivalent circuit of the basic inductive sensor with inductance L (Figure 4.13). The induced voltage U_i over the windings, which is the related to PD current $\boldsymbol{i_{PD}}$, is measured via a resistive burden R_m usually chosen as input impedance (50 Ω) of connected PD measuring device. The acquired PD signal by that method is a voltage pulse with magnitude $\boldsymbol{V_{PD}}$.

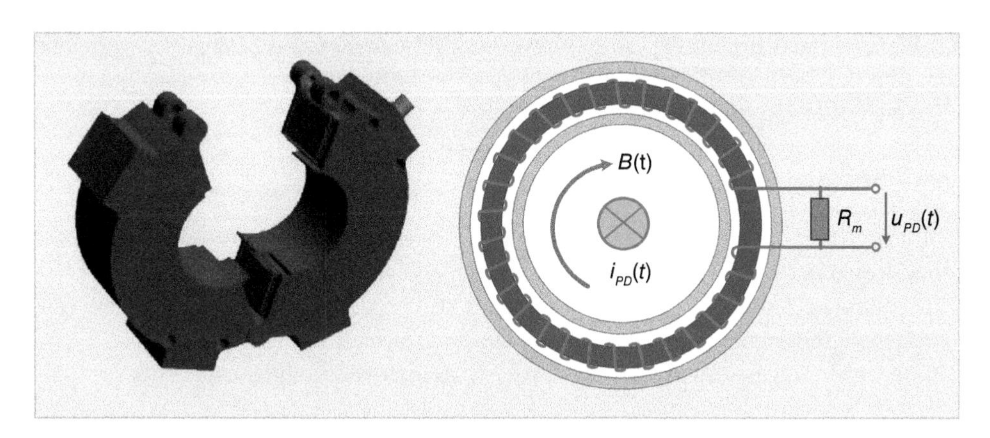

Figure 4.12 Design of a high-frequency current transformer [14].

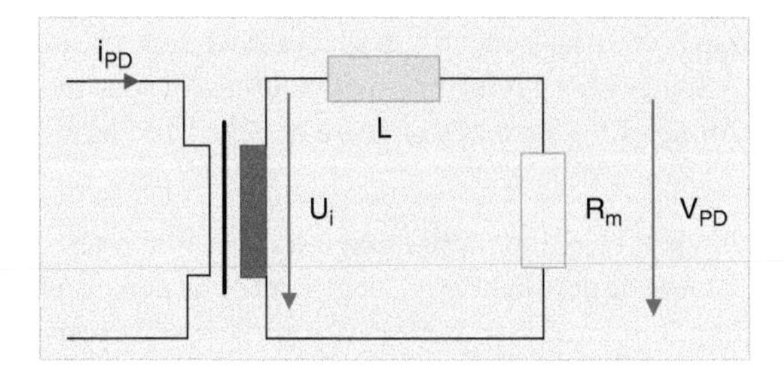

Figure 4.13 Equivalent circuit of a HFCT.

2 The Rogowski-coil based on the same inductive principle but without core is commonly less sensitive.

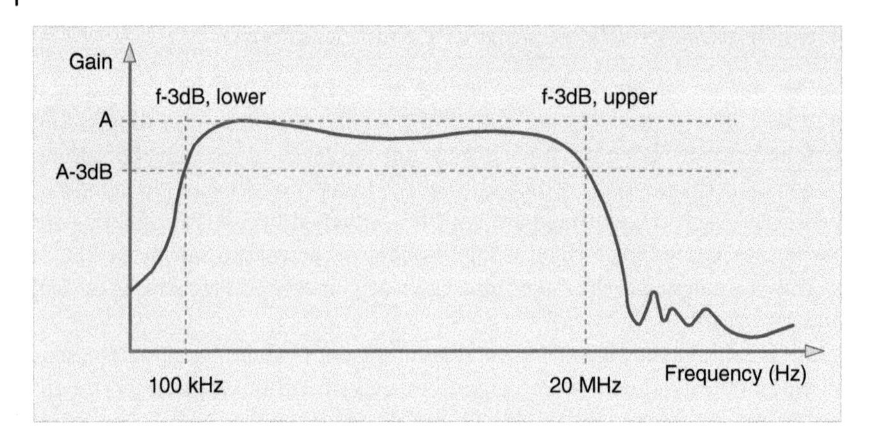

Figure 4.14 Typical frequency behavior of a HFCT-sensor.

A typical frequency characteristic for such inductive sensor is depicted in Figure 4.14. The sensitivity will be determined by the number of turns in relation to R_m, while the lower cutoff frequency ($f_{-3\text{dB, lower}}$) is mainly dependent on the inductance L. An increasing value of L, which means a decreasing value of lower cut-off frequency, may be obtained by higher turns and/or A_L-values[3] of magnetic core.

Characteristic applications for such HFCT sensors are:

- On-line/off-line PD-testing on encapsulated GIS-cable or transformer-cable sealing ends, where no coupling capacitor can be connected.
- On-line/off-line PD-fault-location testing (time domain reflectometry) with optimized low-frequency (50/60 Hz) suppression, even at high operation currents.
- PD testing according to the related standard (IEC 60270, $f > 100\,\text{kHz}$), especially for equipment that cannot be tested ungrounded (transformers, generators).
- PD testing for high-capacitive equipment, such as high-voltage filter capacitors.

One of the advantages of such a sensor is that there are no direct connections to the high-voltage test circuit. This may be demonstrated in the following example of cable testing (Figure 4.15) [15].

It is assumed that a PD current pulse $\boldsymbol{i}_{\mathbf{PD}}$ is starting from an internal cable defect, traveling through the cable insulation or the switchgear and returning to its source. All possible HFCT sensor positions for inductive coupling of PD current signal are depicted and numbered (1–5):

Pos.1: Sensor is in series with an impedance Z bridging high-voltage source. When testing offline Z becomes a coupling capacitor or another cable length, and the sensor is used as coupling device (CD) according to related standard. Performing on-line tests Z is the overall substation impedance that mainly is a complex network of branched

3 The A_L- value is a core-specific property and is expressed by inductance per turns square.

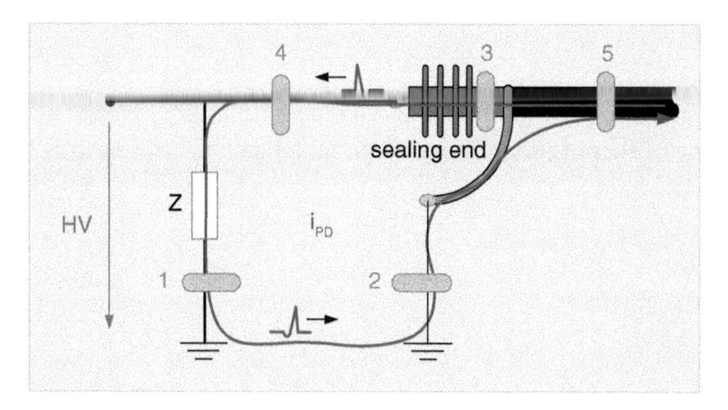

Figure 4.15 Possible positions of HFCT sensors in a PD test circuit.

impedances. Considering that a common ground connection for all impedances could not be identified in most cases. Advantageously the inductive sensor must withstand only the currents flowing through *Z*. Smaller and/or sensitive HFCT sensors can be used.

Pos.2: Sensor is connected to the grounded cable screen. In many cases this is the most convenient measuring position as it is the most sensitive and safest method. In fact, in many cases it is the only possibility to mount the sensor in service. Measuring in this branch the sensor must withstand high reactive currents depending on the cable length and test voltage. Testing on-line additional circulating currents induced by neighbored phases can reach up to several hundreds of amps, which should be considered at choosing HFCT design.

Pos.3: Sensor is enclosing the lower part of the sealing end including the semiconductive layer but just before the cable screen. Alternatively, the sensor can be put near the sealing end enclosing the cable sheath whereas the cable screen is conducted twice through the sensor for canceling out the screen current. In this case, the measured PD signals are superposed with the high-operating currents flowing through the conductor.

Pos.4: Sensor is mounted on high-voltage potential. As in position (3), high reactive and operating currents are passing the sensor. Additionally, this requires the transfer of the detected PD signals to the PD-measuring system located on low potential area. Commonly this will be performed by using optoelectronic data transfer systems.

Pos.5: Sensor is enclosing both the conductor and the screen. When mounting in this position the outgoing and incoming currents theoretically compensate each other, so there is no possibility for highly sensitive measurements in this position.

The decoupling of PD signals by measuring impedance or HFCT sensor described above provide the signal acquisition in the VHF-range of PD spectra. According to Figure 4.3 the PD events have also spectral components up to the UHF range and more. The requirements and/or special properties for signal acquisition under such conditions will be discussed in the following chapters. In the following, the further processing of PD signals from decoupling to characteristic PD parameters will be described.

4.4 Signal Processing

The further processing of PD signals acquired from the high-voltage circuit may be classified by different viewpoints. One of them is the principal type of applied measuring technique, which may be distinguished in

- Full analog processing
- Semi-digital processing
- Full digital processing

4.4.1 Full Analog Processing

The PD signal from the coupling device or other sensors may be amplified or attenuated depending on the signal strength. After that, the signal is filtered by using an analog band-pass filter, which is designed for an integration behavior. The band-pass filter is followed by a peak detector commonly provided by an analog sample–hold circuit. The output signal is proportional to the magnitude of the integrated PD current signal and will be indicated by a needle (pointer) instrument (Figure 4.16). The calibration in units of charge (pC) can be done by adjusting the needle instrument or an additional amplification using a special resistive potentiometer.[4]

The advantages of this setup are the easy and repairable setup without any additional components. There is no direct synchronization with the voltage signal or more suitable data handling like PRPD pattern possible. However, the output (before the sample and hold circuit) can be connected to an oscilloscope and the results can displayed in the *x*-*y* mode (PD and voltage in ellipse- or linear time-based view). The change of the cutoff frequency of the filter is only possible by using different analog filters outside of the device.

4.4.2 Semi-Digital Processing

Following the development of the computer technology the next generation of measuring devices were developed in a semi-digital setup (Figure 4.17). Still the high-frequency PD signal were done by using attenuator and amplifiers as well as an analog band-pass filter.

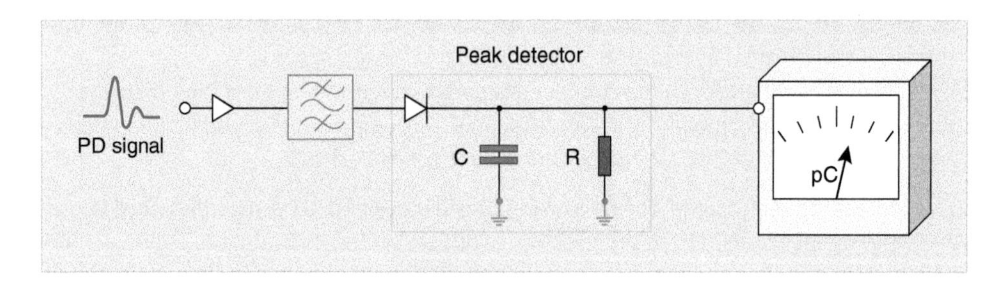

Figure 4.16 Full analogue PD processing (schematically).

4 The complete calibration procedure is described in Section 4.9.

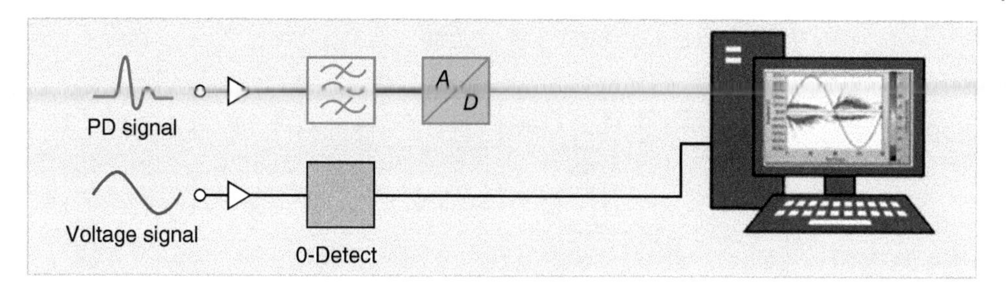

Figure 4.17 Semi digital PD processing (schematically).

The output of the band-pass filter with or without a sample and hold circuit were digitized by using an analog digital converter (ADC) and further processed by a special PC based software. In case of AC testing the AC test voltage may be also digitized by using an additional ADC-unit.

For this reason, the acquisition rate as well also the data handling was optimized to the available technology. The advantage at the implementation is:

- Recognition and storage of every filtered pulse
- Synchronized with test voltage
- Generating of PRPD pattern by using filtered PD pulses and test voltage signal (phase angle)
- Integration of the measuring device directly to remote controls or other operational systems
- Fault location by using additional single shot oscilloscope digitizer system

However, the filter settings were defined by the analog circuit, whereby a change of the cutoff frequencies where possible by using different analog filters and remote-control switches. A recording of each raw pulses is not possible by using this setup.

4.4.3 Full Digital Processing

With further development of analog digital converters and the digital signal processors (DSPs) a full digitalization of the PD pulses and the adequate processing procedures was possible. According to practical experience and to related standard for PD measurement, the applied ADCs should have a sample rate of 10^7, at least. The sample rate is a characteristic parameter of analog-digital converters and is measured in samples. Moreover, for PD fault location of cables even higher sample rates of above 2×10^7 seem to be necessary. However, nowadays PD devices are also available with analog bandwidths up to 20 MHz and sample rates of $>6 \times 10^7$ or even higher.

The PD signal is adapted by attenuator and amplifier to the input dynamic of the analog digital converter and if necessary, a high-pass filter removes components with power frequency caused by a circuit current. After adaption, the PD signal is digitized by using an analog digital converter (ADC) with above-mentioned sampling rates and the PD signal is filtered by using DSP and evaluated by using a peak detection (Figure 4.18). Another concept would be to deliver the raw data information stream to

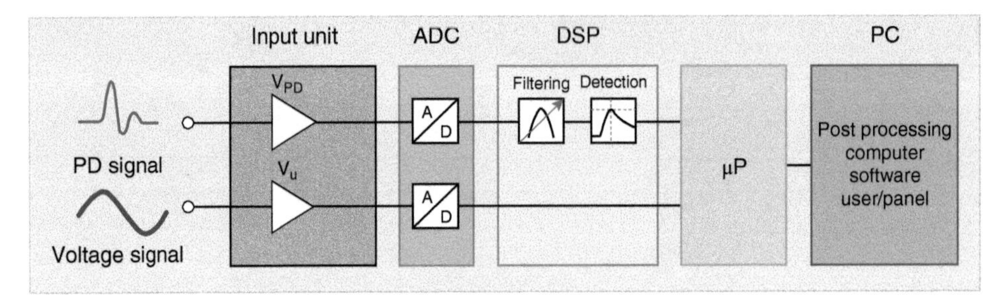

Figure 4.18 Full digital PD processing (schematically).

the PC and provide at the PC the filtering and peak detection. The voltage signal is also adapted, digitized, and provided to the PC.

By using this setup, the original (raw) PD pulses are digitized and a full flexibility of analyzing of the PD signals is possible. The advantages are that the filter settings can be adjusted to the PD as well also to the noise spectra fully digital, if possible, also after recording the PD measurement by using the original raw data measured pulses. Furthermore, if possible, a synchronized multichannel measurement can be used to measure different phases or different locations of a setup at the same time.

The advantage at the implementation is:

- Recognition and storage of every original (raw) PD pulse
- Free setup of filtering of PD pulses
- Multichannel measurement at different phases or locations
- Synchronization with test voltage signal
- Evaluation by using filtered pulse and voltage signal possible PRPD pattern
- Integration of the measuring device directly to remote controls or other operational systems
- PD fault-location

4.5 Measurement Principles

As opposed to the previous classification, the further processing of PD signals may also be distinguished by frequency characteristic of applied measuring technique. As discussed in previous chapters, the main description of partial discharge behavior should be expressed in terms of charge [12, 13]. Therefore, to process PD signals, it is necessary to integrate the detected signal, which is proportional to the PD current. Concerning the frequency spectrum of PD signals this integration can be done either by using a fraction of the whole spectrum in frequency domain or by actively integrating of the complete signal in time domain. In the first case, the signal surpasses a low-pass filter providing a so-called quasi-integration assuming an undistorted current signal of the original discharge. The processing takes place in the frequency domain, whereby two methods may be distinguished – the wide-band and the narrow-band detection both characterized by different frequency limits. The principal behavior of those systems at processing a PD pulse may be demonstrated

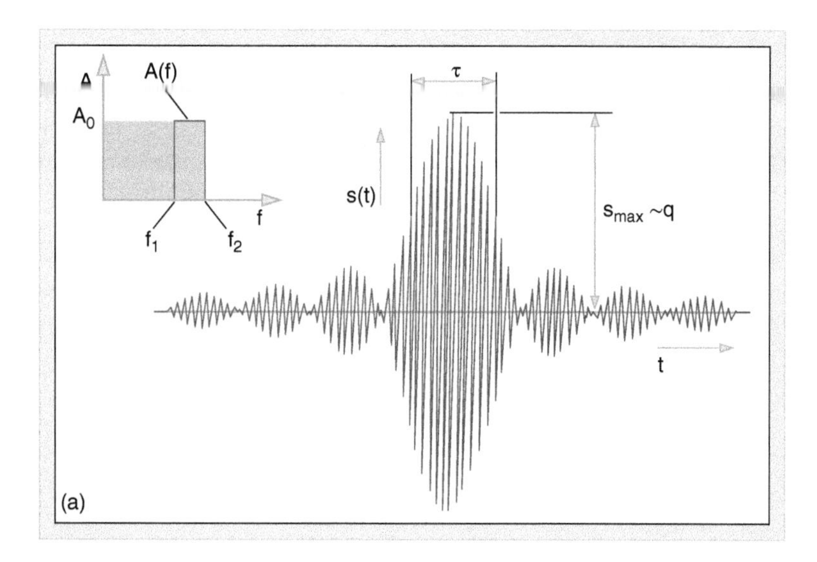

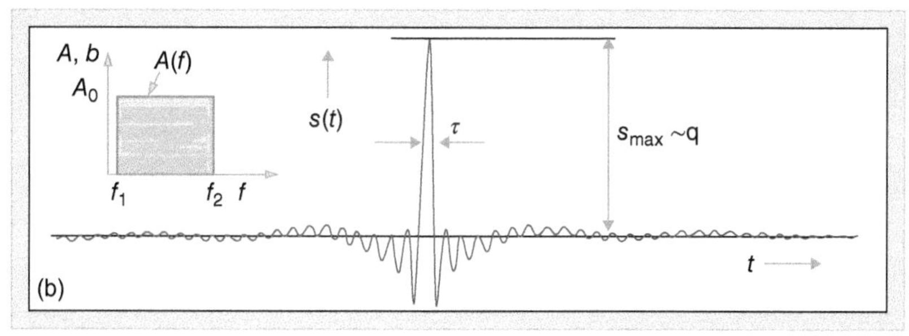

Figure 4.19 System response on a Dirac pulse at systems with different bandwidth as principles for integration of PD signals in the frequency domain (according to [16]). (a) System response for a narrow band system. (b) System response for a wide band system.

theoretically by the response on a Dirac-signal exciting the different systems (Figure 4.19). The system properties are characterized by amplitude spectra *A(f)* with corner values $f_{1,2}$ providing the system's bandwidth Δf, the response function $s(t)$ with the magnitude s_{max} and the average duration τ, [17]. According to system theory the response on the Dirac-signal is different dependent on the bandwidth as can be seen in Figure 4.19.

For transferring these results to the PD measuring, it must be assumed that the PD spectra in the given frequency limits $f_{1,2}$ is constant. Under these conditions, the magnitude of the response signal s_{max} is proportional to the PD charge q.

Thus, with the related standard IEC 60270, the frequency range of PD measurement systems is set for a so-called wide-band detection to a lower cutoff frequency f_1 of 30 kHz up to 100 kHz and an upper cut-off frequency f_2 to a maximum of 1 MHz with a bandwidth Δf of 900 kHz [13]. For the so-called narrow-band detection the limits are given to a small bandwidth Δf between 9 and 30 kHz with a recommended middle frequency of about 30 kHz up to 1 MHz.

In the following, both principles are discussed in more detail.

4.5.1 Narrow-Band Measurement

It should be noticed that the narrow band measurement principle and adequate devices were developed and applied in the early stage of partial discharge testing history [18, 19].

As just mentioned, the narrow-band detection and, therefore, the adequate measuring instruments may be characterized with a small bandwidth Δf between 9 and 30 kHz with a recommended middle frequency of about 30 kHz up to 1 MHz. The affecting on a PD pulse may be demonstrated by an injected (raw) charge pulse of 10 pC with filter setup of a middle frequency of 100 kHz and 1 MHz with a corresponding bandwidth of 9 and 30 kHz (Figure 4.20). As theoretically expected, the raw pulse is triggering an oscillation response with positive and negative peak values, where the polarity of the PD pulse cannot be examined. From Figure 4.20 it follows that in case of higher bandwidth of 30 kHz the oscillation is more damped, so the average duration time of the pulse is smaller than in case of 9 kHz.

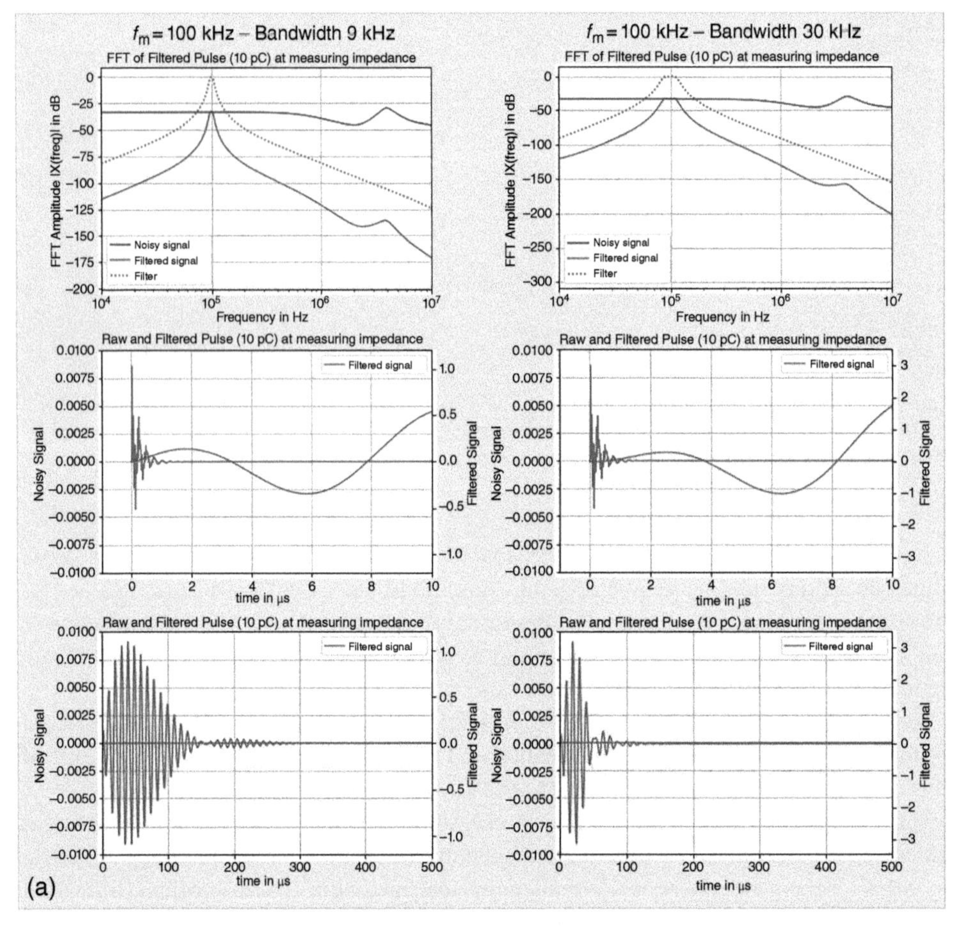

Figure 4.20 Pulse response at narrow-band detectors with different middle frequency and bandwidth (left column: bandwidth 9 kHz, right column: bandwidth 30 kHz). (a) Pulse response in frequency and time domain at middle frequency 100 kHz. (b) Pulse response in frequency and time domain at middle frequency 1000 kHz.

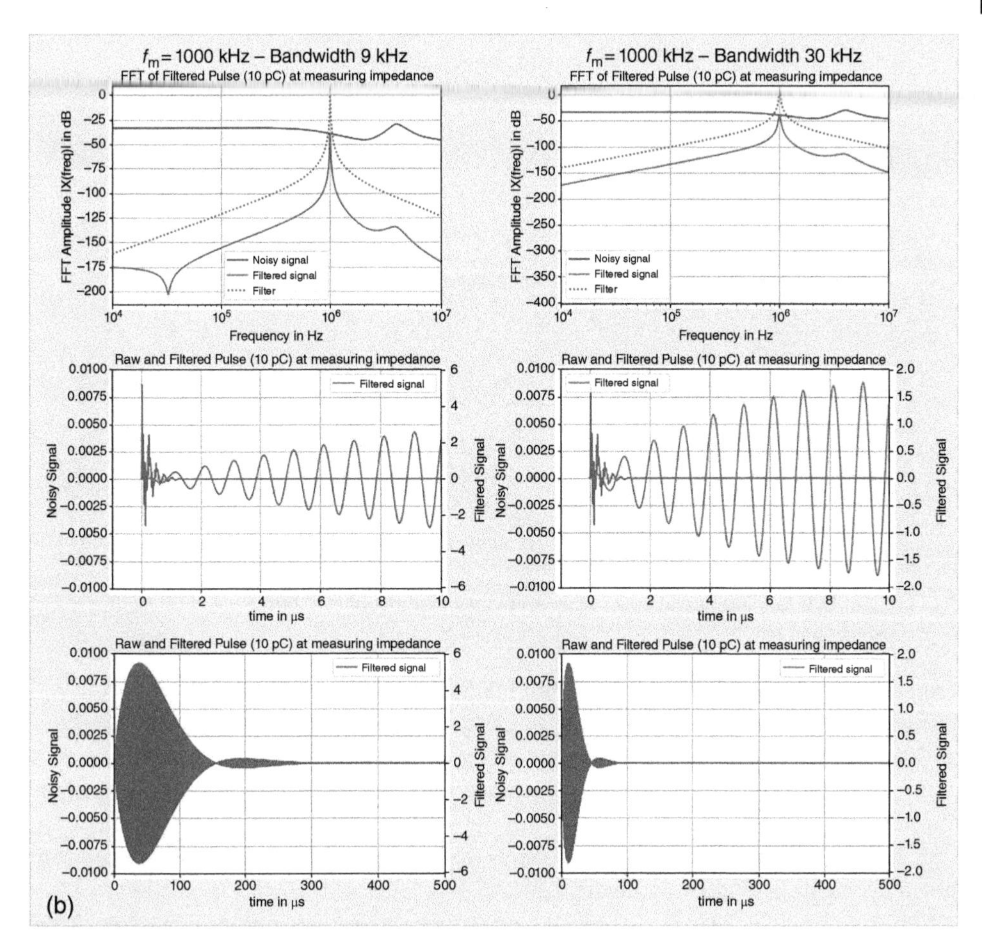

Figure 4.20 (Continued)

This also means that the pulse resolution is improved, which enables the processing of higher rate of detected PD pulses. The peak value of the response envelope represents the PD charge value. The upper limit of the middle frequency of 1000 kHz should only be used for small, noninductive devices under test and small test setups; otherwise, the frequency behavior of the PD pulse provides less signal content in the higher frequency ranges.

In addition to the middle frequencies and the bandwidth, it can be recognized that, depending on these two parameters at a bandwidth of 9 kHz, a damped oscillation with a length of up to 150 μs is triggered. However, if additional (successive) pulses provide an additional trigger to the system, a separation of pulses is impossible, as shown in Figure 4.21, due to the caused overlapping. Furthermore, no polarity can be derived from the filtered signal. As recognized in the amplitude spectra also the transmitted energy of the signal is due to the limited bandwidth quite low, which leads to small amplitudes.

Likewise, the overlapping response indicates that any separation of individual successive pulses seems to be impossible. However, also the amplitude of the filter signal is not reflecting the amplitude of one of the pulses as well as of the superimposed pulses. Therefore,

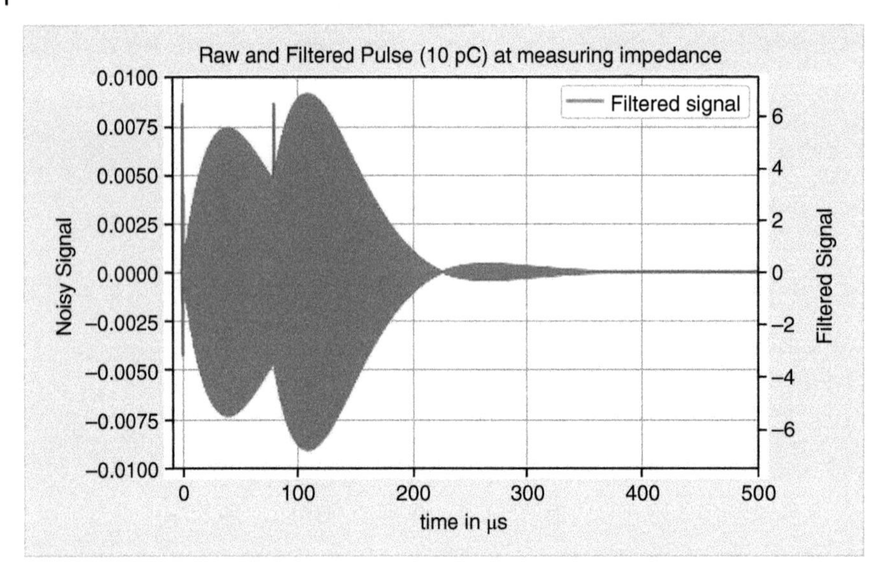

Figure 4.21 Overlapping of two successive PD pulses with the same charge value after 80 μs (f_m = 1000 kHz, bandwidth 9 kHz).

results of the narrow band measurement obtained by the described detection system are strongly dependent on the filter settings, but also on the intensity (pulse rate) of PD behavior. The qualified evaluation of PD processes by applied narrow band measurement systems needs a lot of experience about PD behavior of investigated objects as well as certain knowledge of appropriate filter settings.

An improvement over the previous setup is the application of heterodyne principle, as shown by the block diagram in Figure 4.22 [20].

With that principle the input PD signal is mixed/multiplied with a sinusoidal signal from an oscillator (analogous or digital), and afterward the intermediate signal is filtered according to the corresponding bandwidth. This filtering acts as low-pass filter performing the required quasi-integration, and the demodulator response is reasonably stable for indication of results.

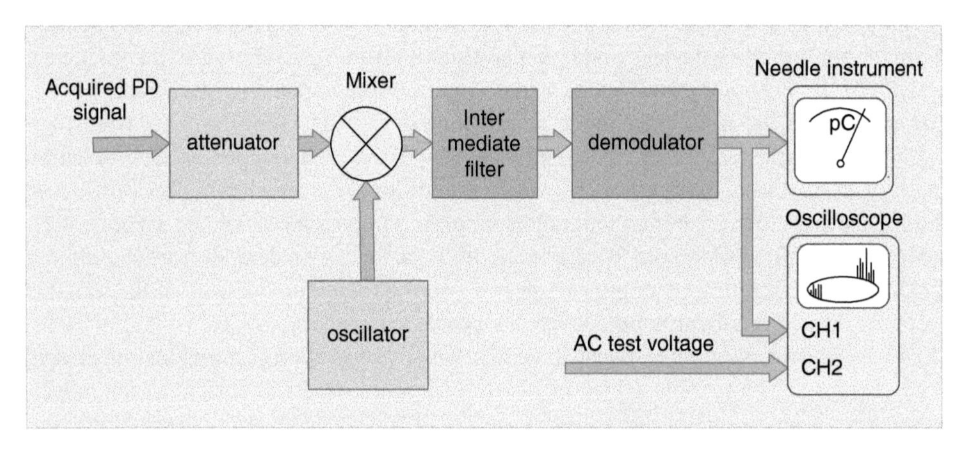

Figure 4.22 Narrow band measurement system based on heterodyne principle.

The "shifting" of the input signal with the mixing frequency is a well-known and comfortable principle in high-frequency technology, as shown in Figure 4.23.

The results of those procedures are shown in the frequency behavior, where after mixing and the possible signal level adaption by using an amplifier, the band-pass filtering performs low-pass filtering (Figure 4.24).

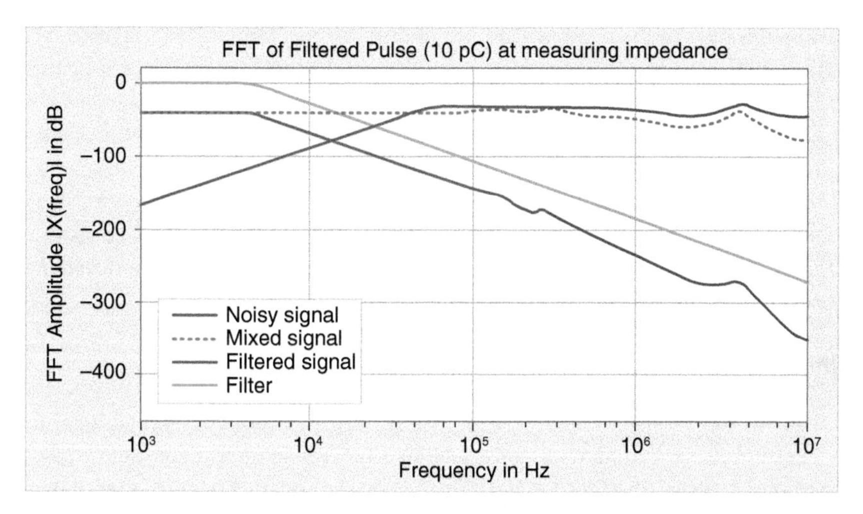

Figure 4.23 Frequency spectra of shifted and origin input signal after filter surpassing.

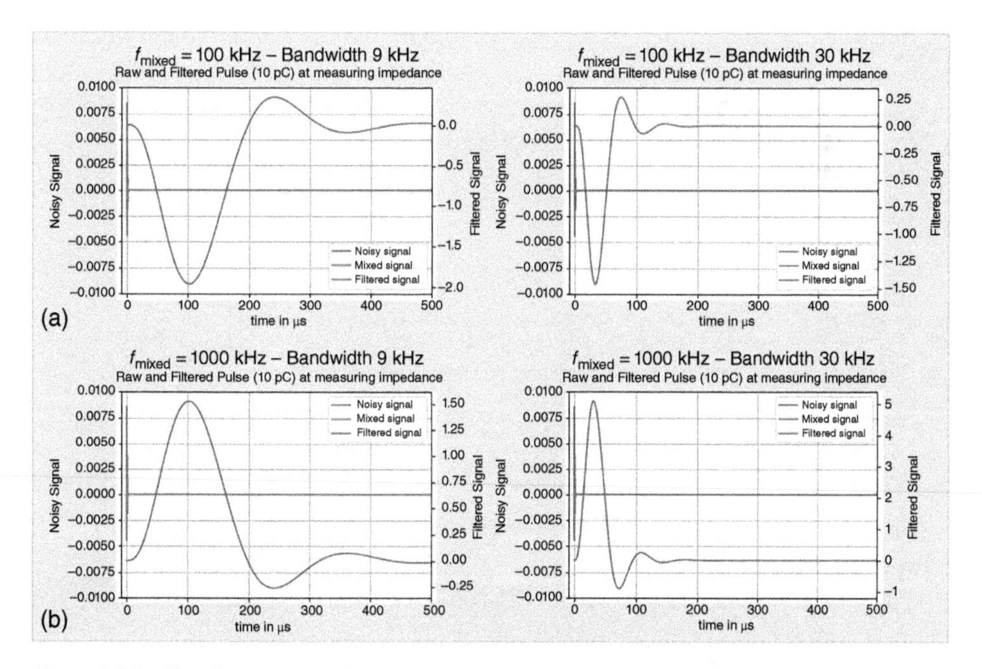

Figure 4.24 Time behavior of injected PD signal mixed at two different options (a) 100 kHz, (b) 1000 kHz – (heterodyne principle)) and after surpassing filters with different bandwidth (left column – bandwidth 9 kHz), right column – bandwidth 30 kHz).

Compared with the previously discussed bandwidth filter method, the heterodyne approach does not provide a significant oscillation in response to the input pulse (Figure 4.24). However, the quality and phase angle of the mixing signal has an influence on the output signal and polarity, but the design of low-pass filter is much easier for analog as well as for digital applications.

Concerning the case of two successive (overlapped) pulses, it can be stated that based on signal theory, the same information resp. pulse energy can be provided by this solution. Also, a separation of a second PD pulse within the filter responds, which was triggered by the first PD pulse, will lead to the same results as already discussed with the bandpass structure (see Figure 4.25).

Despite the discussed problems it should be noticed that one of the main advantages of the narrow-band PD measurement is the ability to evaluate PD signals in a very noisy environment. With the tight bandwidth and adapted filter design, any background noise can be damped in a very effective manner. Note that the discussed narrow-band PD measurement should be distinguished from the radio-interference-voltage measurement (RIV), which was also introduced in the early PD measurement history. This method will be discussed in section 4.5.4.

4.5.2 Wide-Band Measurement

The wide-band measurement is characterized by wider limits of frequency range for the entire measuring circuit compared with the previous one. Per standardized definition, these are as mentioned above 30 kHz $\leq f_1 \leq$ 100 kHz, $f_2 \leq$ 1 MHz and 100 kHz $\leq \Delta f \leq$ 900 kHz, where $f_{1,2}$ is the cutoff frequency and Δf the bandwidth of the applied band-pass filter [13]. Notably, due to this frequency settings, the partial discharge pulse also triggers an oscillation with positive and negative peak values, but due to the significantly higher bandwidth,

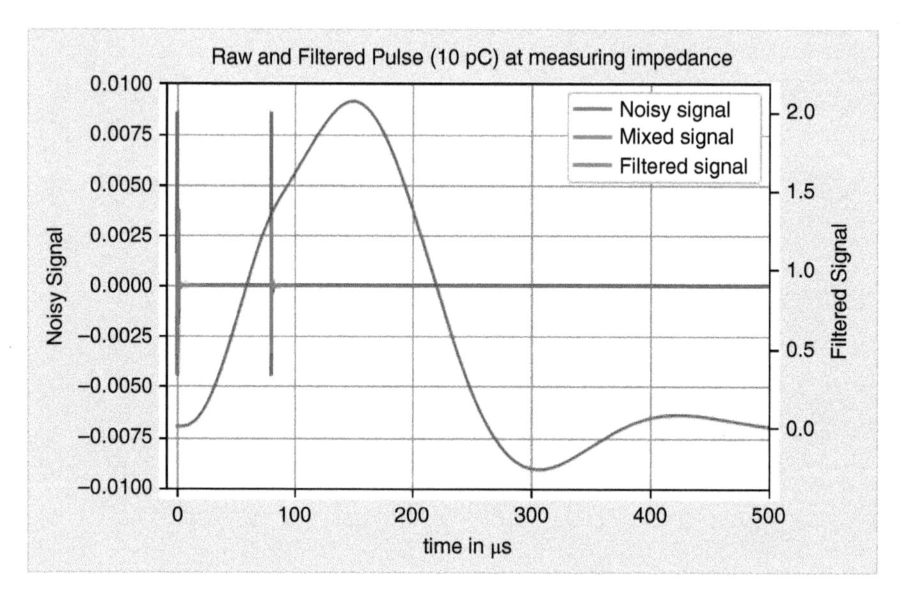

Figure 4.25 Overlapped PD pulse (same charge) after 80 µs (f_m = 1000 kHz, bandwidth 9 kHz).

the pulse polarity can be distinguished. Additionally, the oscillation is damped higher as compared with the narrow band method so that the pulse resolution time is typically only ≈ 5–10 μs. The peak value of the envelope of pulse represents the PD charge value. The limit of f_2 to 1 MHz should only be used for small, noninductive devices under test and test setups of small size; otherwise, the frequency behavior of the apparent PD pulse provides less signal content in the higher frequency ranges.

The principal behavior in frequency and time domain in response to an injected charge pulse under various filter conditions is demonstrated in Figure 4.26.

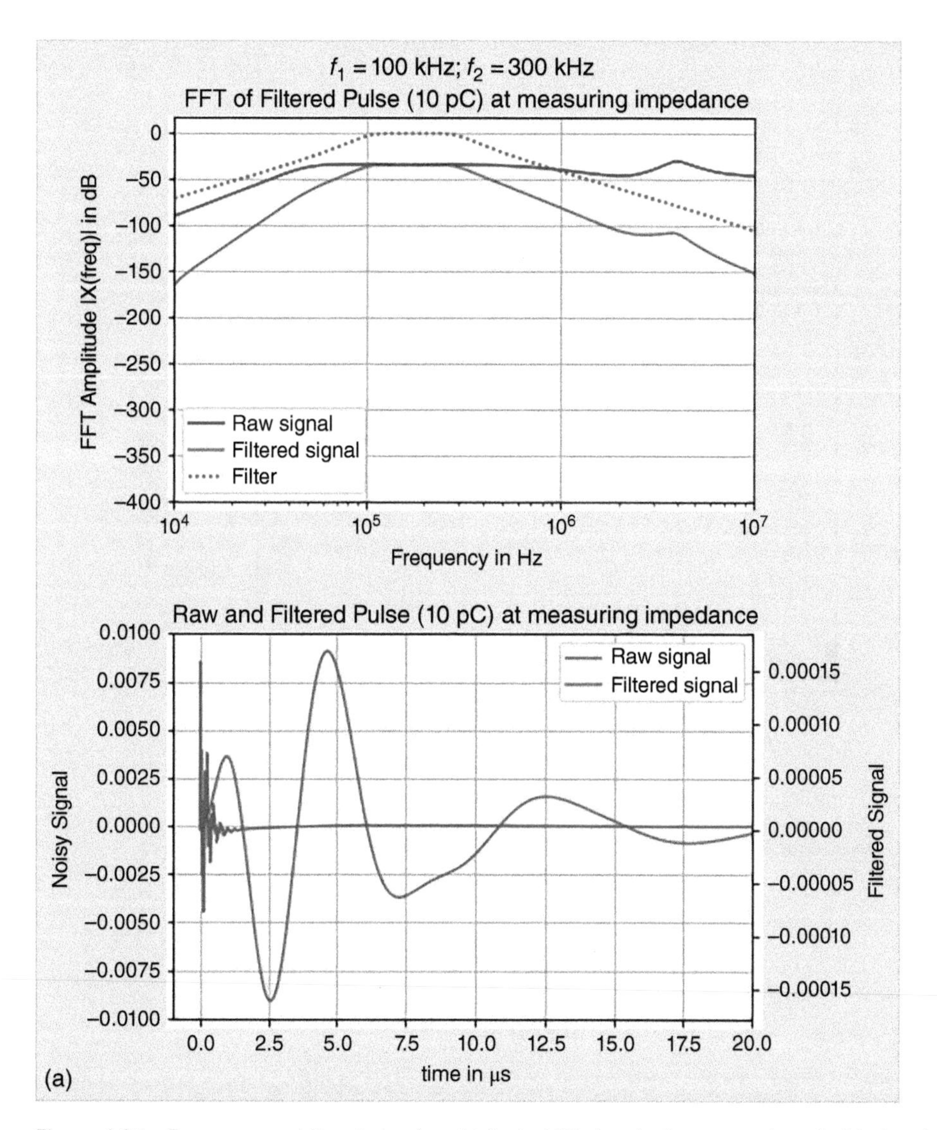

Figure 4.26 Frequency and time behavior of injected PD signal after surpassing of wide-band filter with different upper cutoff frequencies. Filters with (a) upper cutoff frequency 300 kHz; (b) upper cutoff frequency of 600 kHz; and (c) upper cutoff frequency 1000 kHz.

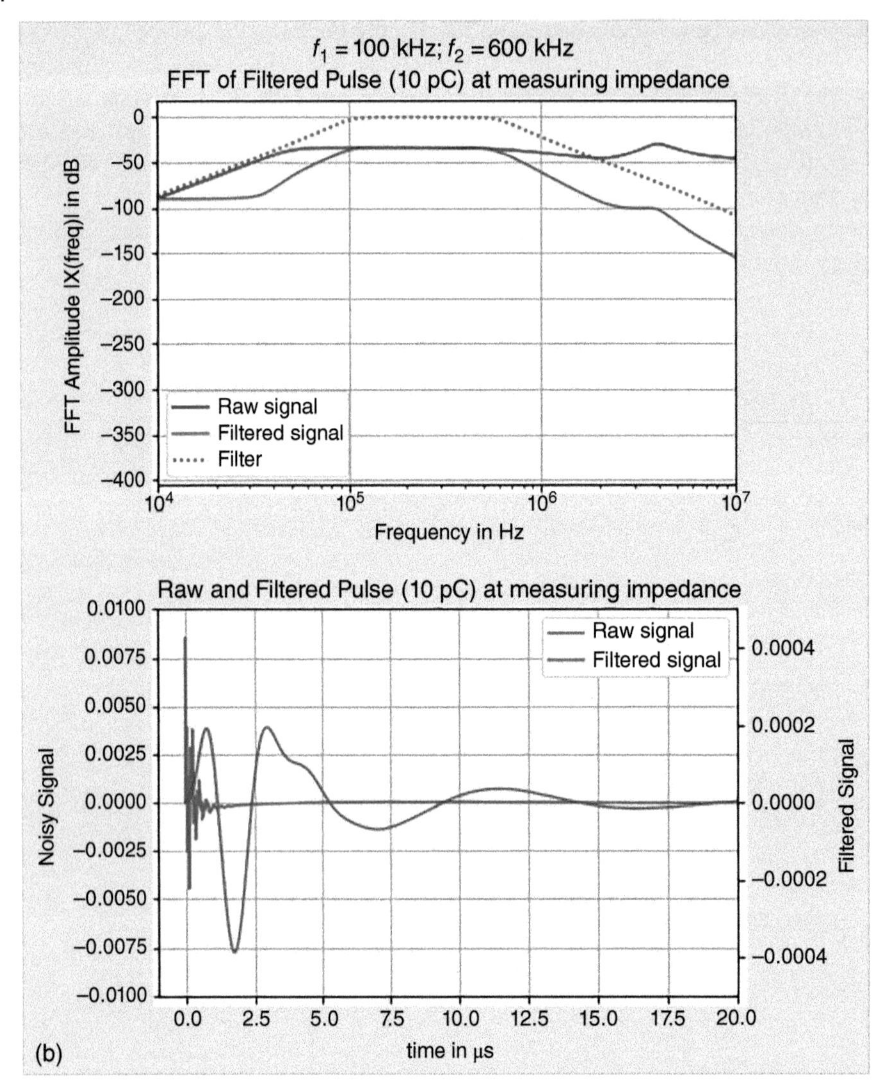

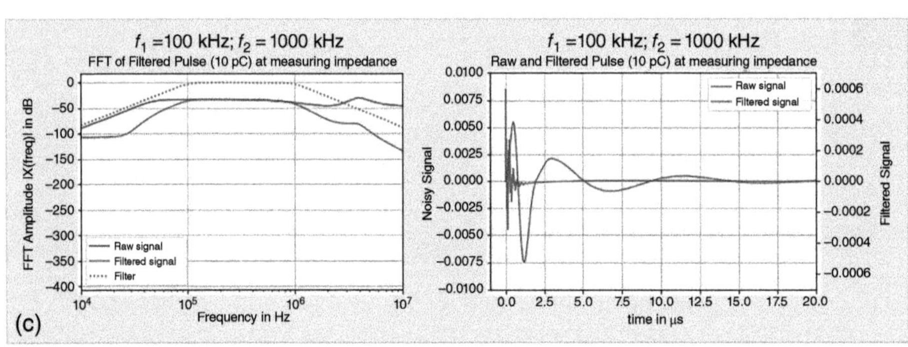

Figure 4.26 (Continued)

In case of successive pulses, no overlapping takes place and the separation of next pulse depends on the filter setting (Figure 4.27). An improvement can be reached at initializing the filter after certain microseconds of the peak value, so the pulse resolution can be increased about three times higher than compared with the nominal behavior of the filter.

4.5.3 Time Domain Integration

As already mentioned, the estimation of the PD charge may also be performed by direct integration of the complete signal. Hereby it must be assumed that it is possible to process the undistorted PD signal. Concerning the previous discussion about frequency limitations of measuring path is in case of direct time integration such a measuring bandwidth required, which enables the undistorted transfer of the PD signal to the processing unit. Compared with the limitations discussed in the previous subchapter the upper corner frequency should significantly exceed the 1 MHz value. Therefore, that principle is often termed as ultra or very high wide-band measurement.

To estimate the PD charge q the current signal i must be integrated according to Eq. (4.2).

$$q = \int_{t_0}^{t_1} i(t) \cdot dt \tag{4.2}$$

with t_0 and t_1 as time parameter of the pulse (starting, ending). For an idealized pulse, the response is unambiguous (Figure 4.28), but under practical conditions the PD pulse is very often superimposed by noise signals (Figure 4.29), which hamper the estimation of clear time values. The amplitude of the integrated signal is proportional to the charge value q.

In practical cases, very often a so-called measuring threshold level for detected signals is applied, which suppresses the noise level. At the same time, however, this method decreases the signal sensitivity of the measuring system. This obstacle can be avoided if the starting time of the current pulse may be determined by using the *energy criterion* [21]. This method provides the integration of the squared value of current pulse by reduction of a constant value (Eq. 4.3). The minimum of that value is identical to the starting point of the pulse (Figure 4.30).

$$y(t) = \int_{t_0}^{t} \left(i^2(t) - \varepsilon\right) \cdot dt \tag{4.3}$$

The energy criterion is quite efficient and also detects the starting point of a PD pulse in heavy noise environment. However, in the illustration, the integral of the pulse over the given time is not affected too much because the noise is white noise (Figure 4.31).

The technical realization of the principle of direct time integration was already made in the early 1970 years, in that time realized by an analog design solution (Figure 4.32), [22]. The PD signal, after amplifying will be integrated by an electronic integrator providing the time-dependent charge pulse signal. The magnitude of that signal evaluated by a corresponding unit is equivalent the pulse charge. Notably, this kind of very-wide-band measurement is also more influenced by disturbances or noise; therefore, special measures for noise suppression seem to be necessary using this technique. It should be also noticed that such PD instruments must have an appropriate control unit, which avoids the overload of

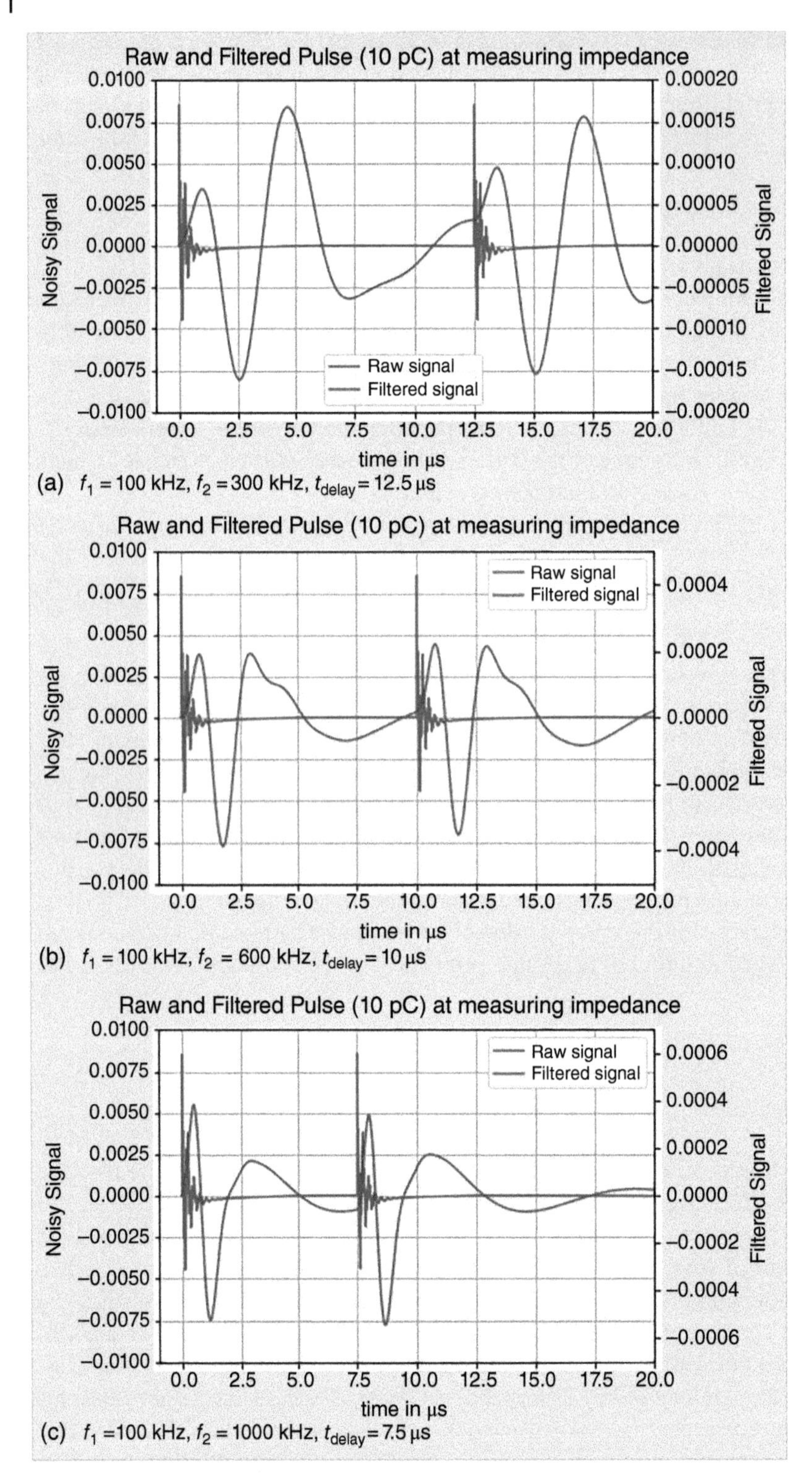

Figure 4.27 Response behavior at two successive injected PD signals at different upper cutoff frequencies f_2 and time slot t_{delay} between pulses: (a) f_2 = 300 kHz, t_{delay} =12.5 μs; (b) f_2 = 600 kHz, t_{delay} =10 μs; (c) f_2 = 1000 kHz, t_{delay} = 7.5 μs]

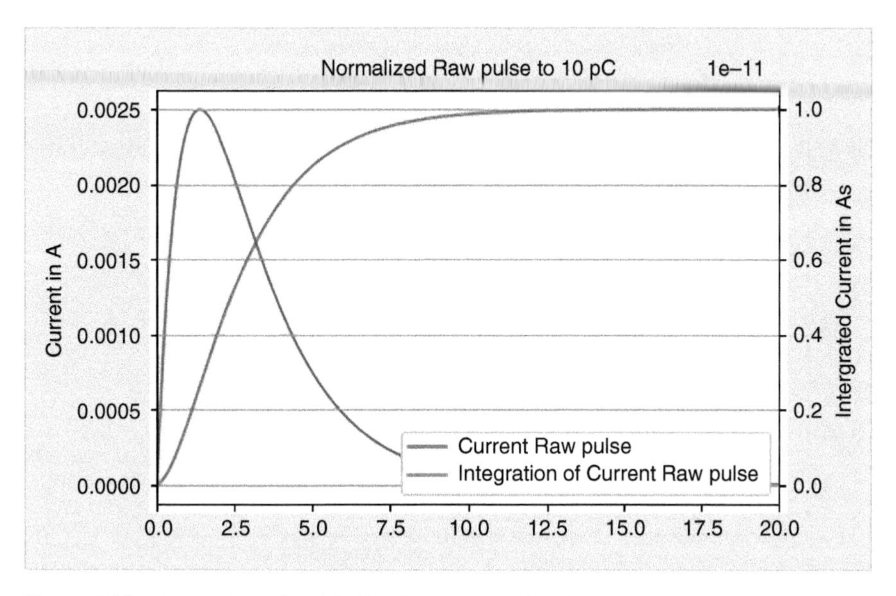

Figure 4.28 Integration of an idealized current (raw) pulse.

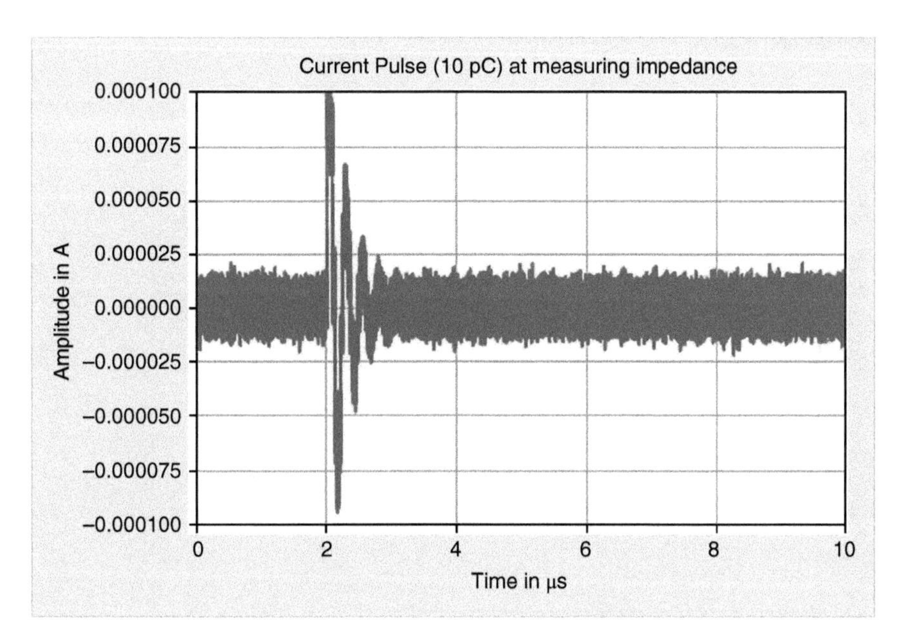

Figure 4.29 PD current pulse superimposed with noise signal.

the electronic integrator when more than one PD pulses are detected. After surpassing the evaluation unit, the elaborated PD signal with respect to its charge is ready for further treatment as by reading instruments or by visualization units.

As mentioned in Section 4.4 with progress in electronic technology, also digital PD measuring systems were available, following the measuring principles discussed. For example,

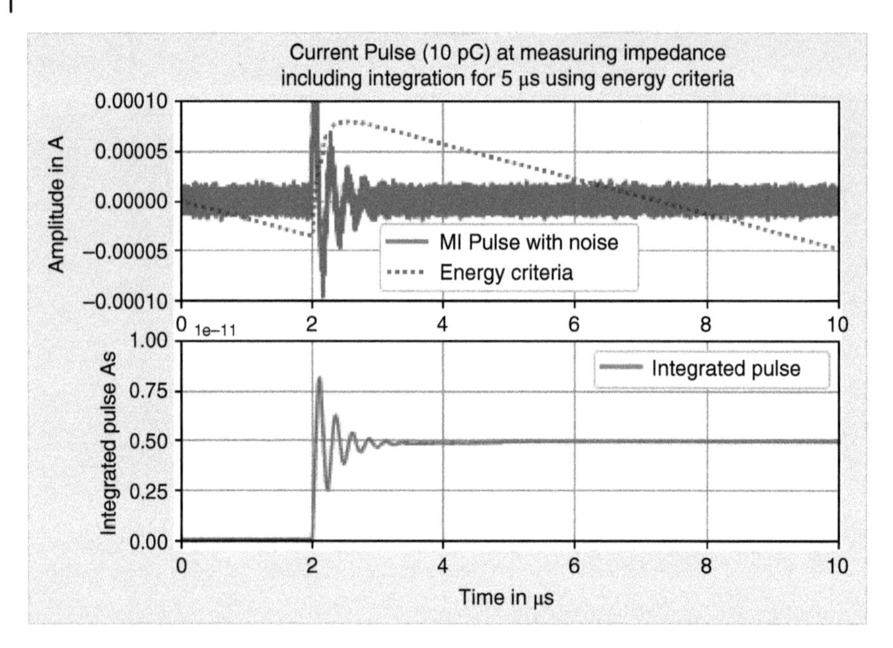

Figure 4.30 Superimposed noised PD signal and processed response signal by using energy criteria for pulse detection.

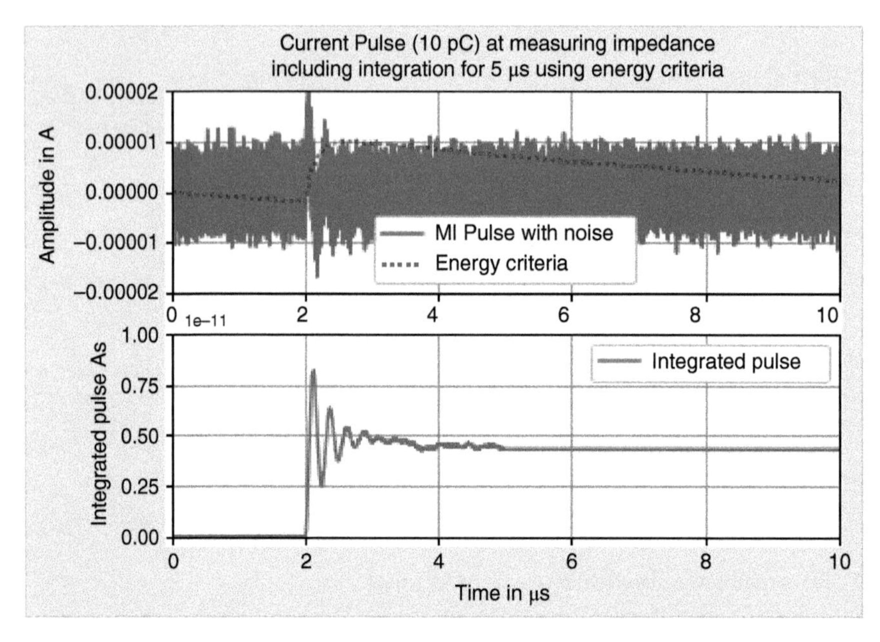

Figure 4.31 Heavy noised PD signal by using energy criteria for pulse detection.

Figure 4.33 shows the block diagram for digital design of a PD measuring system, where the PD signal is after the input unit digitized by an ADC and then processed for further evaluation. Using a computer-based acquisition and evaluation unit significantly enhances and improves the opportunities of evaluation and interpretation of measured PD behavior.

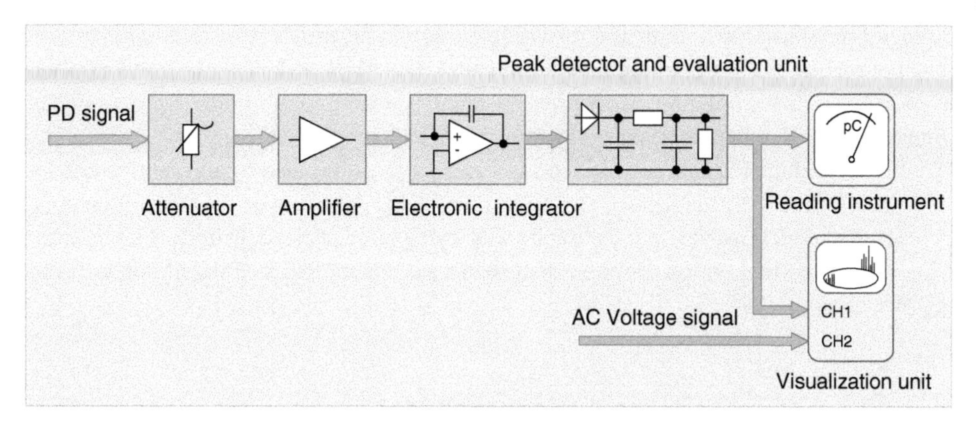

Figure 4.32 Block diagram of an analog PD instrument equipped with an electronic integrator according to IEC [13].

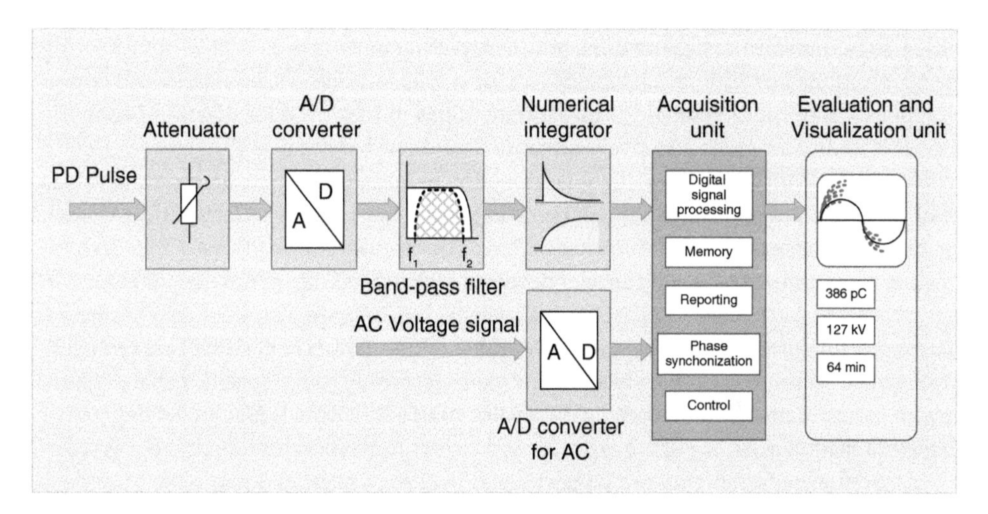

Figure 4.33 Block diagram of an digital PD measuring system according to IEC [13].

4.5.4 Radio Interference Voltage (RIV) Measurement

The radio interference voltage (RIV) measurement was developed to quantify disturbances (interferences) occurring in radio (AM) broadcast transmission. After exploring that these disturbances mainly were caused by (external) partial discharges on overhead lines, the applied measuring technique used for their detection was also used even for PD measurement on other electrical insulation. Based on standardized requirements, the measuring design is very similar to that for narrow-band measurement. Together with a coupling device with adequate high-pass characteristics, an RIV-meter consists of band-pass filter with bandwidth 9 kHz and a quasi-peak detector with corner frequency between 15 kHz and 30 MHz. As already discussed, under certain conditions the measured peak value (envelope) of RIV signal is proportional to the magnitude of charge, but the significant difference to PD measurement devices is the strong dependence on PD pulse rate with respect to the measurable time distance between successive PD pulses. If the pulse rate is low, the

RIV meter provides fluctuation, and the reading instrument provides an oscillation in the range between a minimum and a maximum value. The highest deflection of the reading meter during a 15-second interval of observation will be recorded. Additionally, an acoustic signal can be used, which is observed by the human ear.

Therefore, various PD sources with equivalent charge values lead to different results in RIV measurement. For this reason, an application for PD detecting or PD measurement is not recommended; however, that technique was widely used, especially in North America, and is a measurement option for pulse disturbances often applied in EMC immunity testing procedures [23, 24].

4.5.5 Synchronous Measurement for Multichannel Application

In some practical cases, it is necessary to acquire and process PD signals from more than one input unit of measuring system. This is applied for example at PD measurement in a three-phase system where single PD signals should be acquired synchronously in each phase like in transformers, generators, or other rotating machines as well as at measuring sensors positioned on different positions (e.g. in extended cable systems) [25]. The reason for application of multichannel PD measurement lies in improved localization possibility of defects and/or in more effectively separation of disturbances and PD signals in noisy environments. In all these cases, the synchronizing of measured signals in time and values obtained from different input-channels (sensors) is necessary. The accuracy in time of such synchronization is essentially for further processing and evaluation of the acquired signals, which is commonly provided by special developed vectorized algorithms implemented in appropriate computer-based PD measuring systems [26, 27] (see also Chapter 7). Here we assume that for defect localization in a power cable, some sensors (e.g. HFCT) are arranged. If PD signals occur, they propagate along the cable according to the specific cable parameter with a certain propagation speed v_p. If a value of $v_p \sim 20\,\text{cm/ns}$ is assumed along with a defect localization accuracy of ~ 1 m, the error in synchronization time should be less than ~ 2.5 ns. Similar requirements can be postulated in case of parallel 3 phase measurement with required phase-angle. Therefore, for multichannel synchronization of acquired PD signals, the signal transfer from input units to the measuring system should be provided by special connection links that enable a minimized error in acquisition of temporal parallel PD signals (Figure 4.34).

The following connection methods for synchronous PD measurement are applied:

- Synchronization with fiber optic link and appropriate optical converter
- Synchronization using real time ethernet link via internet/intranet
- Synchronization using global positioning satellite (GPS) signal for input units that are not connected directly to each other

Such PD measurement systems more and more are used in condition monitoring applications for important or system-relevant assets in power energy networks.

Another application for synchronous multichannel system is used for PD measurement at three-phase apparatus. In such cases, measurement issues often arise due to false identification or distinguishing of PD pulses to disturbances caused by EM coupling

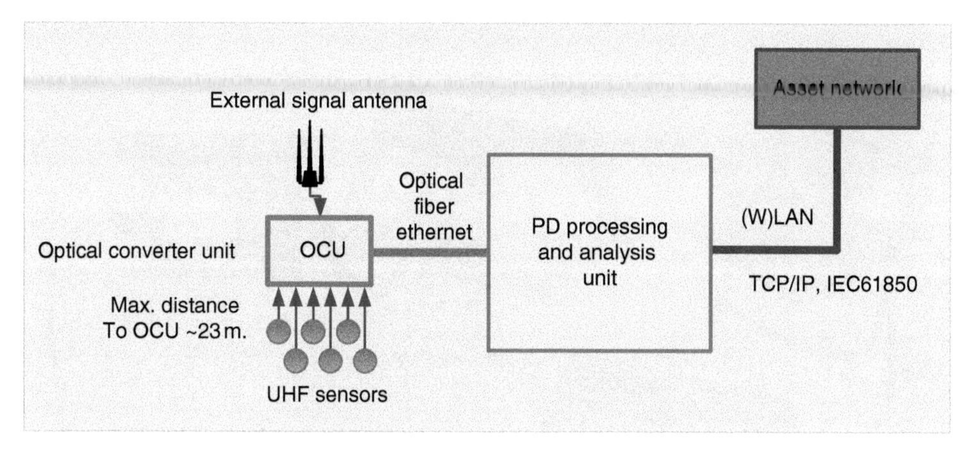

Figure 4.34 Multichannel synchronous PD measurement system (adopted from PDM system [QUALITROL] for transformer monitoring diagnosis) [27] (schematically).

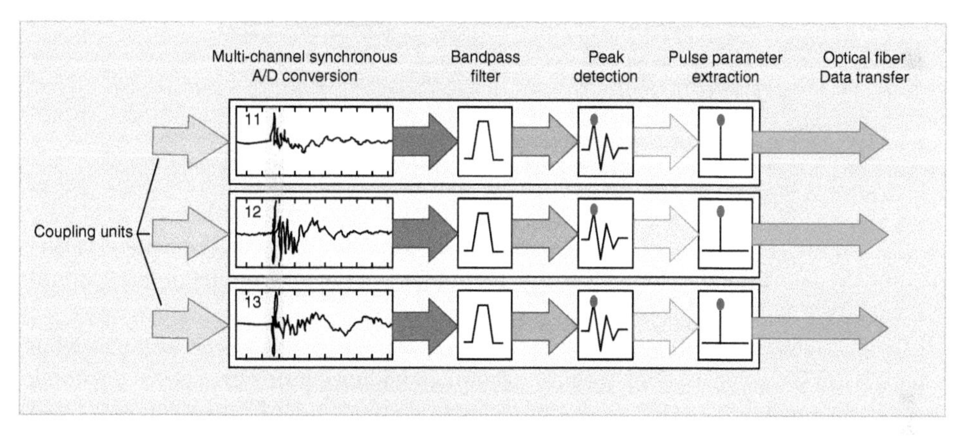

Figure 4.35 Principle of signal elaboration at multichannel synchronous PD measurement system (according to [26], schematically).

between the phases (crosstalk effects). The developed PD system enables the synchronous processing of noisy PD signals acquired in separate coupling units and their evaluation, respectively, visualization by appropriate software [27]. To illustrate that, Figure 4.35 demonstrates the schematic elaboration of PD signals through the internal "measurement path" from signal analog-digital conversion up to pulse parameter identification and possible further data-transfer by optical fibers (e.g. for condition monitoring). Based on the assumption that each pulse source has its own transfer properties while traveling to coupling unit, a separation of different PD pulses as well as interference disturbances might be possible when the synchronization of different input units is provided with a proper value of 1–2 μs, at least.

Another example for ultra-fast synchronous measurement is applied for test of related synchronization software (Figure 4.36). In that test setup a mean clock synchronization error of 1.59 ns at 500 samples is provided, what means a (theoretical) number up to

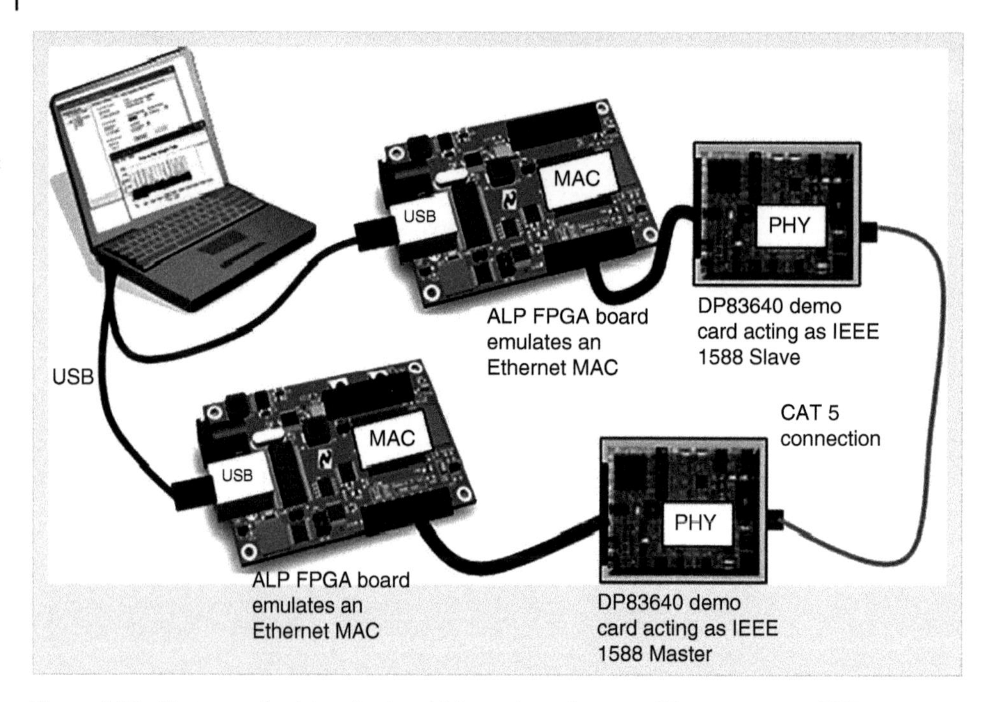

Figure 4.36 Test setup for (ultra-fast) multichannel synchronous PD measurement [28].

500 input-units may be processed. By using more samples, the synchronization can be enhanced and for the data acquisition only the stability of the local oscillator is important. A resynchronization can be done after a certain time span.

If the measurement units are not directly connected, GPS or global navigation satellite systems (GNSS) may be used for synchronization, providing a time signal based on atomic clocks in the satellite. The GPS system provides an accuracy limit of 30 ns compared with the Galileo GNSS system provided in spring 2022 an UTC Time Dissemination Service Accuracy of 5.1 ns [29]. However, by using the difference time evaluation of two or more GNSS satellites a better accuracy is achievable.

4.6 Noise Suppression and Reduction

4.6.1 Introduction

Based on the measurement principles mainly the quasi-integration within the frequency band of the PD current signal is used. In practical measurement cases also several sources of noise occur having a similar frequency behavior, which might lead to an overlapping and influencing of measurement result. Additionally, the PD signals for several application are quite small in comparison to the noise signals, which lead to a full overlap and misleading result interpretation afterwards. For this reason, in the following based of an overview of selected noise sources several de-noising methods are discussed.

4.6.2 Noise Sources

The disturbances and EM noise may be caused by several sources (Figure 4.37). The "coupling" to or the "entering" into the PD test circuit might be different dependent on the frequency characteristic of the source. According to that, the disturbances can also be distinguished by their propagation way – namely, *conducted* and/or *radiated coupled* disturbances, which require adequate noise suppression methods.

4.6.2.1 Main Sources of Conducted Coupled Noise

Typical *conducted coupled* noise sources are caused by high-voltage supply unit mainly consisting of *power supply,* variable transformers (variac), and the high-voltage transformer itself:

- *Power supply.* The power supply feeding the high voltage test circuit may deliver noise in the frequency range of partial discharges via the conducting leads. This can be happened by mains plug power network or other electronic devices or motors that are using the same network areas.
- *Variable transformers (Variac).* A variable transformer (variac) for setting the primary voltage for the high voltage transformer uses usually moveable taps which are sliding over the coil. During the moving of taps small electrical sparks occur which leads to disturbances of the PD measurement. Especially for proper estimation of partial discharge inception and extinction voltage (PDIV, PDEV) by increasing or decreasing of test voltage these disturbances should be avoided. Also, EM disturbances arise when using modern electronic units like PWM converters and should be suppressed (see also Chapter 8).

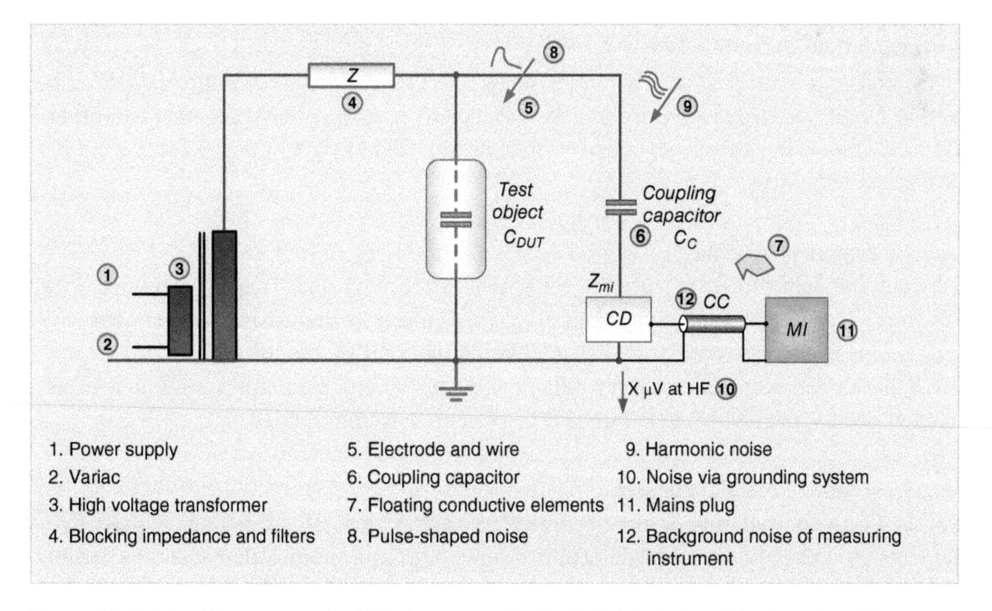

Figure 4.37 Possible sources for disturbances and noise in PD test circuit (schematically).

- *High-voltage transformer.* The high-voltage transformer provides the test voltage for PD measurement and might have different constructive design. The noise might be arisen dependent on design type, so like classical oil-paper insulated transformer with metallic steel tank resp. insulating tubes or dry type with resin-cast insulation. Due to insulation or similar defects inside or outside of the transformer also partial discharges might be occur, which must be suppressed because they are not caused by the test object and lead to misinterpretation of PD results.

In case of using a high-voltage-resonant test set the applied voltage transformer provides a comparable lower high voltage than the nominal value of the set, because the set-high-voltage is generated by serial resonance of inductance with the capacitance of the test object. However, the partial discharge issues are also related to both main parts of the set – the voltage transformer as well as the applied inductance.

4.6.2.2 Blocking Impedance and Filters

The high-voltage blocking impedance (mainly an inductance) is used to suppress (filter) the conductive coupled noise from the high-voltage supply. In principle, it is commonly designed as low-pass filter with a suitable damping behavior for the expected noise frequencies. The easiest way of damping the noise is using a blocking impedance, which provides the filter behavior in conjunction with the used test capacitances and stray capacitances. The blocking impedance (inductance) and/or the filter (combination of inductance, resistor, and capacitors) must be partial discharge free itself.

4.6.2.3 Electrodes and Wire

Depending on the value of test voltage, a PD test circuit also consists of various connection leads and related connection points. These must not generate additional partial discharges (e.g. corona discharges). Therefore, these elements should have a higher PD inception voltage (PDIV) than the applied PD test voltage, which required their proper electrical design (e.g. sufficient wire resp. electrode radius, surface quality and so on). Additionally, the connection quality of leads expressed by low connection resistance must be provided; otherwise, small sparking effects (so-called contact noise) might be generated, affecting the PD measurement results.

4.6.2.4 Coupling Capacitor

The coupling capacitor necessary for capacitive decoupling and basically designed with polymeric or oil-paper insulation material may also generate disturbances by an unproper design. This is often the result of voids or delamination within the insulation.

In light of the discussed "noise capability" of the test circuit elements, it should be noted that a pretesting of "disturbance-free" PD test circuit is recommended.

4.6.2.5 Floating Potential Elements

If in the immediate vicinity or directly around the area of test circuit metallic (conductive) elements are located, it must be determined how to ground them. Otherwise, at a certain value of high-voltage some disturbances may be generated, mainly by capacitive coupling to the high-voltage parts (electrodes). Especially in case of AC testing, these discharges occur in both half cycles of test voltage, which affect the PD measurement in a serious way. A similar situation may arise at DC testing when insulated parts out of the test circuit are charged by electric field or moving particles. Also, here a preliminary testing of test circuit is recommended.

4.6.2.6 Pulse-Shaped and Harmonic Noise

The source of pulse-shaped noise can be widely spread (e.g. fluorescent tube lamps, gas discharge lamps, inverter for motors, electronic devices, moving forklifts, electrical sparking of tram or trains from the neighborhood, brushfire in motors, and so on). The source can be closed to the test circuit or even outside of the test room. That resulting noise (mainly from its very high frequency parts) may be *radiative coupled* by EM waves through ambient air or the connections of the high-voltage circuit, the measurement devices, or even the grounding wires. In that case, these elements act like an antenna for the disturbances.

In the same manner, harmonic noise caused by radio signals of communication devices might affect the PD measurement. Disturbances with frequency in kHz range by broadcast transmission have been reduced over the last decades, but at the same time there has been a significant increase in mobile and satellite communication, and the related noise signals in GHz range especially affect PD measurement for UHF measurement applications. The suppression of those disturbances needs special filtering and shielding measures.

4.6.2.7 Noise via Grounding System or Wire Loops

All the described noise sources can also couple into the grounding system if the grounding system is provided as mesh. Therefore, it is necessary to use a star-formed grounding system connecting separate groundings at a common point, including other grounded devices from the safety, control or measuring units, or else the grounding systems must have a high frequency decoupling. Moreover, the ground connection leads should also be of low inductance, which is best accomplished using Cu or Al foil.

If wire loops are placed incorrectly, those loops could act as inductive coupled noise by interfering voltages induced into the loop area. This can be minimized by reducing loop cross over as much as possible.

4.6.2.8 Mains Plug and Background Noise of the Measurement Instrument

If the measurement instrument is connected directly to the mains plug and the power unit does not provide enough filter capacity, noise from the mains plug can be transferred to the measurement device. However, battery-powered systems or suitably designed measuring instruments will provide a solution for it.

The background noise may be caused by physical properties of the device components or surrounding conditions, such as thermal noise or white noise. That could affect the PD measurement, especially at very low partial discharge levels and/or high preamplification of PD signals. That could be avoided only by adequate selection of responsible components and must be considered already at design of measuring device.

4.6.3 Denoising Methods

After describing various noise sources, which affect partial discharge measurement, possible methods for effective denoising will be discussed.

4.6.3.1 Shielding

To reduce the impact of *radiated* coupled disturbances shielded test rooms or entire shielded test laboratories are commonly used. Note that a proper design of shielding mostly together with the grounding system is one of the main challenges when related laboratories are planned and erected [30]. Without going into detail, the shielding efficiency and thus the reduction efficiency is in first order dependent on the frequency spectra of disturbances to be suppressed. For performing standardized PD measurement, a shielding effect for frequencies up to 1 MHz seems to be sufficient, what means a damping (shield) factor ~100 dB up to ~5 MHz [31]. With the method of shielding, it is possible to reduce the noise from any radiated sources. Additionally, it is important that all the connection lines (feeder, measuring cables that go into or out of the shielded room) must be included in the shielding concept (e.g. provided with filtering equipment at the shielded laboratory walls; otherwise the noise will be transmitted into the shielded room).

4.6.3.2 Filters

One of the common methods to reduce *conducted coupled* noise is to use analog or digital filters located in the measuring signal path along that the noisy PD signal will be transmitted. The filters are mainly designed as low-pass, high-pass, band-pass, or notch filters or a combination of these. Analog filters are set up as classical LC-filters with a damping characteristic of up to 30–60 dB if the design and the screening is proper performed [32]. However, also for digital devices analog filters have a positive impact, because with analog filters the signal energy of the noise can be filtered before the presignal adaption and the A/D conversion (see also Section 4.4). If the noise amplitude is greater or in the same range as the partial discharge signal, the analog filter will provide a significant effect for increasing the signal-to-noise ratio (SNR). The digital filters are used after the A/D conversion implemented in appropriate FPGA or DSP-arrays (on-line) or offline at a PC. At PD measurement systems, various types of digital filters are applied – infinite impulse response (IIR) or finite impulse response (FIR) filters with different properties for signal acquisition. So IIR-digital filters uses the input signal and previous samples of the output signal with some stability issues for single pulse events, while FIR-digital filters are more stable for single pulse events but need for implementation in FPGA or DSP arrays much more resources [33]. With digital filters, tunable bandpass-filters might be effectively realized, enabling the PD measurement system to be adapted to the noisy environment (e.g. to select a suitable frequency range for measurement with low noise content and high SNR-value) (Figure 4.38).

For PD wide-band measurement, the band-pass filter might be combined with an adapted notch filter to optimize the measurable frequency range with high SNR-value (Figure 4.39). The notch filter is used for blocking (damping) of high noise components within the transmitted signal, which is already filtered by band pass (measuring impedance).

4.6.3.3 Balanced Bridge Measurements

As mentioned in Section 4.2 external EM noises may be eliminated or reduced by using of so-called balanced PD bridge method (Figure 4.40). The tunable measuring impedances Z_{m1}, Z_{m2} are located in both branches (test object, reference object). Tuning the measuring impedances to balance the bridge, common mode disturbances appearing at the

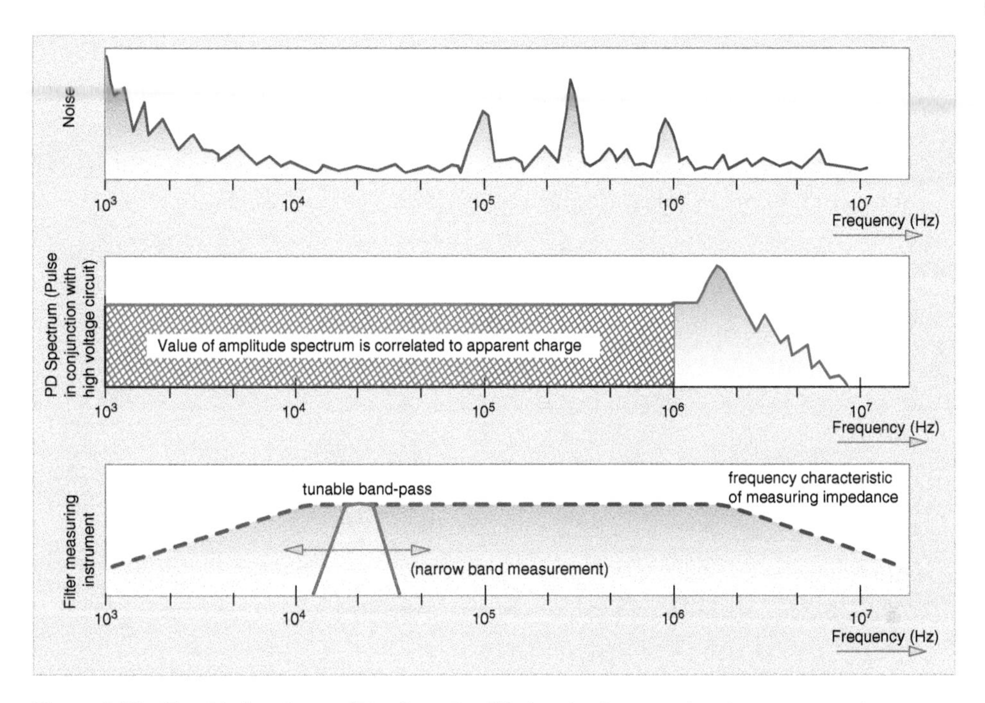

Figure 4.38 Tunable band-pass filter for noisy PD signals at narrow band measurement.

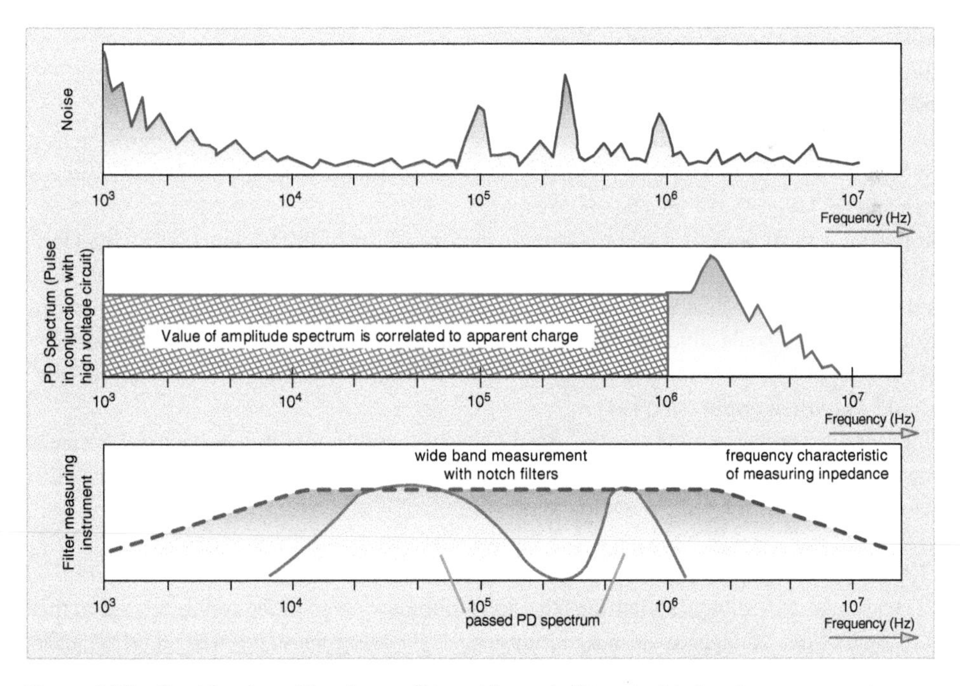

Figure 4.39 Combination of band-pass filter with notch filter at wide band measurement.

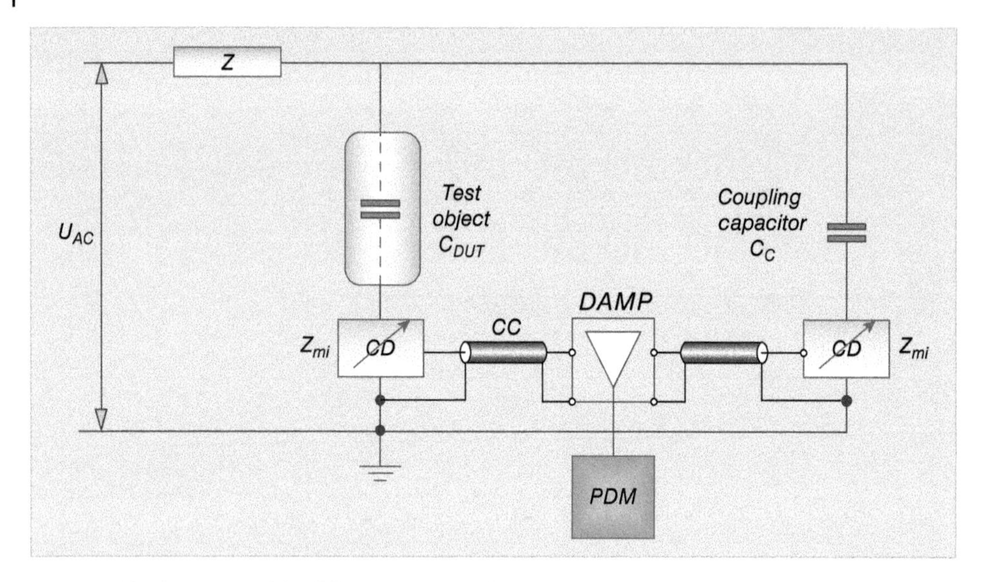

Figure 4.40 Balanced bridge PD measurement for noise suppression.

high-voltage side of both branches might be suppressed by the differential amplifier DAMP. In that case, only the PD signal originating in the test object appears at the output of the differential amplifier and might be measured by the PD measuring system (PDM). To obtain a high sensitivity by high common mode rejection of DAMP, the bridge branches should be designed as symmetrical as possible.

Therefore, as reference object a complementary PD-free test object is recommended. Despite the benefits of the balanced bridge for noise suppression, this approach is not generally employed in practice due to a certain design challenge because both bridge-branches must have an equivalent frequency response over the full bandwidth used for the PD signal processing [31].

Moreover, it must be considered that the travel time of the high-frequency noisy interfering signal along the complementary PD-free test object branch is equal to that noisy PD signal traveling along the test object branch. It was investigated that a small deviation in ns-range of both signals (like jitter effect in electronics) might lead to contra-acting results with even higher noise than before (e.g. by superposition of noise signals); therefore, this method has limited application [34].

The noise suppression methods discussed above are commonly independent of applied PD test voltage. The following sections describe suppressing methods introduced and developed especially for AC-PD testing applications.

4.6.3.4 Windowing

At PD testing at AC-voltage, sometimes the noise signals are dependent on or related to the phase angle of test voltage, as shown in Figure 4.41. By using the windowing method, that noise might be suppressed by a "measuring window," which should be positioned in amplitude and phase at the same time-slot at which the noise signals occur. The positioning of that window should be performed by detecting of noise signals using the PRPD pattern or

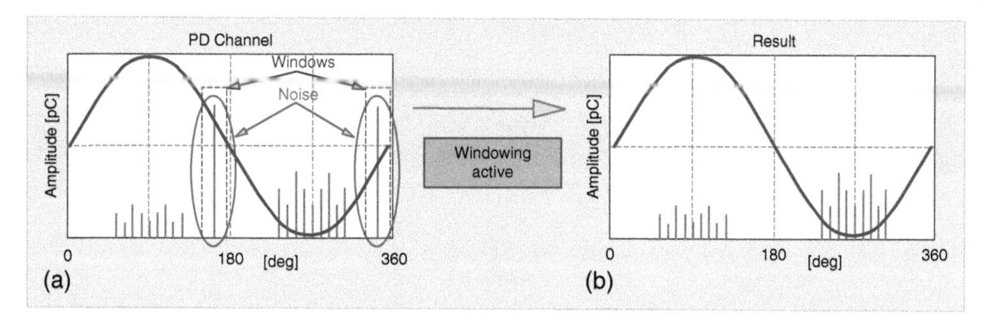

Figure 4.41 PD pulse diagram with (a) noise and related window positions and (b) PD pulse diagram with activated windows.

in the pulse view. By activating the window as a "blocking" time-slot in which no measuring is possible, the disturbances are not considered for further processing or evaluating the PD results. However, if noise signals and PD pulses occur in the same time, the latter one could not be measured and, therefore, the whole PD behavior cannot be evaluated. Therefore, in the related standard the application of such deactivated measurement time-slots is limited [13].

Note that moving windows are also applicable, which recognize that the noise is moving over the phase and so the activated window is moving with the noise over the phase, too.

4.6.3.5 Gating

For stochastic noise sources, which are not strongly related to an AC phase angle, the gating method is a well-proofed noise suppression possibility [31]. With this method, the noise signal is measured with an additional channel (e.g. gating channel or full measuring channel dependent on the applied system performance and acting like a trigger (gating) signal). If that gating signal exceeds a certain threshold, the "gating logic" will be activated and the PD processing unit does not recognize the measured signal as PD (Figure 4.42). The couplers for the gating channel can be provided by capacitive, inductive sensors, or UHF antennas. In case of known noise sources (e.g. electronic converters or similar), the gating signal might also be derived directly from the "noise-producing" device.

It should be noticed that in some cases, even the *inverse-gating* method is applied. In that method the gating channel receives usable PD signals, such as by using UHF methods [36].

In such cases, the processing PD channel recognizes these signals not as noise but as recordable signals, and the PD pulses will be measured and evaluated. This can be used for PD measurement in highly noise environment by applied UHF coupling and requires a related logic software unit in the PD system for recognizing PD events of concern.

4.6.3.6 Clustering

Due to the recent advancements in digital signal processing and fast electronic components, very promising denoising software tools have been developed, mostly based on a cluster separation approach implemented in appropriate PD measuring systems. These

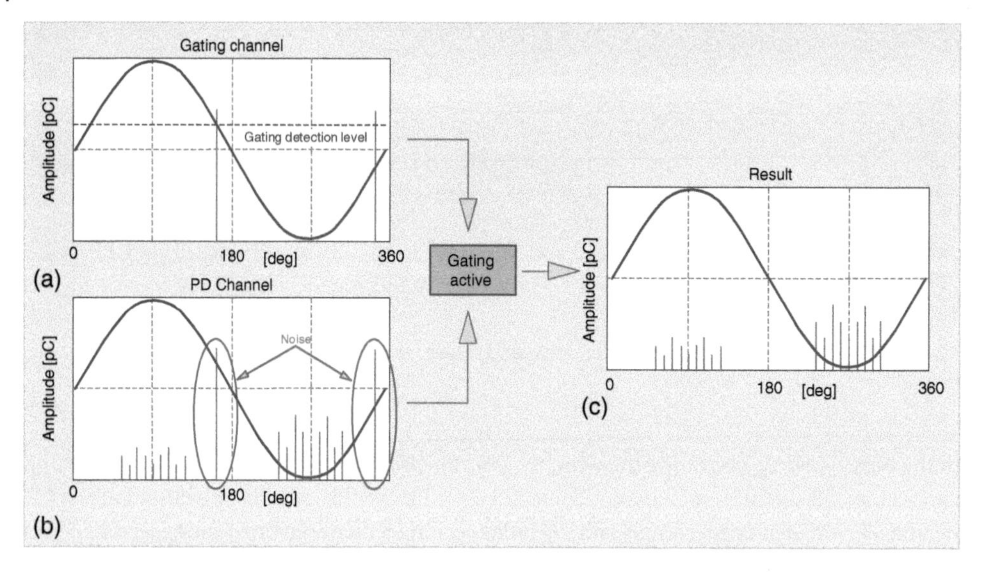

Figure 4.42 PD pulse diagram with (a) gating signals, (b) noise, and (c) PD pulse without noise at active gating.

clustering techniques mainly based on waveform of the acquired noisy PD signals and on the assumption that each PD source as well as noisy signal have a characteristic signature in time and/or frequency domain. With appropriate mathematical procedures or special algorithms equal or similar pulse signatures are collected to a data cluster and presented mostly in visible or image form such as by star diagrams or classical phase-resolved PD pattern (PRPD), see also next section. The visibility of those clusters enables the recognizing of PD pulses and noise signals as well as in some cases also the identification of different PD sources.

As an example, such visualization in kind of star diagram using synchronous multichannel PD measurements at PD testing of a power transformer enables the unambiguous recognition of external noise signals (Figure 4.43), [35]. It is seen that the detected noise signals caused by external corona pulses with similar signal intensity occurring in each of three phases are recognized by blue data cluster, while the colored cluster in phase L_1 indicates a single PD event. That might be used even for further processing of the identified PD pulse cluster by various post-processing methods e.g. PRPD pattern or statistical analyzing procedures.

That processing and evaluation method is usable mainly for magnitude of PD signals leading to a so-called three-phase-amplitude-relation-diagram (3PARD), but also for the travel time of signals from PD origin to terminals of the investigated test object with respect to the connected coupling units. In that case, a so-called three-phase-time-relation-diagram (3PTRD) might be established and used for noise separation. For more details in obtaining and treatment of such diagrams related to PD measurement on three-phase apparatus, see [26]. Note that accuracy of such applications is dependent on the number of acquired data points (cluster) – the more PD pulses are detected, the better might the source recognition be with regard to their separation from noise signals.

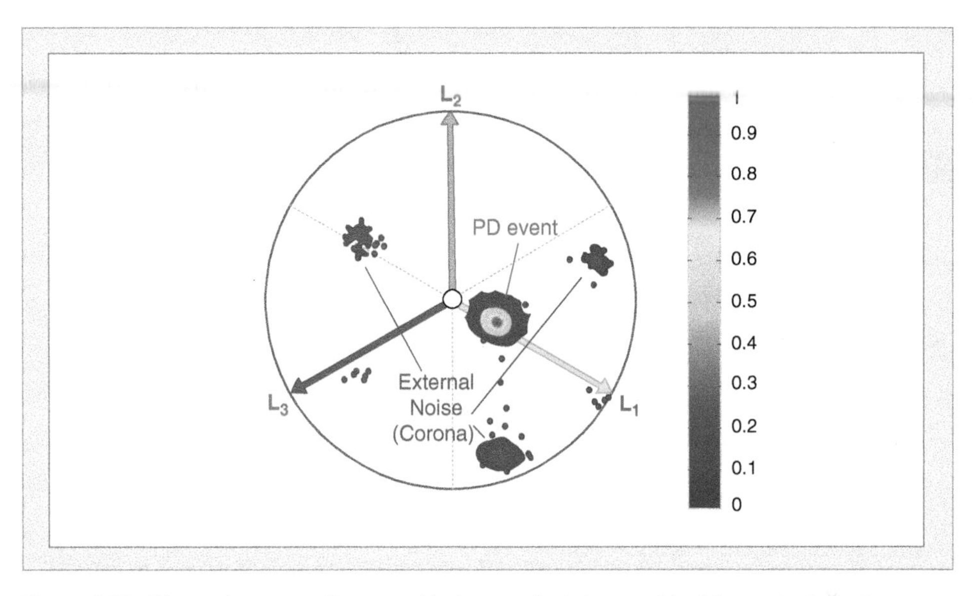

Figure 4.43 Three-phase star diagram with three typical clusters (blue) for each phase due to external noises as well as a single cluster (colored), indicating that a PD defect is located in phase L_1 of the investigated power transformer (according to [35]).

Based on the experience with the 3PARD-method, the multispectral PD measurement was introduced [37]. At this method, also termed as three-frequency-related-diagram (3CFRD), only one coupling or acquisition unit is necessary. Each acquired noisy PD pulse will be simultaneously filtered by three band-passes equipped with tunable center-frequencies and bandwidth (Figure 4.44). After passing filters, the corresponding signal response will be vectorially added with a 120° shift again and visualized in a star diagram (Figure 4.45). If the amplitude spectrum of noisy PD signal provides a constant (flat) value, the result will be visible in the zero crossing of the related diagram; otherwise, if the

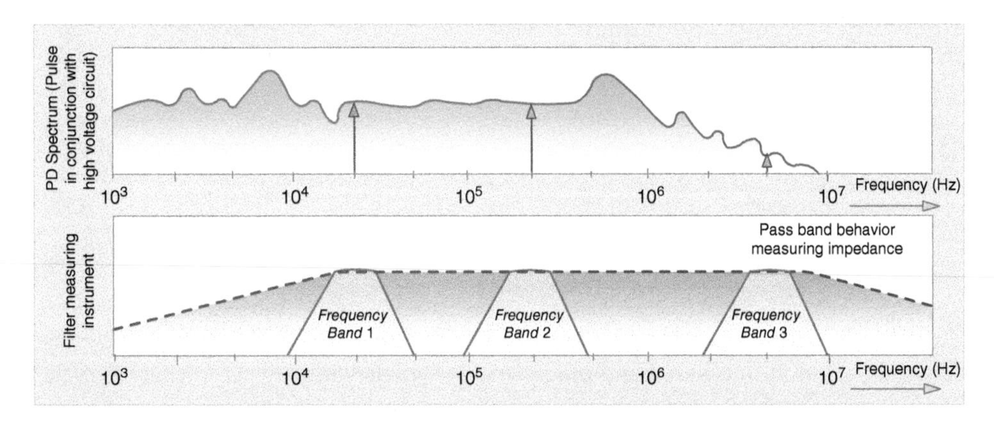

Figure 4.44 Amplitude spectrum of a noisy PD signal at simultaneous filtering with three band-pass filter (schematically).

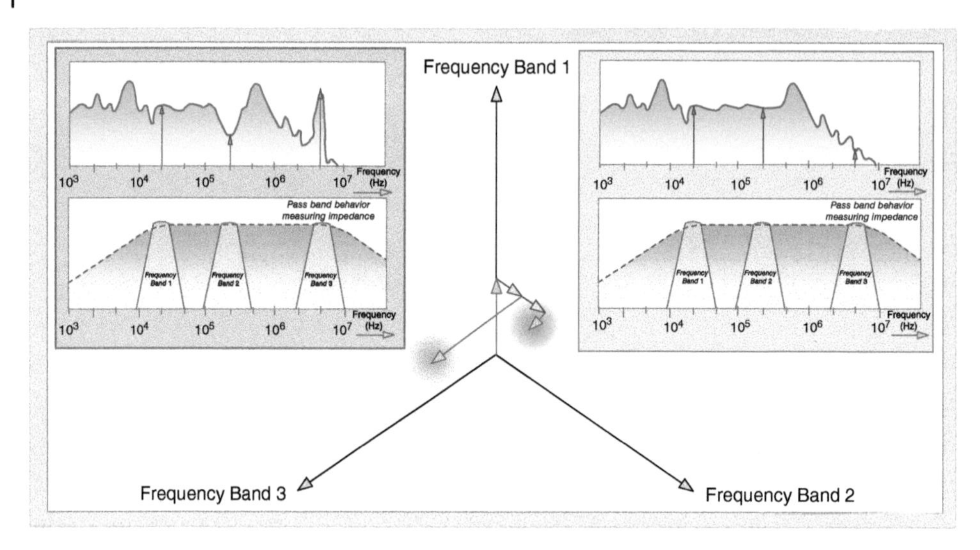

Figure 4.45 Separation of two PD signals with different amplitude frequency spectra by the 3CFRD method.

amplitude spectra will not be constant, the results lead to different points in the related view. This enables the separation of PD pulses and noise signals as well as the distinguishing of PD pulses caused by different origin PD sources.

Another approach based on the cluster separation is the decomposition of the acquired PD pulse waveforms. For this purpose, characteristic PD pulse parameters are used in either the frequency or the time domain, such as the rise and decay time as well as the pulse width [38, 39]. Further approaches provide noisy PD pulse decomposition by using wavelet-transformation tools, which enable a separation of detected PD signals superimposed with disturbances [40, 41]. One recent development is the application of methods originally used in the field of image retrieval and classification from the natural language processing [42].

It should be noticed that successfully using all these advanced noise suppression or reduction methods is possible only with deep knowledge about fundamentals of PD measurement as well as operation principle and functionality of the applied denoising tools. In other words, this cannot be simply replaced by an automatic expert system (ES) but in the future might need an experienced test engineer who is familiar with PD measurement technique features.

4.7 Visualization and Interpretation of PD Events

4.7.1 Introduction

The visualization of measured and processed PD signals with respect to characteristic PD parameters enables a compromised overview of gathered results and, at the same time, a valuable support for identification of origin PD sources. In many cases interpretation of the visualized PD behavior leads to decisions for further treatment of tested objects. The historical and recent visualization options are mainly focused on applications for PD testing at AC voltage, it might be separated in classical and advanced computerized features strongly accompanied with the progress in development of modern digitized PD measuring systems.

4.7.2 Classical Methods

One of the first examples for visualizing is the measured apparent charge dependent on time and voltage realized by an oscilloscope connected with the output of a PD measuring device. The visualization was provided by charge behavior and AC test voltage over time (mainly one period) or as so-called Lissajous-figure with charge dependent on voltage (Figure 4.46).

The classical Lissajous technique has further been modified in order to display the complete PD behavior during a single cycle of AC test voltage. This was provided by an elliptical time base of oscilloscope superimposed with the pulse charge train occur in the related AC cycle (Figure 4.47).

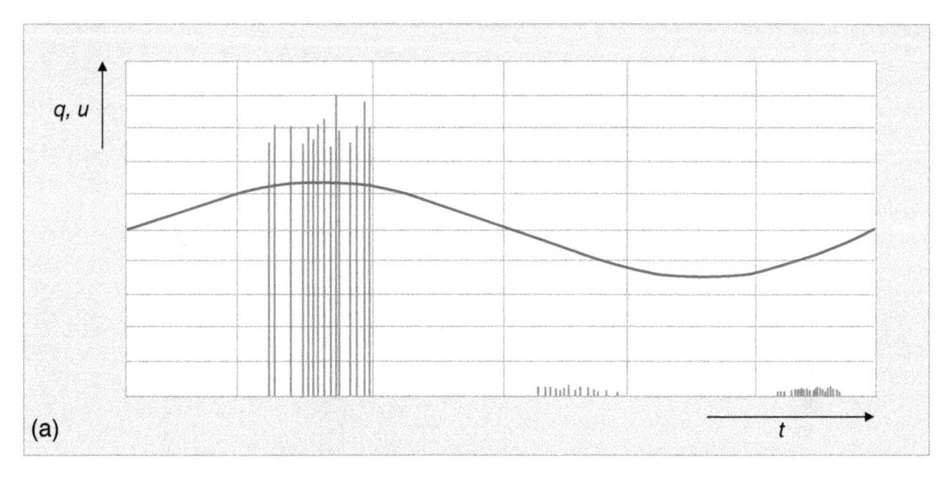

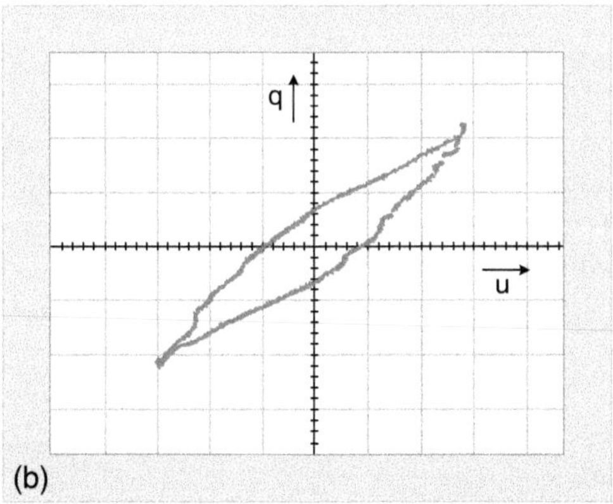

Figure 4.46 Visualization by oscilloscope for measured PD charge and AC test voltage. (a) Charge pulses and AC voltage (one period) over time. (b) Lissajous-figure for $q(u)$ dependence.

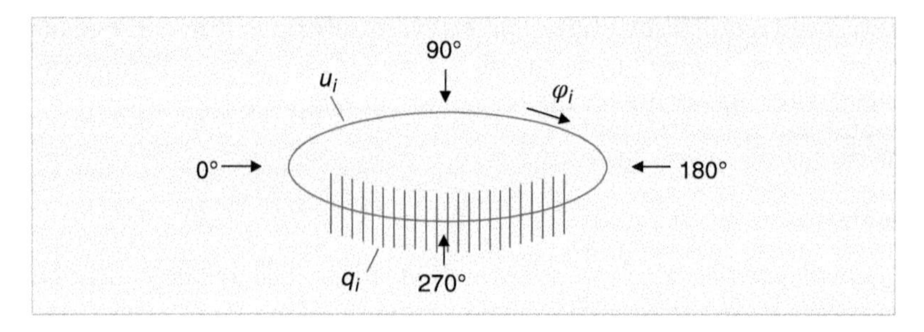

Figure 4.47 Improved visualization for measured PD charge q_i and AC test voltage u_i.

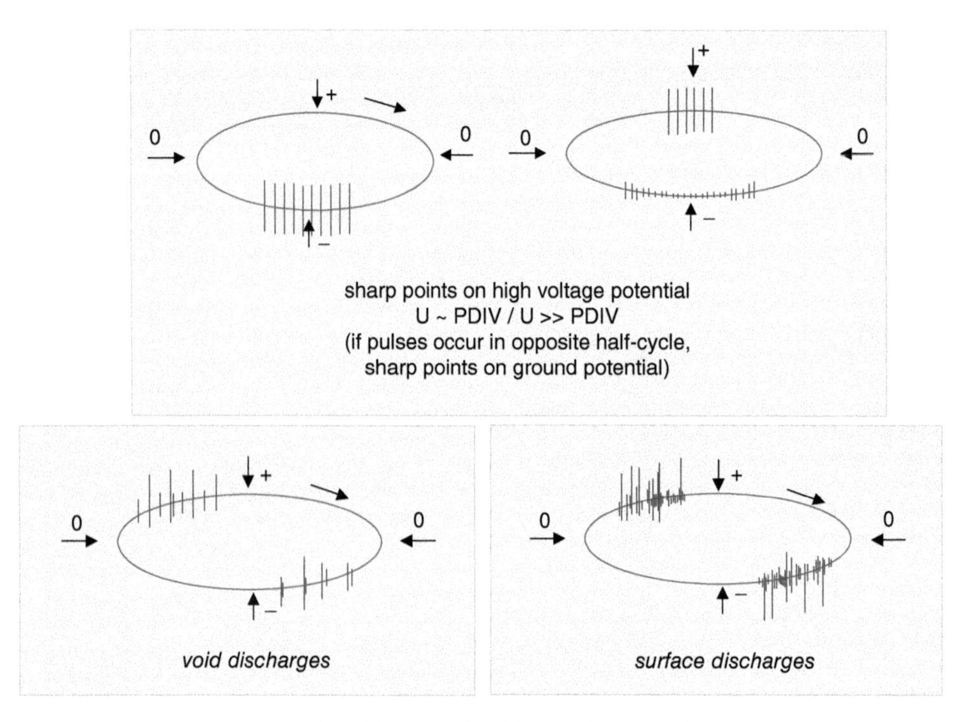

Figure 4.48 PD patterns of typical PD sources in elliptical display mode.

That visualization of measured PD events was used for extended recognition of various PD sources accompanied by improved PD measuring technique [43, 44]. To support the interpretation of measured PD behavior in PD testing of practical equipment, some characteristic patterns of typical PD sources were published as shown in Figure 4.48 [45, 46].

The use of such phase-resolved PD patterns, later also recorded with a linear time-base, became common practice in interpretation of practical PD measurement results.

It should also be noticed that the recording of measured charge dependencies over voltage or over time by using of *X*-*Y*-(*t*)-plotter[5] was a valuable support for PD interpretation in basic investigations, as depicted in Figure 4.49.

5 In that case the *Y*-input is connected with the PD measuring device, the *X*-input either with an appropriate value of test voltage or internal time-based.

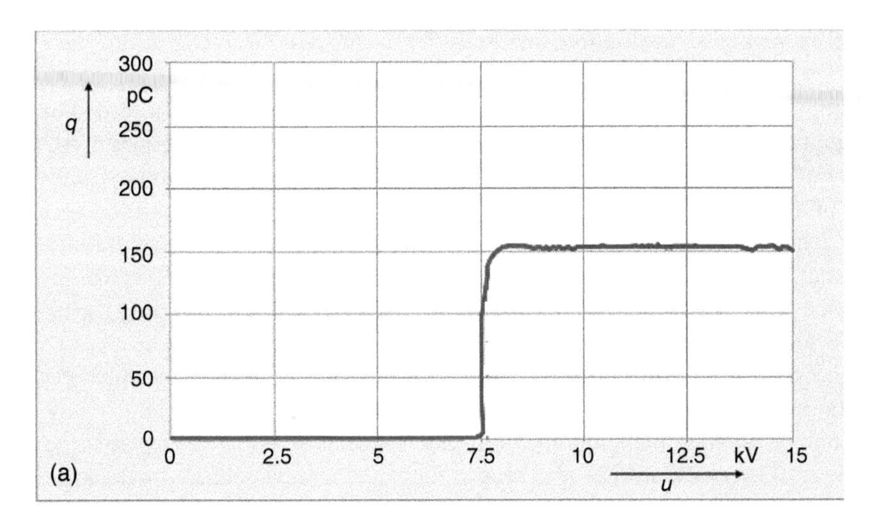

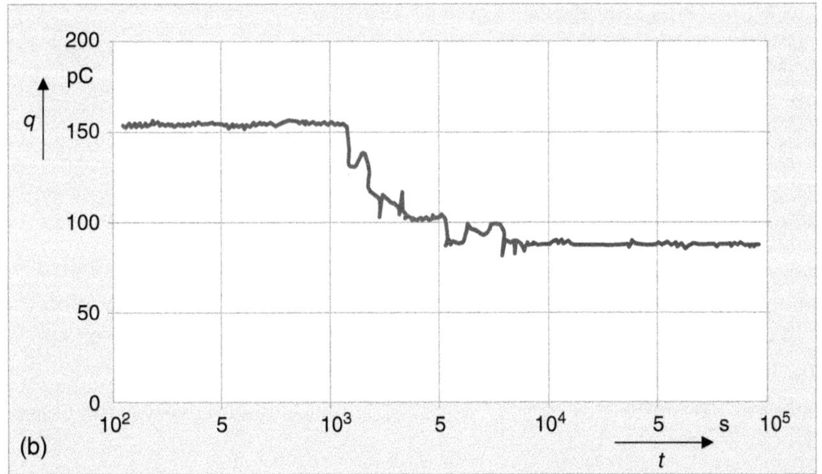

Figure 4.49 Records by applied *X-Y-(t)* plotter for visualization of results at PD measurement on point-to-plane arrangement (a) Charge behavior dependent on test voltage; *q*(*u*)-plot, (b) Charge behavior dependent on stressing time; *q*(*t*)-plot.

4.7.3 Advanced Methods

Note that the described methods are based on PD measurement performed with analog measurement devices. Visibility is currently determined using advanced digital PD measuring systems characterized by acquiring and storing a parameter-vector (q_i, u_i, t_i, φ_i) for each measured pulse, with:

q_i – Pulse charge of the individual PD current pulse
u_i – Instantaneous value of the applied test voltage
t_i – Instant of PD occurrence
φ_i – Phase angle at instant of PD occurrence[6]

6 In case of test voltage other than AC, that parameter is missing.

Based on this advanced evaluation, methods for improved PD recognition were developed as e.g. the *cumulative phase-resolved PD pattern* or φ-q-n PD pattern [47, 48]. At this method similar to an impulse-height analyzer principle the pulse number n related to its charge magnitude will be counted over a certain measuring time (mainly periods of AC test voltage) but will be displayed over about one period of the test voltage only (Figure 4.50), [49]. The information about pulse number n is provided by the colored point-cloud where each color is related to a certain pulse number. This information submits an improved view of the PD intensity. To support the interpretation of practical PD measurement as provided for previous applications (see Figure 4.48), an overview regarding typical and basic PD sources was published [47–49], as illustrated in Figure 4.51.

Modern digital computer-based PD measuring systems provide the storage of acquired PD data in real-time mode in the computer memory and after performing the measurement procedure the option of recalling again to visualize it in a *replay mode*. Using the replay mode makes possible a 3D presentation that goes beyond the traditional two-dimensional graphs shown in Figure 4.51, which makes it even easier to discriminate EM interferences from real PD events, as well as to distinguish different PD sources (Figure 4.52), [31].

That kind of postprocessing was also used to evaluate statistical PD behavior, in particular [50]:

- Standard deviation
- Skewness
- Kurtosis
- Cross-correlation

The PD behavior expressed by such statistical operators might be established with PD fingerprints that in certain cases enable the improved identification and classification of typical PD sources [50]. To create characteristic PD fingerprints, the following statistical parameters are commonly displayed, as shown in Figure 4.53:

Hn (phi): Number of PD pulses H occurring within each phase window versus the phase angle phi.

Hq (phi) peak: Peak values of PD pulses occurring within each phase window versus the phase angle.

Hq (phi) mean: Mean values of PD pulses occurring within each phase window versus the phase angle phi. This quantity is deduced from the total charge amount within each phase window divided by the pulse number occurring in this phase window.

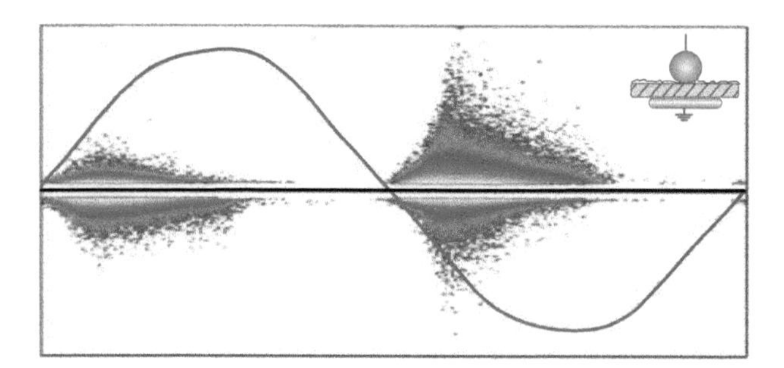

Figure 4.50 φ-q-n – pattern for PD behavior of surface discharges (acc. to [49]).

(a)

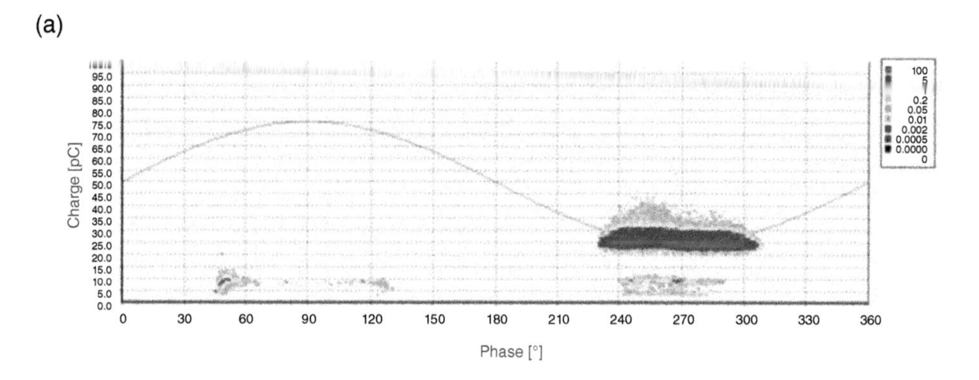

(b)

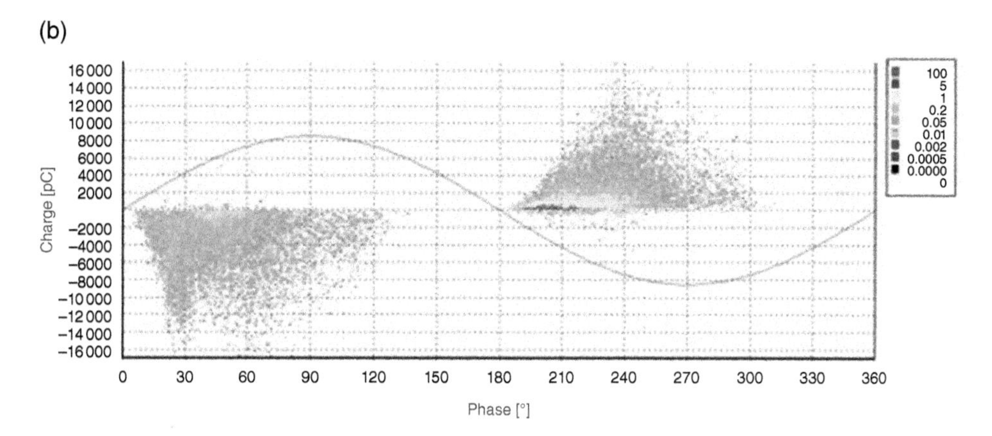

(c)

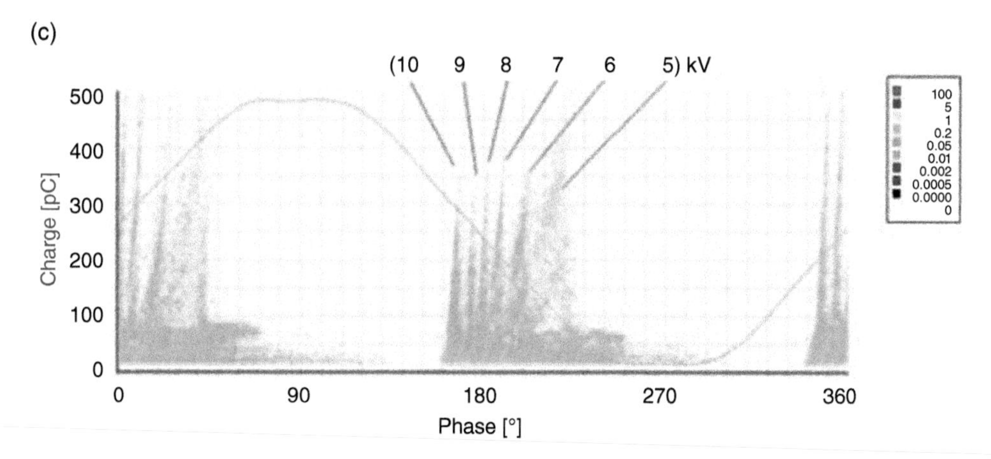

Figure 4.51 Overview about φ-q-n – pattern of PD behavior for various PD sources (a) sharp point in air on earth potential, (b) surface discharges, (c) internal [void] discharges at different test voltage magnitude).

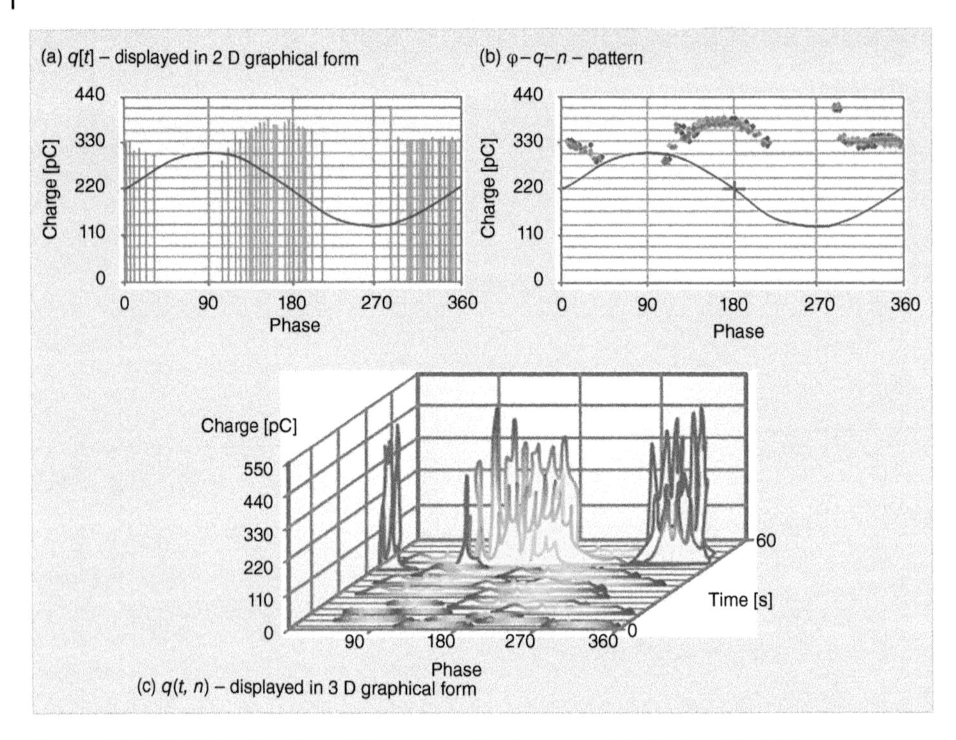

Figure 4.52 PD behavior of metallic parts on floating potential displayed in 3D form.

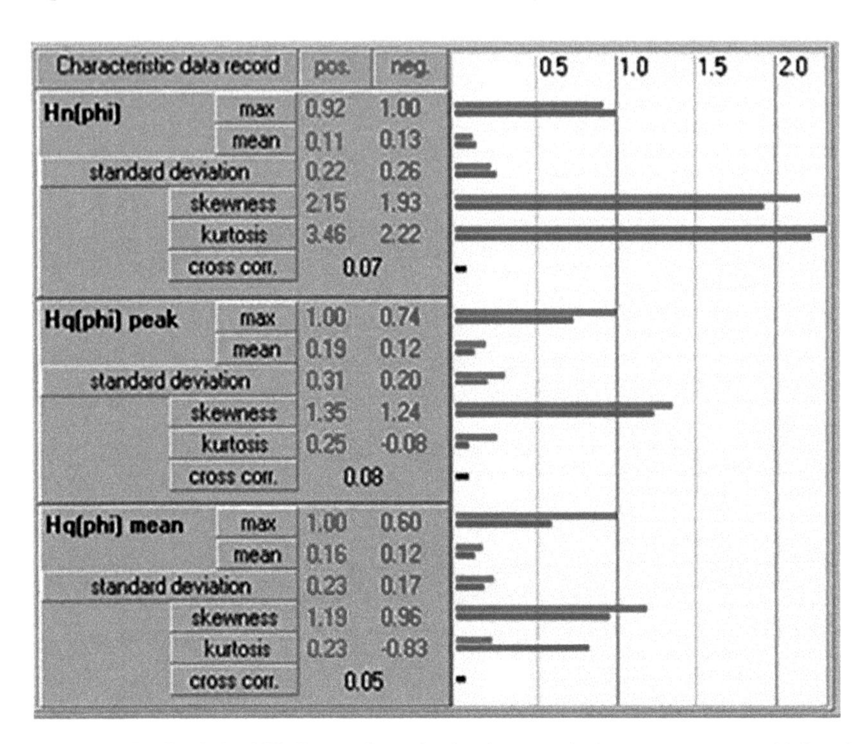

Figure 4.53 Statistical PD fingerprints obtained at practical measurement on high-voltage cable terminations.

To illustrate that for an insulation defect within a high-voltage cable termination the described collection of PD fingerprints is shown in Figure 4.53.

Based on statistical PD fingerprints established for various types of failures and implemented in the postprocessing algorithms in the measuring system, a comparison between those values and the measured values obtained in practical measurements might be led to improved interpretation of gained results as illustrated in Figure 4.54. However, it can be seen that the measured PD intensity cannot be unambiguously identified by only one typical PD source. Here, dependent also on the test conditions, three various PD sources have an “identification probability” of >90%. That means the evidence of an experienced engineer familiar with PD measurement technique seems to be necessary for clearer interpretation. Nevertheless, these PD fingerprints obtained in defined time intervals within condition-based maintenance procedure may be applied as valuable assessment tool for evaluation of insulation quality and, in case of its changing, for related decisions.

4.7.4 Pulse Sequence Analysis

Another prospective approach adopted for PD pattern recognition is the analyzing of PD pulse sequences [51]. In opposition to the φ-q-n – pattern no transformation of PD data in one cycle of test voltage takes place and the correlations between consecutive pulses are evaluated and visualized. Considering that each measured and processed PD pulse might be characterized by its magnitude q_i, instantaneous voltage u_i and the phase angle φ_i or time t_i of occurrence respectively, differences of these values between three subsequent PD pulses might be measured and depicted in characteristic charts using special developed algorithms as illustrated in Figure 4.55a. The obtained chart for that (basic) point-to-plane

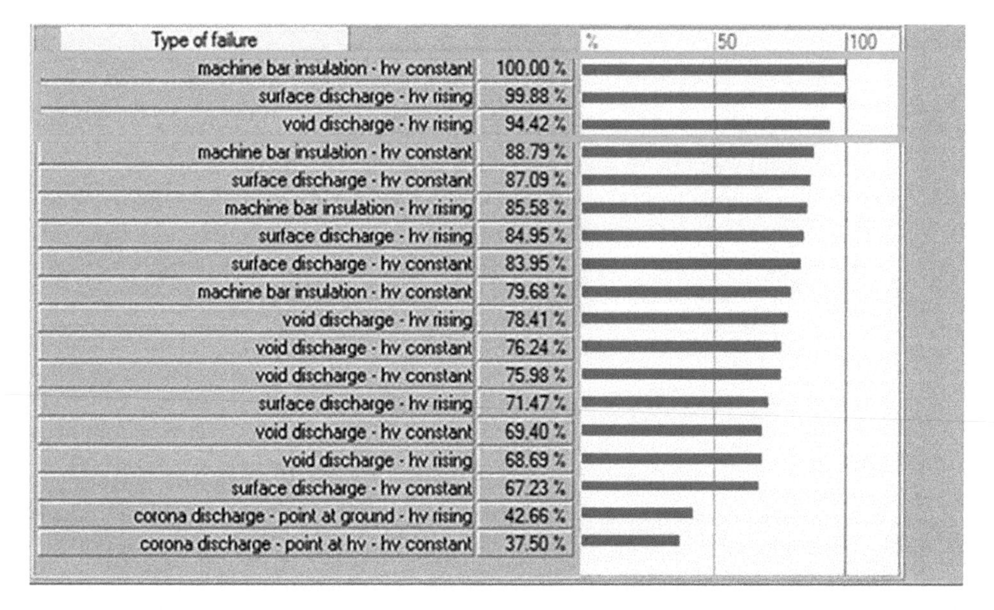

Figure 4.54 Statistical PD fingerprints of practical measurement results in comparison with (hidden implemented) reference data (green box - identification of measured results with reference data >90%).

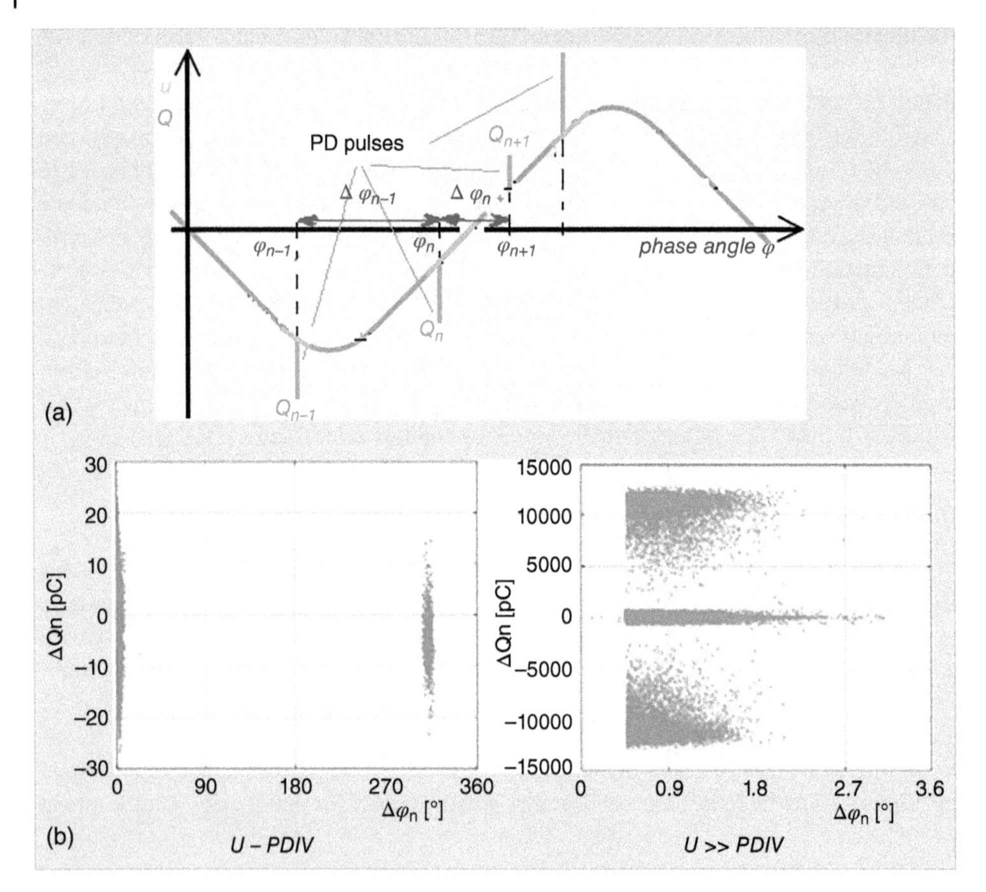

Figure 4.55 Visualization of PSA charts. (a) principle of PSA for q-φ dependencies; (b) practical example for point-to-plane arrangement at different test voltage.

arrangement shows for a test voltage near the inception voltage (U ~ PDIV) with TRICHEL discharges significant other performance than for higher test voltages (U>>PDIV) where TRICHEL - and STREAMER discharges occur (Figure 4.55b). These visualization technique for evaluation of characteristics of consecutive PD pulses seems to be very prospective, but further application and in particular cumulated experience in PD testing need to be necessary for valid establishing and common introducing in PD diagnostic procedures.

4.8 Artificial Intelligence and Expert Systems

4.8.1 Introduction

Computer technology and software solutions progress rapidly and, therefore, also in field of PD diagnosis new methods and procedures should be considered for application. The common aim of all these innovative methods should be an improvement in interpretating obtained PD results and, finally, the message about the harmfulness of detected PD signals.

The following sections provide a short review of the state-of-knowledge in artificial intelligence (AI) methods as well as discussion of expert systems under the aspects whether a successful application of such methods in PD diagnosis might be expected in near future.

AI is now quite commonly applied in various fields of engineering, business, medicine, and other domains due to the rapid advancement of computer technology (and sensors, respectively). AI has been developing since 1950 to enable a computer program to imitate human intelligence to solve given problems. Usually, AI is divided into three levels (i.e. AI, machine learning (ML), and deep learning (DL)). ML is defined as the subset of AI, which teaches the computer to learn and decides with minimal human intervention, its efficiency relies predominantly on data. ML needs to understand the features that represent data. DL is the subset of ML that performs feature extraction of the original data and then carries out identification, classification, and localization. In DL, the algorithm uses several layers to define high-level features based on the input dataset by which it does not need to understand the best feature that represents the data. DL architectures also deal with heuristics and empirical results. Deep-learning neuron network has been very successful in image classification, speech recognition, and natural language processing. Typically, ML provides excellent performances on a small/medium data set, whereas DL offers excellent performance on big data. AI technology with a computer program is used to develop an ES to emulate the judgment of expertise for the decision-making of complex problems.

The expert system uses knowledge and inference procedures to implement complex heuristic tasks based on sophisticated logical deduction. ES relies on the accumulated knowledge information and the ability of algorithms integrated into this system. The ES needs and is flexible to be improved its performance over time with the new input data to become more reliable and accurate. The ES will provide the recommended optimal decision, describe a differential diagnosis, and/or recommend contingency plans. Since about 1990, AI techniques have been developed for PD analysis and diagnosis. Then, AI and ES applications for PD diagnostic techniques continued to evolve and have recently developed by leaps and bounds due to the great advances in computer technologies and other related areas. Automated identifying of simultaneously activated multiple PD sources, including insulation integrity estimation due to the existence of PDs, is a common goal for maintaining the operation of high-voltage apparatus more accurately and efficiently.

4.8.2 Artificial Intelligence and Artificial Neural Networks

AI is a branch of computer science that has been developed since 1940. AI aims to enable the programmed machine to have the cognitive ability to think and imitate the actions of a human. Generally, AI operation is based on the collection of rules facilitating to development of the expert system. There are six aspects that AI would have the following capabilities– natural language processing, knowledge representation, automated reasoning, machine learning, computer vision, and robotics. The first four aspects primarily supported the PD diagnostic system. The last two aspects support the field of robotics [52]. AI can perform multiple tasks with too complicated and time-consuming. Therefore, AI is used to develop a comprehensive PD diagnosis system by which all applicable machine learning algorithms, especially artificial neuron networks (ANNs), have played a crucial role. AI is very promising as a part of the PD diagnosis system to classify PD signals out of strong noise, categorize the types of PD from multiple PD sources, and locate the multiple PD sources with extreme speed.

An artificial neural network (ANN) is a machine learning type that simulates the learning mechanism of biological organisms. ANNs are also developed to support the replication of human brain processing, even though the human brain is far more complicated and many of its cognitive functions are unknown [53]. The ANN is composed of a connected network of neurons. ANN would have the following abilities: learning the relationships between input and output data; handling noisy, imperfect, or incomplete data; dealing with the complex, higher-order functions, and nonlinear interactions among the input variables; and parallel carrying on of numerous operations in a short time.

The parts of a neuron's cell are illustrated in Figure 4.56a. Each neuron comprises a cell body or soma, the dendrites, the axon, and the synapse. The neurons are connected by axons and dendrites, by which the connection junction is called a synapse. The dendrites branch receives the input signals from other neurons and passes these signals onto the cell body; these signals are processed and transferred to the axon as an output channel to nearby neurons [54]. Signals are propagated from neuron to neuron by a complicated electrochemical reaction. The strengths of synaptic connections are responsible for the stimulation pattern to form the basis for learning in living organisms.

In artificial neural networks, the simple neuron, as illustrated in Figure 4.56b, consists of several inputs, the input function, the activation function, and several outputs. Each input to a neuron is scaled with a weight affecting the input function. The learning process occurs if the weights are adjusted. Then the activation or transfer function could be just a threshold, or a nonlinear function is applied to send out the outputs. In the training process, the training data containing examples of input–output pairs of the function is implemented. According to the biological neuron, the cell body, axon, and synapses are identical to the ANN's node, weight, and activation.

Such an artificial neuron can be written as a mathematical function (Eq. 4.4):

$$a_j = g\left(in_j\right) = g\left(\sum_{i=0}^{n} w_{i,j} a_i\right) \tag{4.4}$$

where

a_j = Output activation of unit j
a_i = Output activation of unit i
$w_{i,j}$ = Weight on the link from unit i to this unit
$g(z)$ = Activation function
in_j = Weighted sum of its inputs

There are various kinds of activation functions applied for neurons; for example, activation function R, activation function S (sigmoid function), activation function tanh ($g(n)$), and activation function $2/\pi$ arctan($g(n)$). Each activation function has unique characteristics, as described in [55].

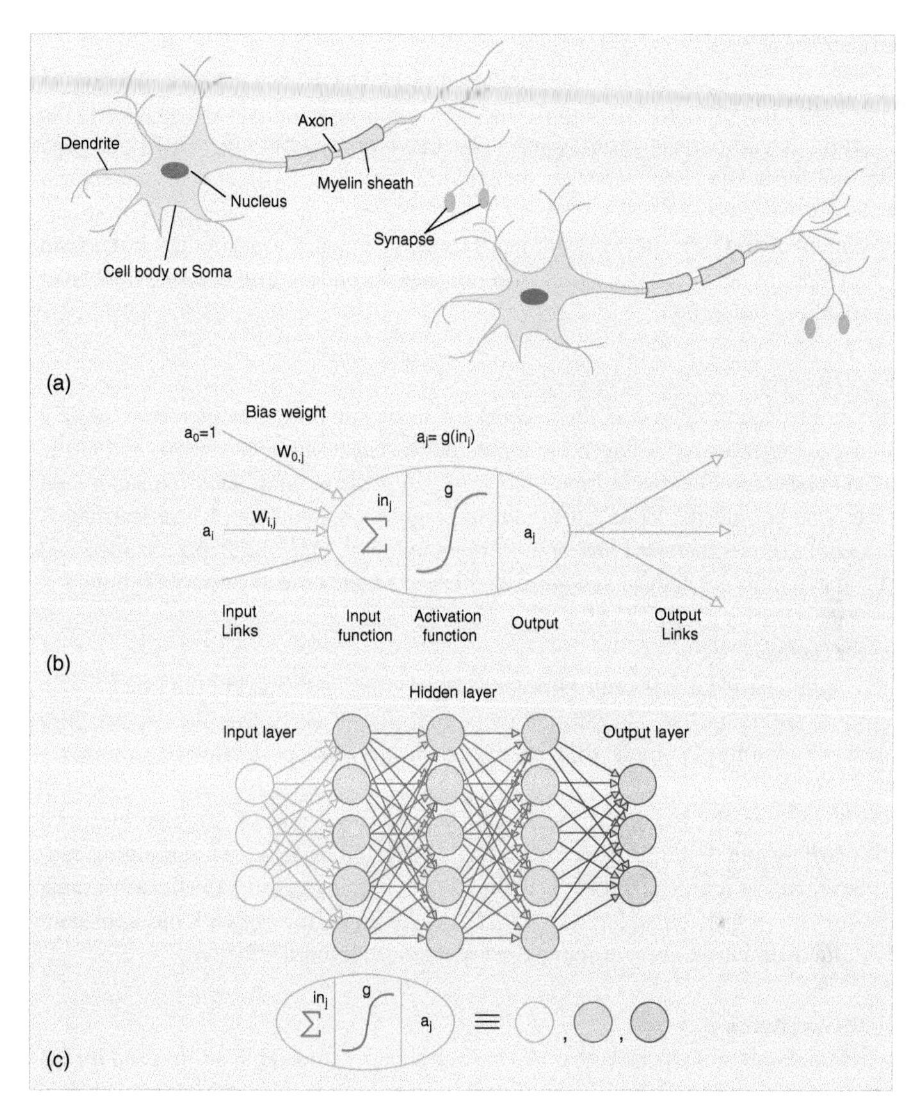

Figure 4.56 Biological and artificial neuron networks (a) biological neuron, (b) artificial neuron, (c) artificial neural networks, according to [52]).

4.8.2.1 Learning Process

The learning process is a significant procedure for modifying the weights and biases of the ANN. The aim of performing the learning process is to train the network to perform the given tasks of the ANN. Machine learning can be mainly classified into three types:

1) Supervised learning
2) Unsupervised learning
3) Reinforcement learning

Supervised Learning

In the supervised learning process, a set of training data is provided. Each output is determined by what its desired response is to the input. Once the inputs are applied to the network, the network output is compared to the targets. The network weights and biases are then adjusted to minimize the error between the network output and the targets. Supervised learning can also be referred to as classification. Deep learning models are based on supervised learning involving large networks of neurons that can learn to recognize patterns in the data set. Different network layers learn different aspects of data, and then the final layer predicts the desired outcome.

Unsupervised Learning

In the unsupervised learning process, the network learns the important features from the data set, exploiting the relationship between the input and the clustering. The weights and biases are modified in response to network input. The target outputs are unavailable. Unsupervised learning can also be referred to as clustering. In some cases, semi-supervised learning is acknowledged. In semi-supervised learning (a combination of supervised and unsupervised learning), some outputs are determined what their expected response to the given inputs.

Reinforcement Learning

For reinforcement learning, it is pretty similar to supervised learning, but the exact expectation output is unknown. The algorithm only gives feedback about the success or failure of the answer (providing a grade that measures the network performance over some sequence of inputs).

Machine learning methods can also be classified based on the type of data that they deal with static learning and dynamic learning. A single snapshot of data with remaining constant properties will be learned in static learning. For dynamic learning, the network must deal with the continuously changing data with time; therefore, the network has to be continuously trained, or within every suitable time window, to remain effective.

4.8.2.2 ANN Architecture

This part describes common principles of ANN Architectures and applications used for PD classification problems. The described ANN architectures selected are from both classical ANN and the ANN applied for deep learning. For classical ANNs, the backpropagation neural network (BPNN) and the probabilistic neural network (PNN) are briefly summarized. Additionally, the convolution neural network (CNN) and the recurrent neural network (RNN) are discussed.

4.8.2.3 Common Principles

Classical ANNs can be separated in BPNN and the PNN:

- *Backpropagation neural network (BPNN).* The BPNN structure is composed of the input layer, output layer, and multiple hidden layers, as exhibited in Figure 4.57a. When input data is fed to BPNN, it propagates through the network layer by layer until the output is produced as the response of the network. Then, the network output is compared to the expected output for that input pattern. The error is passed back from the output layer to

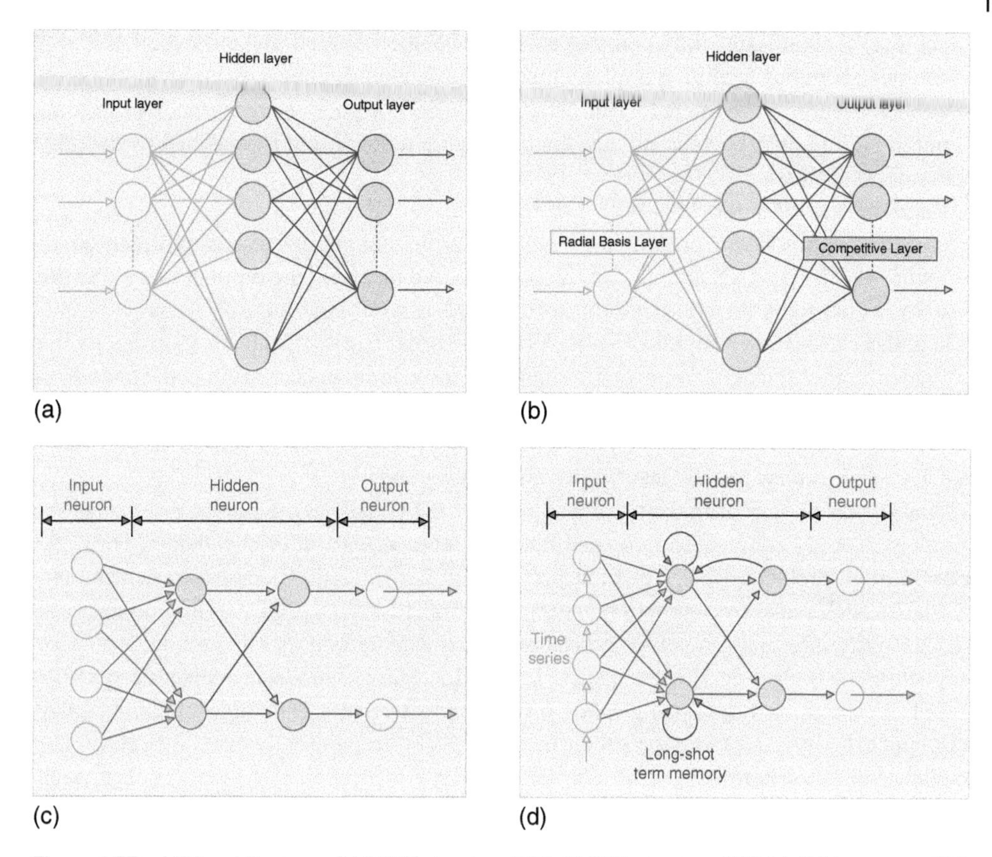

Figure 4.57 ANN architectures: (a) BPNN structure [55], (b) PNN structure [55], (c) CNN structure [56], (d) RNN structure [56].

the input layer, with the weights adjusted as the error backpropagates. The weights are adjusted until the actual response of the network moves closer to the expected response.

- *Probabilistic neural network (PNN).* The PNN structure is composed of the input layer, output layer, and multiple hidden layers, as demonstrated in Figure 4.57b. When input data is fed to PNN, the first layer of the hidden layer calculates distances from the input vector to the training input vectors. It then generates a vector indicating how close the input is to a training input. In the second layer, these contributions are summed for each class of inputs to generate as its net output a vector (a0, a1) of probabilities. A maximum of these probabilities is then chosen by a complete transfer function on the output of the second layer, and a 1 will be produced for that class, whereas a 0 will be generated for the other class at the output layers.

For deep learning ANNs also the convolution neuron network and the recurrent neuron network are applied.

- *Convolution neural network (CNN).* The convolutional neural network or CNN is based on convolution operation. CNN is widely used for object classification and detection. The CNN structure comprises the input layer, output layer, and multiple hidden layers, as shown in Figure 4.57c. The hidden layers consist of the convolutional, activation, pooling, dropout, and fully connected layers. Feature extractions and evolution from simple to complex forms are implemented in the convolution layer. In contrast, the non-existence features are filtered in the activation layer. The pooling layer executes the computations and preserves the suitable feature size then the output is predicted by the fully connected layer.

- *Recurrent neural networks (RNN).* The recurrent neural network (RNN) structure contains the input layer, output layer, and multiple hidden layers, as represented in Figure 4.57d. The RNN architecture is similar to that of CNN. The input data is fed to the RNN sequentially for computation. In the hidden layer, each neuron feeds its output signal back to its own inputs of all the other neurons. The self-feedback loop causes the hidden state of the neural network to change after the input is carried out in the sequence. The output is from the output layer.

4.8.2.4 Applications for PD Classification and Localization

Various types of ANN architecture are applied for PD classification and localization problems. Each ANN architecture provided different advantages and disadvantages. For a PD classification problem, BPNN, counter propagation network (CPN), cascaded neuron network (CCNN), radial basis function neuron network (RBFNN), extension neuron network (EXNN), ensemble neuron network, PNN can be utilized. Moreover, other methods are also applied to a PD classification problem – i.e. Hidden Markov models (HMM), fuzzy logic-based classifiers, self-organizing maps (SOM), inductive inference algorithms, adaptive resonance theory (ART), support vector machine (SVM), rough set theory (RST), cerebellar model articulation controller (CMAC), and sparse representation classifier. Details of each technique can be found in [57–60]. For PD localization problems, after receiving the PD signals from the given sensors, such as UHF sensors and/or acoustic emission sensors, the time of arrival (TOA) method or the direction of arrival (DOA) method needs to be executed. For the TOA method, the times of a PD signal – for example, an EM wave or acoustic signal, reaching each sensor are different depending on the distance between the position of sensors and the PD source. A numerical calculation can be applied to obtain the location of the PD source. For the DOA method, the direction of arrival of the incoming signal in either three or two dimensions can be calculated by applying the phased array theory (PAT) [61]. ANN is combined with a different method to localize the PD sources – for example, K–NN regression connected with time domain feature extraction and correlation-based feature selection [62], Fast recurrent-convolution neural network combined with SVM [63], wavelet kernel-based convolution neural network (W-CNN) [64], deep neural network [65], and multioutput convolutional recurrent neural network (CRNN) [66].

4.8.2.5 Basic Principles of PD Recognition

Input Signals

Typically, in AC systems, the phase-resolved partial discharge (PRPD) patterns and the PD pulse currents are used as valuable diagnostic parameters. The PRPD pattern visualizes the occurrence of PD activities in reference to the phase of AC voltage. For the simple PRPD patterns, the PD types – i.e. corona discharge, surface discharge, and internal discharge – can be classified by PD experts. Likewise, the apparent noise may be separated from the PD pulse current without considerable effort. However, the detected PRPD patterns or PD pulse signals from onsite PD measurements are always

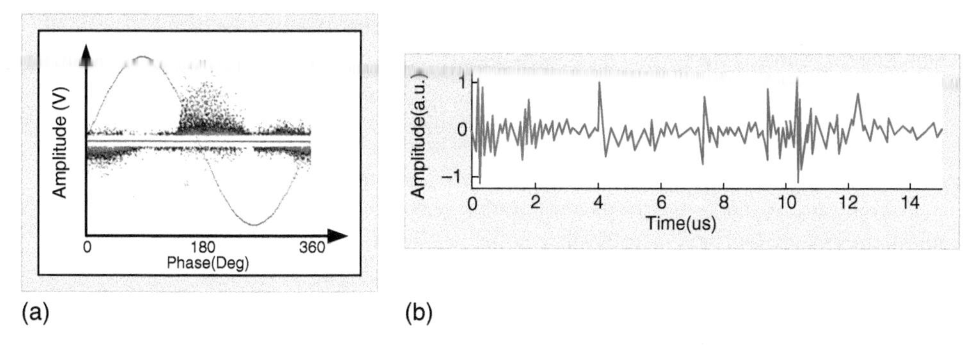

Figure 4.58 Examples for typical input signals (a) PRPD pattern superimposed with noise signals [67], (b) PD pulse current superimposed with noise signals [68].

superimposed with noise signals. A mix of PD signals from different PD sources is also found. Therefore, it is a complex task that needs a big effort to evaluate PD signals and classify PD types obtained from real testing, especially in a robust noise environment. Examples of the PRPD pattern and the PD pulse current superimposed with noise signals are manifested in Figure 4.58.

Normally, the PD measuring system detects both PD signals combined with noise signals, which causes difficulty in PD identification, classification, and localization. It is difficult to identify PRPDs or PD signals with noise interference or even when multiple PD signals from different PD sources are superimposed. To avoid the noise/disturbance problem, avoiding the acquisition of noise signals in the PD measurement process is preferred, then the noise reduction algorithm is applied in a post-processing phase. Commonly, the PD spectrum differs from the noise spectrum; therefore, it can be applied for discrimination between PD sources and noise. White noise, narrow band noise, and impulsive noise exist in onsite PD testing. White noise may be generated from the operation of the equipment, grounding system etc., the narrow band noise may come from the radio communication, and the impulsive noise may be caused by the corona discharge, the switching operation, and so on. As summarized in Section 4.6, various denoising techniques can be performed.

Feature Extraction

Feature extraction is a vital part of a PD classification and localization problem due to the high dimensionality of PD measurement data. The features could be statistical parameters, image processing features, frequency/time features, and fractal features.

The feature extraction is processed once original input PD signals, such as PRPD patterns or PD pulses, are obtained. The input variables will be determined. However, some input variables may be correlated with other input variables, so-called multicollinearity, which negatively impacts the performance of the PD analysis and diagnosis model. To mitigate such a problem, which leads to increasing the processing speed and reducing the storage required memory, a dimensional reduction algorithm is needed.

Dimension Reduction Algorithms

The dimension reduction algorithm is designed to transform the original data from the high dimensional feature space to a new informative space with lower dimensionality by which the redundant and ineffective information is removed while keeping a high portion of information [69]. Dimensionality reduction is applied to reduce the number of input variables, which helps to mitigate overfitting and, consequently, to reduce the complexity of the model.

Generally, the dimension reduction algorithm can be divided into two main groups:

- *Linear technique.* The original data will map onto a low-dimensional feature space with a selected linear transformation by which it will preserve some information of interest. The examples of linear techniques applied success for PD classification are principle component analysis (PCA) and Fischer discriminant analysis (FDA) [70].
- *Nonlinear technique.* This technique can be divided into two subgroups, i.e. global and local group techniques. In the global group techniques, the global properties will be preserved when the original data is mapped onto a low-dimensional feature space. Examples of global nonlinear techniques are kernel PCA, kernel FDA, metric multidimensional scaling, stochastic proximity embedding, Isomap, and stochastic neighbor embedding. For the local group technique, the small neighborhood properties around each sample data will be preserved in the same as the global properties of the data manifold will also be kept. An example of the local nonlinear technique is a locally linear embedding.

PD Variable Feature Extraction

PD variable feature extraction is performed to obtain the relevant input feature from the original PD test data. The newly established PD parameters are used for facilitating PD classification and localization.

Newly established PD parameters may be from the basic and derived PD quantities according to IEC 60270 obtained from the PRPD patterns. The parameters of the PD pulse shape, such as the front and tail times and the pulse duration, provide time domain parameters, whereas the peak power frequency, the median frequency, and the spectral entropy derived from the PD signal in the frequency domain provide significant support for PD classification and localization problems [71]. Various PD features can be extracted from the input PD signalize, the statistical parameters, features from image processing, frequency, and time feature extraction, Weibull features, fractal features, cross wavelet spectrum, wavelet coefficients, features from two-pass slit window scheme, features from autocorrelation techniques, and chaotic characteristics [58]. The statistical PD parameters, PD pulse shape parameter, and wavelet transform are introduced in the next section.

Statistical PD Parameters

As already described, PD quantities useful for PD diagnostic can be divided into two groups, i.e. the basic quantities such as PD intensity expressed by apparent charge q, PDIV, PDEV, and the derived quantities such as q-V charts, PRPD patterns, PD magnitude-intensity mappings, and several other plots. The definition of statistical parameters related to the PD quantities (e.g. mean value, variance, skewness, kurtosis, and cross-correlation factors), and already discussed in Section 4.7 are formulated and written as in Eqs. (4.5)–(4.12) in Table 4.1.

More details involved with the statistical parameters can be found in [58, 72, 73].

PD Pulse Shape Parameters

The PD pulse shape is quite unique and strong depending on the nature and location of PD sources as well as the measurement circuit. The feature extraction that is respected to the

Table 4.1 Parameters of statistical features used for PD evaluation.

Parameters	Function	Equation
Mean	The average value of all pulses existing in both half cycles	$$\mu = \frac{\sum_{i=1}^{N} x_i}{N} \quad (4.5)$$ N is the number of test samples. x is the charge amplitude of each test sample.
Variance	The value explaining the spread between numbers in a data set	$$\sigma^2 = \frac{\sum_{i=1}^{N} (x_i - \mu)}{N-1} \quad (4.6)$$
Kurtosis	The sharpness of the distribution	$$k_\mu = \frac{\Sigma (x_i - \mu)^4}{\sigma^4} - 3 \quad (4.7)$$
Skewness	The symmetry of the distribution	$$S_k = \frac{\Sigma (x_i - \mu)^3}{\sigma^3} \quad (4.8)$$
Cross-correlation factor (CC)	The difference in the shape of distributions	$$CC = \frac{\Sigma xy - \frac{\Sigma x \cdot \Sigma y}{n}}{\sqrt{\left[\Sigma x^2 - \frac{(\Sigma x)^2}{n}\right]\left[\Sigma y^2 - \frac{(\Sigma y)^2}{n}\right]}} \quad (4.9)$$ x is the mean discharge magnitude in the positive half of the voltage cycle. y is the mean discharge magnitude in the negative half of the voltage cycle. n is the number of phase positions per half cycle.
Discharge asymmetry	The value explaining the quotient of the mean discharge level of the Hnq(Φ)	$$D_A = \left(Q_n^- / N^-\right) / \left(Q_n^+ / N^+\right) \quad (4.10)$$
Phase asymmetry (Φ_A)	The value explaining the difference in the maximum pulse of the Hnq(Φ)	$$\phi_A = \phi_{-\max} / \phi_{+\max} \quad (4.11)$$
Modified cross-correlation factor	The differences between discharge patterns in the positive and negative halves of the voltage cycles	$$MCC = D_A \cdot \phi_A \cdot CC \quad (4.12)$$

PD pulse shape has the potential to support PD classification and localization. The separation of multiple PD sources and noises using the newly explored PD pulse shape parameters by applying the clustering technique is very popular nowadays. This technique is based on the transformation of the time series of the input PD pulse into the time sub-series, corresponding to pulses with a similar waveform. The equivalent duration of the waveform and the equivalent bandwidth of the spectrum of each pulse are plotted on a time-frequency (TF) map. Applying this technique, the standard deviation of the normalized signal for the time domain (σ_T) and for the frequency domain (σ_F) are calculated based on the first normalize of the obtained $s(t)$ PD signal as expressed in Eqs. (4.13)–(4.15) [74]

$$\tilde{s}(t) = \frac{s(t)}{\sqrt{\int_0^T s(t)^2 \, dt}} \tag{4.13}$$

$$\sigma_T = \sqrt{\int_0^T (t - t_0)^2 \, \tilde{s}(t)^2 \, dt} \tag{4.14}$$

$$\sigma_F = \sqrt{\int_0^\infty f^2 \left|\tilde{S}(f)\right|^2 df} \tag{4.15}$$

where f stands for the frequency, $\tilde{S}(f)$ represents the Fast Fourier Transform (FFT) of $\tilde{s}(t)$, and t_o represents the gravity center of the normalized signal. The formulation of t_o is expressed in Eq. (4.16).

$$t_0 = \int_0^T t \, \tilde{s}(t)^2 \, dt \tag{4.16}$$

Wavelet Transformation

Wavelet transformation (WT) is developed to overcome the drawbacks of Fourier Transformation (FT) application for PD denoising. Obviously, FT provides the frequency components of a periodic signal but not the information concerning where in the time domain the change in frequency occurs [75]. The short-time FT (STFT) tries to solve such a problem by introducing a sliding window, extracting a small portion of the signal, and then performing FT [68]. However, the region in the very low-frequency component cannot be detected on the spectrum, aka a resolution problem [68, 76]. To conquer the aforementioned problem, WT is introduced by utilizing a large time window to extract low frequencies and a smaller time window for high frequencies [68].

Wavelet is a small wave-type signal, which satisfies

$$\int_{-\infty}^{\infty} \psi(t) dt = 0 \tag{4.17}$$

$$\int_{-\infty}^{\infty} \left|\psi(t)\right|^2 dt < \infty. \tag{4.18}$$

where $\psi(t)$ is the wavelet basis function.

A signal can be transformed into wavelet coefficients as:

$$\psi_{a,b}(t) = \frac{1}{\sqrt{|a|}}\psi\left(\frac{t-b}{a}\right). \tag{4.19}$$

where a is the scaling parameter, and b is the translation parameter. The scaling parameter is for compressing and stretching the mother wavelet, as expressed in Eq. (4.19), which is a specific shape function whose shape remains the same for a given wavelet, only changing its nonzero width and position. The translation factor is for shifting the mother wavelet along the time axis. The factor $\frac{1}{\sqrt{|a|}}$ is used to ensure that each scaled wavelet function has the same energy as the wavelet basis function [68, 75]. The continuous WT (CWT) is defined as

$$W_{a,b} = \int_{-\infty}^{\infty} X(t)\frac{1}{\sqrt{|a|}}\psi^*\left(\frac{t-b}{a}\right)dt \tag{4.20}$$

where $X(t)$ is the function of a signal and ψ^* is the complex conjugate of the mother wavelet [77]. Performing the CWT, determining wavelet coefficients at every scale is time-consuming and generating much data. Therefore, the discrete WT (DWT), the discretized version of the CWT, is proposed. DWT is defined as:

$$W_{j,k} = \sum_{n\in Z} X(n)2^{(-j/2)}\psi\left(2^{-j}n-k\right). \tag{4.21}$$

where $X(n)$ is a discrete function of a signal, j and k are integers. The inverse DWT (IDWT) can be defined as:

$$X(n) = \sum_{j\in Z}\sum_{k\in Z} W_{j,k}\psi_{j,k}(n) \tag{4.22}$$

In the implementation of DWT, a signal goes through a series of low-pass filters and high-pass filters and is decomposed into a number of approximation and detail coefficients [77].

The application of wavelet transform for PD denoising requires three steps [78, 79]:

1) Apply the DWT for signal decomposition by selecting a mother wavelet, selecting a level N, and computing the wavelet decomposition coefficients of the signals at levels from 1 to N.
2) Select threshold detail coefficients for each level from 1 to N by which a selected threshold, a soft or hard threshold, is applied to the detail coefficients.
3) Apply the IDWT for the reconstruction of the signal by using original approximation coefficients and modified detail coefficients from levels 1 to N.

An example of wavelet transform application for PD denoising is shown in Figure 4.59.

Applying DWT provides an extent degree of satisfaction in extracting PD pulses from noisy backgrounds, which is useful for PD classification; however, in a strong noise environment, DWT may lose its effectiveness and does not usually reproduce PD magnitude and pulse shape accurately. These features may affect PD classification and diagnosis [80]. Different techniques

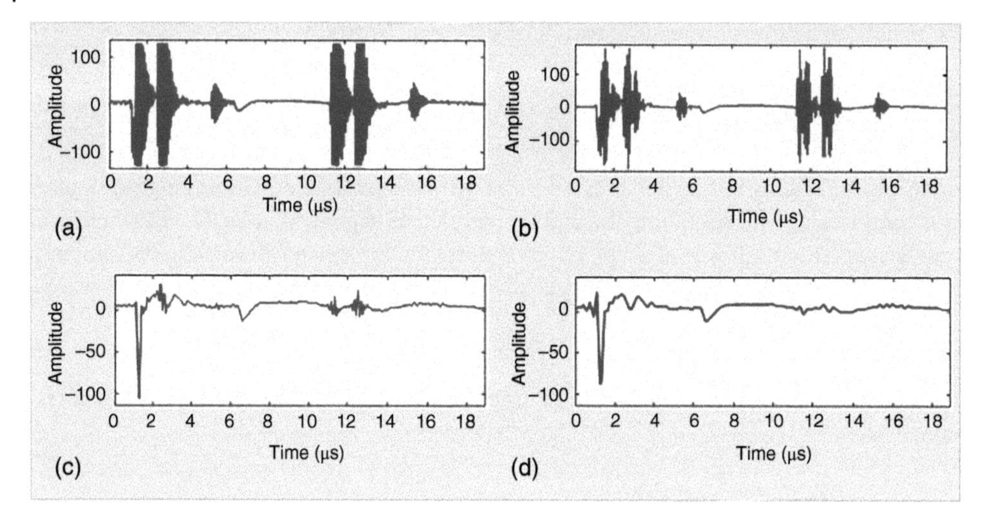

Figure 4.59 Denoising with the threshold of four levels [68]. Where (a) original contaminated PD with noise impulses (b)–(d) the reconstructed signal after one, three, and four levels of decomposition, respectively.

based on DWT have therefore been developed, such as the wavelet packet transform (WPT) [81], DWT with PCA application [82], and the second-generation WT (SGWT) [83] to enhance the competency of traditional DWT for PD classification and diagnosis.

For more detailed information on the advantages and disadvantages of some feature extraction methods, see [83]. As an example the separation results of multisource PD and noise signals by using different methods at PD testing of hydrogenerator are illustrated (Figures 4.60 and 4.61), revealing two slot discharge sources in the tested object. In Figure 4.60 the separated results of multiple PD sources are depicted using the TF map [67], and in Figure 4.61 is the applying of PC cloud demonstrated. In both figures we see the PD clustering performance that separates the PD sources (i.e. corona discharge, surface discharge, and internal discharge), taking place in the investigated insulation system.

Picture Recognition

The big picture of the PD recognition system is composed of PD input signals such as PRPDs or PD pulse recorded, PD noise reduction or data cleaning process, the feature extraction process, and ANN model application with classifier and locator algorithm, as displayed in Figure 4.62.

4.8.3 Expert System

4.8.3.1 Introduction

Expert systems apply artificial technology to an intelligent computer program mainly to solve complex problems that demand significant human expertise as a knowledge base. Generally, an expert system is comprised of the knowledge base, the inference engine, and the user interface (Figure 4.63). Typically, an expert system requires a knowledge base

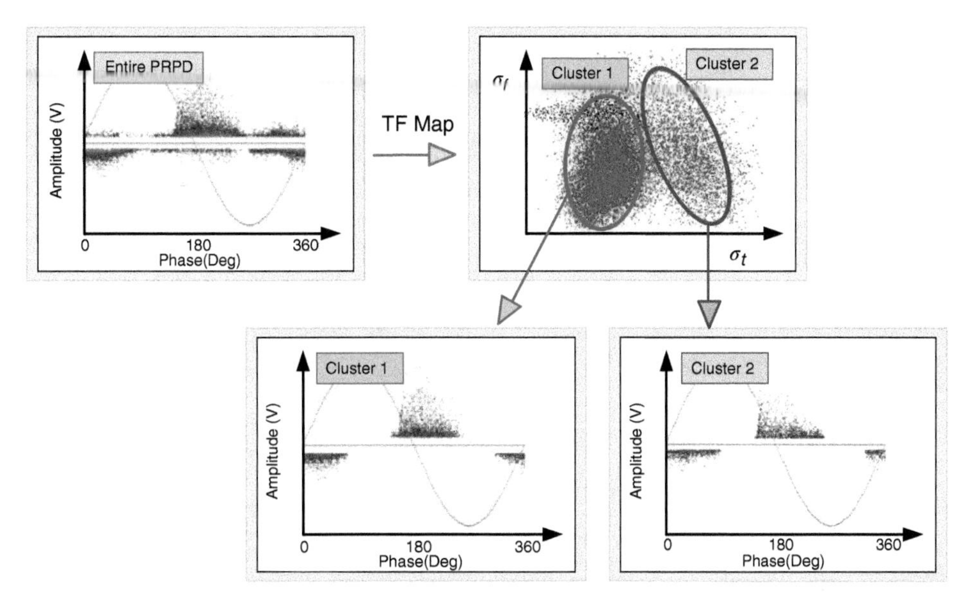

Figure 4.60 Graphic illustration of separated results for multiple PD sources using the TF map.

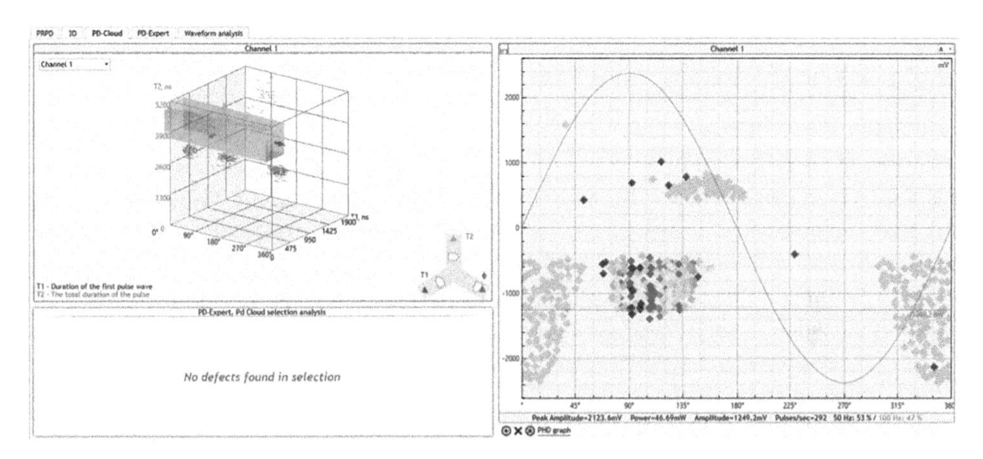

Figure 4.61 Graphic illustration of separated results of multiple PD sources using the PD cloud.

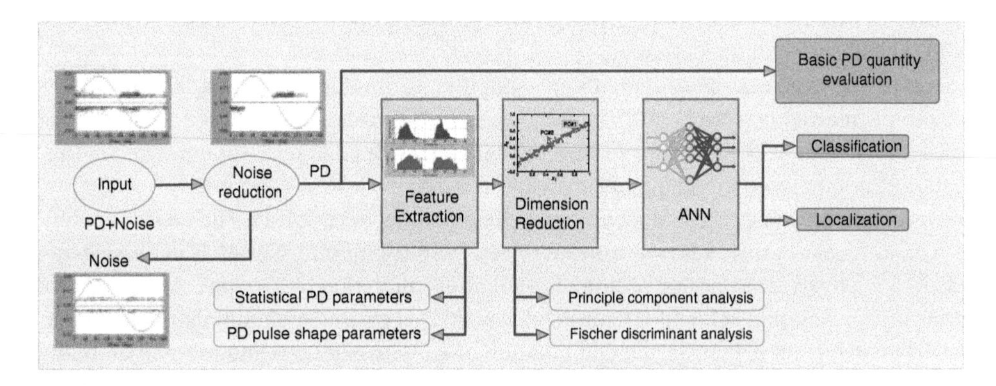

Figure 4.62 Big picture of a PD recognition system (schematically).

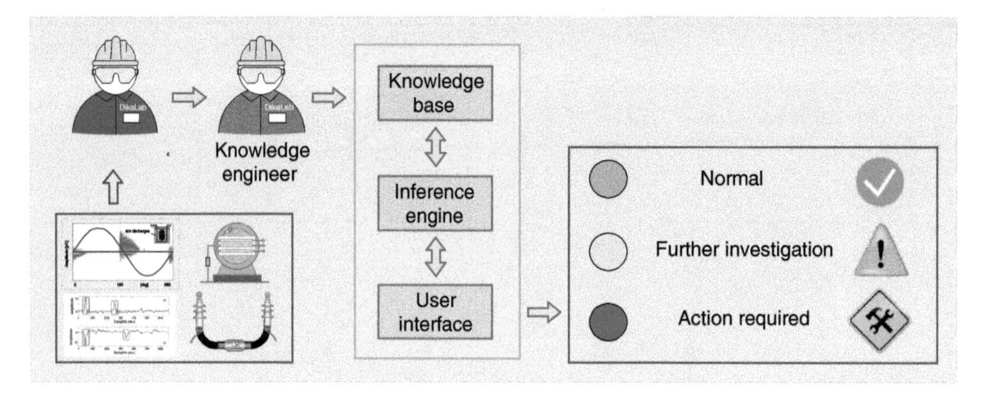

Figure 4.63 The PD expert system (schematically).

comprising accumulated experiences and a set of rules for handling each problem described in the program. Elaborated expert systems can be enhanced with additionally gathered knowledge. The information obtained from human expertise is well organized and stored in the knowledge base. The knowledge base of an expert system should accumulate more knowledge from multiple experts to prevent a bias or skewing of any one expert for decision-making. In the inference engine, complex heuristic tasks are carried on. A rules-based system is implemented by which it will map known information from the knowledge base to a set of rules and make decisions based on those inputs. An explanation module is also included in the inference engine to elucidate the acquisition of a system conclusion. To perform logical deduction of an expert system, the conclusion should be drawn from the existing facts in conjunction with the knowledge base using different types of rules. The end users can interact with the expert system in the user interface to get an answer. Usually, various types of solutions are obtained from the expert system – for example, the suggestion, the demonstration, the diagnosis, the description, the interpretation, the prediction, the alternative option suggestion, and a conclusion.

4.8.3.2 Application for PD Diagnostic

The application of an expert system for PD classification and localization problems is very helpful for maintaining the operation of high-voltage apparatus in satisfactory efficiency as expected. The expert system integrates the merit of AI technology with the capability of sensors for this task. It can be applied to general problems and even very complex problems of PD classification and localization.

Due to the dramatic developments in computer technology, including the explosive growth of Internet of Things (IoT) technologies, AI has been widely applied in the industrial sectors. AI is also applied for PD test result analysis and diagnosis, as described in this section. Various types of AI algorithms can be used, but each AI approach has diverse accuracy and efficiency levels depending on the problems to be solved. The breakthroughs of AI and ES have provided opportunities to deal with useful information from large datasets in a relatively short period of time. In [84], the summation of the application of AI in terms of conventional ML and DL for PD diagnostics is given. To obtain higher efficiency for PD analysis and diagnosis, the combination of AI with Big Data and IoT will dramatically improve PD analysis and diagnosis for supporting the next era of humans.

4.9 Calibration

4.9.1 Calibration of PD Measuring Circuit

Calibration must be carried out in the completed PD test circuit, including the test object, but with the high-voltage source turned off. Each change of circuit components in position or type requires a new calibration [31, 85–87].

A calibrator produces unipolar step voltage pulses of an amplitude U_0 in series with a capacitor C_0 so that each calibrator pulse injected into the test circuit has a charge with a magnitude

$$Q_0 = C_0 \times U_0 \tag{4.23}$$

The unipolar step voltage pulses have the following conditions:

- Determination of the calibrator charge on all settings with an uncertainty of ±5% or 1 pC
- Determination of the rise of the voltage steps with an uncertainty of ±10%
- Determination of the pulse frequency with an uncertainty of 1%

The aim of the calibrator is to determine the scale factor (S_f) as the ratio between the values of (correct) apparent charge (Q_a) and the reading value (M_c) indicated by the measuring instrument. As the input pulse magnitude is reduced by the bandwidth of the PD instrument and the frequency response of the measuring circuit, the measured charge does not consider this fact. Therefore, a calibrator pulse must be carried out, which results in a reference response. This response will be compared with the indicated charge value. A calibration pulse with a value (Q_0) results at the measuring instrument a reading value (M_c). The scale factor (S_f) is calculated from this as:

$$S_f = Q_0 / M_c \tag{4.24}$$

For practical PD testing, a measured pulse charge is indicated by the measuring instrument with the reading value (M_p). With determined scale factor (S_f) the apparent charge (Q_a) generated in the test object may be calculated with:

$$Q_a = S_f \times M_p \tag{4.25}$$

Figure 4.64a shows the calibration procedure to receive the scale factor, and Figure 4.64b represents a calibration procedure for high-voltage apparatus.

To get correct results, the output waveform of the calibrator pulse is specified in the standard [85] (Figure 4.65):

- Rise time $t_r \leq 60$ ns
- Time to steady state $t_s \leq 200$ ns
- Step voltage duration $t_d \geq 5$ µs
- Absolute voltage deviation $U \leq 0.03\ U_0$
- Fall time $t_{fall} \geq 100$ µs

Figure 4.66 shows a practical setup to measure the PD calibration pulse and the measured PD charge pulse (10 pC) obtained from the integration of the measured current pulse.

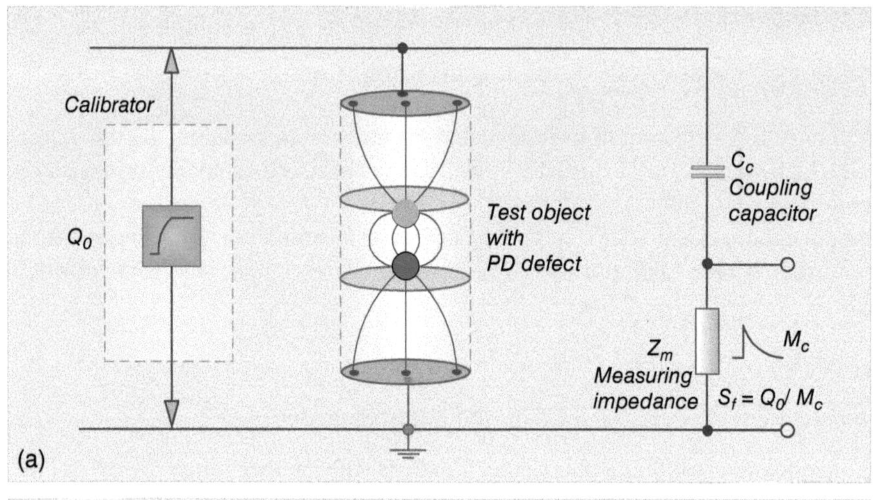

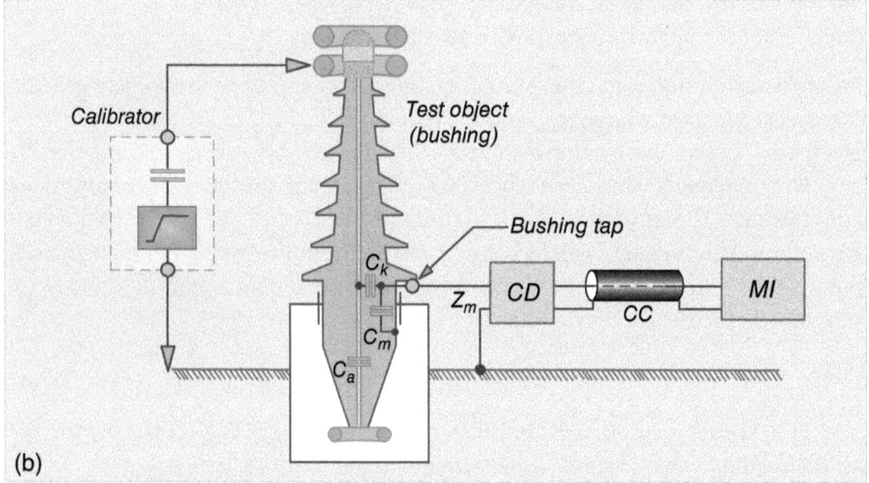

Figure 4.64 Principles of the calibration procedures. (a) Equivalent circuit for calibration. (b) Connection of calibrator to the high-voltage circuit at practical example (bushing).

The connection of calibration generator with the test object within the high-voltage test circuit is shown in Figure 4.67. Principally, the calibration should be made by injecting pulses across the terminals of the test object (Figure 4.67a,b). In the case of high extended test objects, the calibrator should be located close to the high-voltage terminals of the test object, as stray capacitances could cause errors. If the test object must be grounded, the coupling capacitor is in series with the CD; in the other case, CD is in series with the test object.

For test objects represented by a lumped capacitance, the calibrator capacitance should not be greater than 200 pF. For test objects such as power cables exceeding a length of 200 m, the calibrator capacitance C_0 should not be greater than 1 nF. The example of PD pulse calibration for the PD measurement of a 22 kV underground cable with a joint is shown in Figure 4.68.

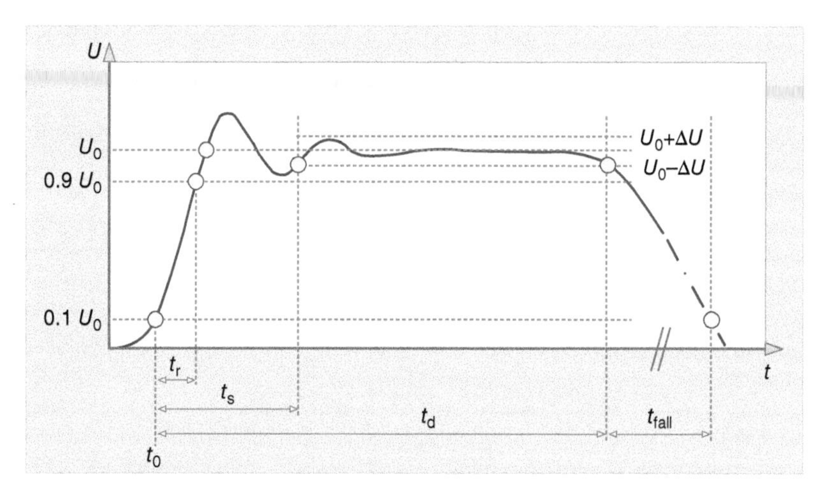

Figure 4.65 Output waveform of the calibrator pulse.

U_0	Step voltage magnitude	t_d	Step voltage duration/decay time
t_0	Origin of the step voltage	$(t_d - t_s)$	Steady-state duration
t_r	Rise time of the step voltage	ΔU	Absolute voltage deviation from U_0
t_s	Time to steady state	t_{fall}	Time interval until next voltage step

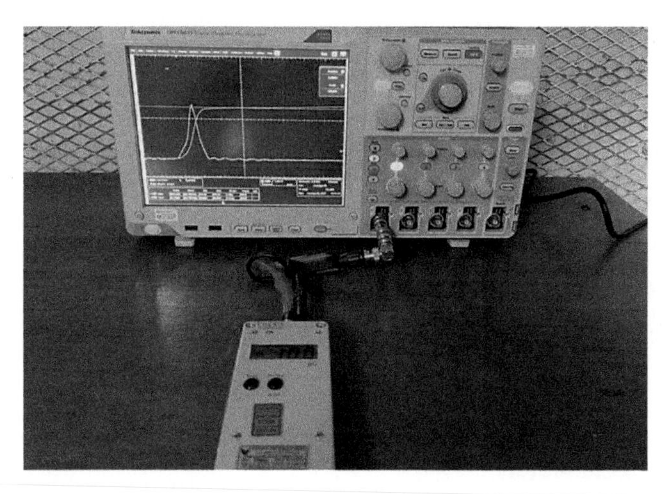

Figure 4.66 Calibration pulses of a commercial calibrator and integration of the measured current (Q_0 = 10 pC).

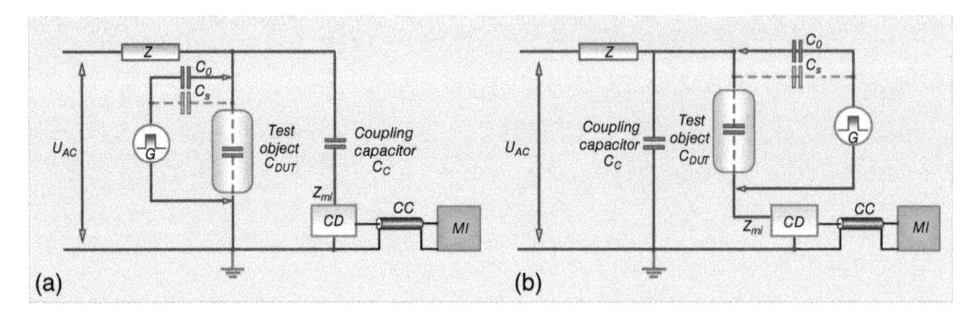

Figure 4.67 Alternative calibration procedures. (a) CD in series with coupling capacitor. (b) CD in series with the test object.

U_{AC}	Voltage supply	C_{DUT}	Test object
Z_{mi}	Input impedance of the measuring system	C_C	Coupling capacitor
CC	Connecting cable	CD	Coupling device
C_0	Injection capacitor	MI	Measuring instrument
Z	Blocking impedance or filter	G	Step voltage generator
C_S	Stray capacitance		

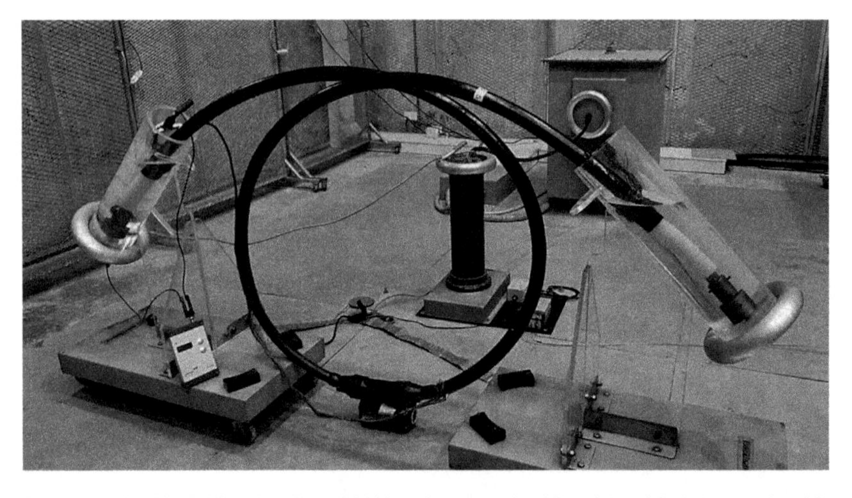

Figure 4.68 PD Calibration for a 22 kV underground cable with cable joint testing (Q_0 = 10 pC).

4.9.2 Performance Test of PD Calibrators

The performance test is the process of determining the stability and accuracy of a measuring device. As the characteristics of a calibrator can change with time, periodical tests of this device should be made at regular time intervals or after repair. The following characteristics of a calibrator are used for the performance test:

- Variable charge magnitude
- Variable time delay between two pulses
- Bipolar pulses

- A series of calibrator pulses with a known number of equal charge magnitude and repetitive frequency
- There are different methods used for this performance test. All these methods are detailed described in IEC 60270 [85].
- *Numerical integration.* The charge from the PD calibrator is determined by numerical integration of the current. The voltage at the input of a calibrated digital oscilloscope with a bandwidth not less than 50 MHz can be measured when the output terminals of the calibrator are connected by a noninductive resistor (Figure 4.69). The value of R_m should be selected between 50 and 200 Ω. The accuracy of the result is related to the accuracy of the integration procedure and the accuracy of the value of R_m.
- *Passive integration method.* The charge of the calibrator can also be determined by measuring the transient voltage appearing across a noninductive integrating capacitor, as shown in Figure 4.70. To ensure that the uncertainty of the measurement is below 3%, the capacitance parallel to the input of the oscilloscope should be selected not below 10 nF including the capacitance of the connecting cables and the input capacitance of the oscilloscope. Repeated measurements could be needed, but in any case, 10 independent measurements are sufficient.
- *Active integration method.* The charge of the calibrator can also be measured using an active integrator (Figure 4.71). Active integration is performed by a fast operational amplifier used in integration mode. The active integrator gain needs to be calibrated by a reference calibrator. The output voltage from the integrator is measured, and the gain of the system is used to calculate charge. The stability of the integrated pulse shall be included in the uncertainty. Repeated measurements could be needed, but 10 independent measurements are sufficient.

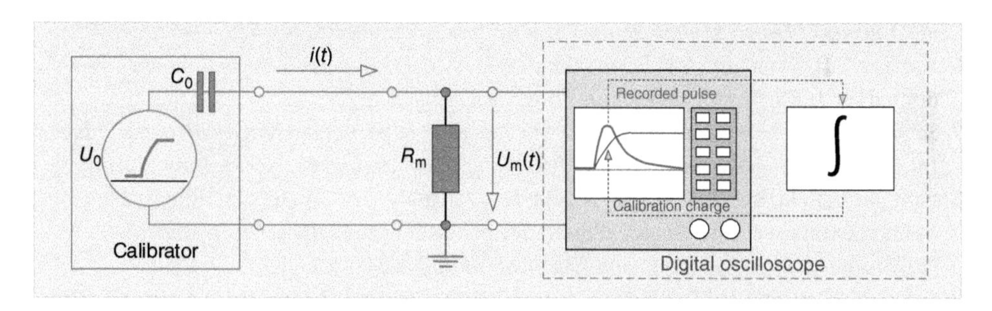

Figure 4.69 Performance test using the numerical integration method.

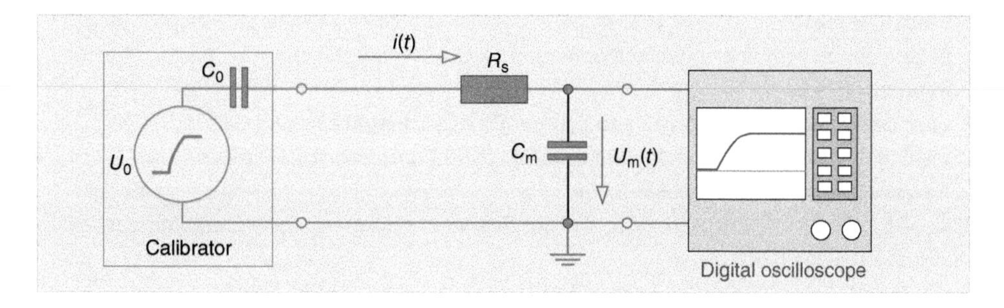

Figure 4.70 Performance test of calibrators using the passive integration method.

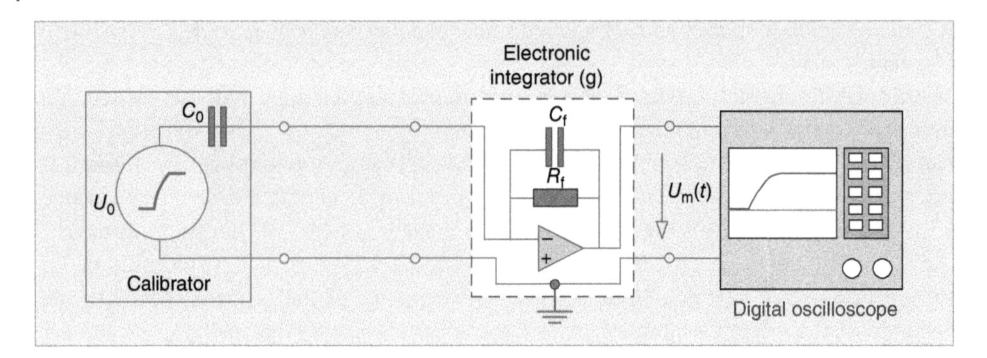

Figure 4.71 Performance test of calibrators using the active integration method.

References

1. IEC 60270: 2000 (2000). *High-Voltage Test Techniques – Partial Discharge Measurements*. Genève: International Electrotechnical Commission.
2. Wichmann, A. (1963). Alterungsmerkmale von Isolierungen elektrischer Maschinen und Folgerungen für den Wicklungsaufbau (in German). *ETZ-A* 84: 827–835.
3. Dakin, T.W. and Malinaric, P.J. (1960). A capacitive bridge method for measuring integrated corona, charge transfer and power loss per cycle. *AIEE Transactions on Power and Apparatus Systems* 79: 648–653.
4. IEC Publ. 599 (1999). *Interpretation of the Analysis of Gases in Transformers and Other Oil Filled Electrical Equipment in Service*. Genève: International Electrotechnical Commission.
5. Fallou, B. (1975). *Detection of and Research for the Characteristics of an Incipient Fault from the Analysis of Dissolved Gases in the Oil of an Insulation*, Electra No. 42e, 32–52. CIGRE.
6. Kreuger, F.H., Morshuis, P.H.F., and Sonneveld, W.A. (1988). Optical detection of surface discharges. *IEEE Transactions on Electrical Insulation* 23 (3).
7. Schwarz, R., Muhr, M., and Pack, S. (2001). *Detection of partial discharge with optical fibres*. In: *12th International Symposium on High Voltage Engineering*. Bangalore, India: ISH.
8. Fard, M.A., Reid, A.J., and Hepburn, D.M. (2017). Analysis of HVDC superimposed harmonic voltage effects on partial discharge behavior in solid dielectric media. *IEEE ransactions on Dielectrics and Electrical Insulation* 24 (1): 7–16.
9. Sonerud, B., Bengtsson, T., Blennow, J. et al. (2009). Dielectric heating in insulating materials subjected to voltage waveforms with high harmonic content. *IEEE Transactions on Dielectrics and Electrical Insulation* 16 (4): 926–933.
10. Saadati, H., Seifert, J., Werle, P. et al. (2015). *Partial discharge analysis under hybrid AC/DC field stress*. In: *19th International Symposium on High Voltage Engineering*. Pilsen, Czech Republic: ISH.
11. Florkowski, M., Kuniewski, M., and Zydron, P. (2020). Partial discharges in HVDC insulation with superimposed AC harmonics. *IEEE Transactions on Dielectrics and Electrical Insulation* 27 (6): 1906–1914.
12. CIGRE_WG D1.33 (2008). *Guide for Partial Discharge Measurements in Compliance to IEC 20670*. Paris: e-CIGRE.
13. IEC 60270: 2015 (2015). *High-Voltage Test Techniques – Partial Discharge Measurements*. Genève: International Electrotechnical Commission.

14 McLyman, C.W.T. (2004). *Transformer and Inductor Design Handbook*. New York: M. Dekker (Inc.).

15 Kornhuber, S., Jani, K., Boltze, M. et al. (2012). Partial discharge (PD) detection on medium voltage switchgears using non-invasive sensing methods. *IEEE Intern. Conference on Condition Monitoring (CMD)*. Paper A-174, Bali, Indonesia.

16 Schon, K. (1986). Konzept der Impulsladungsmessung bei Teilentladungsprüfungen (in German). *ETZ-Archiv* 8: 319–324.

17 Küpfmüller, K. (1974). *Die Systemtheorie der elektrischen Nachrichtenübertragung (in German)*. Stuttgart: Hirzel-Verlag.

18 NEMA (1940). *Methods of Measurement of Radio Noise*. Edison Electric Institute, NEMA, Radio Manufacturers Association.

19 NEMA (1964). *Methods of measurement of radio influence voltage (RIV) of High-Voltage Apparatus. NEMA* 107.

20 Graf, R.F. (1999). *Modern Dictionary of Electronics*. USA: Newnes.

21 Hussein, R., Shaban, K.B., and El-Hag, A.H. (2016). Energy conservation-based thresholding for effective wavelet denoising of partial discharge signals. *IET Science, Measurement and Technology* 10 (7): 813–822.

22 Lemke, E. (1979). *A new method for PD measurement on polyethylene insulated power cables*. In: *3rd International Symposium on High Voltage Engineering*, paper 43.13e. Milano, Italy: ISH.

23 CISPR 18-2 (1986). *Radio Interference Characteristics of Overhead Power Lines and High-Voltage Equipment – Part 2: Methods of Measurement and Procedure for Determining Limits*, 1e. IEC.

24 CISPR 16-1 (1993). *Specification for Radio Disturbance and Immunity Measuring Apparatus and Methods – Part 1: Radio Disturbance and Immunity Measuring Apparatus*, 1e. IEC.

25 Weissenberg, W., Farid, F., Plath, R. et al. (2004). *On-site PD Detection at Cross-Bonding Links of HV Cables*. Paris, France, Paper B1–102: CIGRE Session.

26 Koltunowicz, W. and Plath, R. (2008). Synchronous Multi-channel PD measurement. *IEEE Transactions on Dielectrics and Electrical Insulation* 15 (6): 1715–1723.

27 Qualitrol. (2023). Qualitrol 609 PDM transformer and GIS partial discharge monitor. https://www.qualitrolcorp.com/products/partial-discharge-monitors/transformer-and-gis-partial-discharge-monitors/qualitrol-609/ (accessed 7 June 2023).

28 Texas Instruments. (2013). Texas Instruments application report SNLA098A rev. 2013.

29 EUSPA. Galileo performance reports Q2–2022. https://www.gsc-europa.eu/news/galileo-performance-reports-of-q2-2022-available-for-download (accessed 7 June 2023).

30 Hylten-Cavallius, N. (1988). *High-Voltage Laboratory Planning*. Basel (Suisse): Emil Haefely Publ.

31 Hauschild, W. and Lemke, E. (2019). *High-Voltage Test and Measuring Techniques*, 2e. Switzerland AG: Springer Nature.

32 Schwab, A. (1969). *Hochspannungsmeßtechnik*, 1e. Springer-Verlag.

33 Shenoi, B.A. (2006). *Introduction to Digital Signal Processing and Filter Design*, 1e. Wiley.

34 Haller, R. and Martinek, P. (2014). *Partial Discharge Measurement at High Voltage Test Circuit Having Static Frequency Converters*. Santa Fe, NM (USA): IEEE International Power Modulator and High Voltage Conference (IPMHVC) paper 113.

35 Plath, K.-D., Plath, R., Emanuel, H. et al. (2002). *Synchronous Three-Phase PD Measurement On-Site and in the laboratory*, vol. paper 11. Betriebsmittel, Berlin: (VDE) ETG-Fachtagung Diagnostik elektrischer (in German).

36 Siegel, M., Beltle, M., Tenbohlen, S. et al. (2017). Application of UHF sensors for PD measurement at power transformers. *IEEE Transactions on Dielectrics and Electrical Insulation* 24 (1): 331–339.

37 Schaper, S., Kalkner, W., and Plath, R. (2004). *Synchronous Muli-channel PD Measurement on Power Transformers at Variable Middle Frequency (in German).* Betriebsmittel, Köln: (VDE) ETG-Fachtagung Diagnostik elektrischer.

38 Peier, D. and Engel, K. (1995). *Computer-aided Evaluation of PD Measurement Results (in German).*, 69–78. Betriebsmittel, Esslingen (Germany): (VDE) ETG-Fachtagung Diagnostik elektrischer.

39 Álvarez, F., Ortego, J., Garnacho, F. et al. (2016). A clustering technique for partial discharge and noise sources identification in power cables by means of waveform parameters. *IEEE Transactions on Dielectrics and Electrical Insulation* 23: 469–481.

40 Jang, J.-K., Kim, S.-H., Lee, Y.-S. et al. (1999). Classification of partial discharge electrical signals using wavelet transforms. In: *IEEE 13th Intern. Conf. on Dielectric Liquids*, 552–555. Nara, Japan: ICDL.

41 Lalitha, E.M. and Satish, L. (2000). Wavelet analysis for classification of multi-source PD patterns. *IEEE Transactions on Dielectrics and Electrical Insulation* 7: 40–47.

42 Winkelmann, E., Shevchenko, J., Steiner, C. et al. (2022). Monitoring of partial discharges in HVDC power cables. *IEEE Electrical Insulation Magazine* 38 (1): 7–18.

43 Mole, G. (1954). *The E.R.A. Portable Discharge Detector.* Paris, France: CIGRE Session, paper 105.

44 Kreuger, F.H. (1964). *Discharge Detection in High Voltage Equipment.* London: Temple Press.

45 CIGRE WG 23-01 (1969). *Recognition of discharges. Electra* 11.

46 König, D. and Rao, N.Y. (1993). *Partial Discharges in Electrical Power Apparatus.* Berlin: VDE Verlag GmbH.

47 Fruth, B.A. and Gross, D.W. (1994). Phase resolving partial discharge pattern acquisition and spectrum analysis. *Proceedings of the ICPDAM*, 2 pp. 578–581, Brisbane, Australia.

48 Kranz, H.G. (2000). Fundamentals in computer aided PD processing, PD pattern recognition and automated diagnosis in GIS. *IEEE Transactions on Dielectrics and Electrical Insulation* 7 (1): 12–20.

49 Gross, D.W. and Fruth, B.A. (1998). *Characteristics of Phase Resolved Partial Discharge Pattern in Spherical Voids*, 412–415. Atlanta, USA: CEIDP.

50 Gulski, E. (1991). Computer-aided recognition of partial discharges using statistical tools. PhD thesis. Delft University Press.

51 Hoof, M. and Patsch, R. (1994). Analyzing partial discharge pulse sequences: A new approach to investigate degradation phenomena. In: *Symposium on Electrical Insulation (ISEI)*, 327–331. Pittsburgh, USA: IEEE Intern.

52 Russell, S. and Norvig, P. (2021). *Artificial Intelligence: A Modern Approach*, 4e. London, UK: Pearson.

53 Shanmuganathan, S. and Samarasinghe, S. (2016). *Artificial Neural Network Modelling*, 1e. Switzerland: Springer.

54 Lin Du, K. and Swamy, M.N.S. (2014). *Neural Networks and Statistical Learning*, 1e. London, UK: Springer.

55 Galushkin, A.I. (2007). *Neural Networks Theory*, 1e. New York, Berlin Heidelberg: Springer.

56 Roy, N.B. and Bhattacharya, K. (2020). *Application of Signal Processing Tools and Artificial Neural Network in Diagnosis of Power System Faults*, 1e. Boca Raton, UK: CRC Press.
57 Liu, A.C.C., Law, O.M.K., and Law, I. (2022). *Understanding Artificial Intelligence*, 1e. Hoboken, New Jersey: Wiley.
58 Raymond, W.J.K., Illias, H.A., Bakar, A.H.A., and Mokhlis, H. (2015). Partial discharge classifications: review of recent progress. *Measurement* 68: 164–181.
59 Pattanadech, N., Nimsanong, P., Potivejkul, S. et al. (2015). Generalized regression networks for partial discharge classification. In: *18th ICEMS*, 1176. Pattaya City, Thailand: IEEE.
60 Pattanadech, N., Nimsanong, P., Potivejkul, S. et al. (2015). Partial discharge classification using probabilistic neural network model. In: *18th ICEMS*, 1176–1180. Pattaya City, Thailand: IEEE.
61 Long, J., Wang, X., Zhou, W. et al. (2021). A comprehensive review of signal processing and machine learning technologies for UHF PD detection and diagnosis (I): preprocessing and localization approaches. *IEEE Access* 9.
62 Iorkyase, E.T., Tachtatzis, C., Glover, I.A., and Atkinson, R.C. (2019). RF-based location of partial discharge sources, using received signal features. *High Voltage* 4 (1): 28, 69876–32, 69904.
63 Samaitis, V., Mažeika, L., Jankauskas, A., and Rekuviene, R. (2020). Detection and localization of partial discharge in connectors of air power lines by means of ultrasonic measurements and artificial intelligence models. *Sensors* 21 (1): 20.
64 Ganguly, B., Chaudhuri, S., Biswas, S. et al. (2021). Wavelet kernel-based convolutional neural network for localization of partial discharge sources within a power apparatus. *IEEE Transactions on Industrial Informatics* 17 (3): 1831–1841.
65 Liu, J., Hu, Y., Peng, H., and Zaki, A.U.M. (2020). Data-driven method using DNN for PD location in substations. *IET Science, Measurement and Technology* 14 (3): 314–321.
66 Yeo, J., Jin, H., Mor, A.R. et al. Localisation of partial discharage in power cables through multi-output convolutional recurrent neural network and feature extraction. *IEEE Transactions on Power Delivery* https://doi.org/10.1109/TPWRD.2022.3183588.
67 Tozzi, M., Salsi, A., Busi, M. et al. (2011). Permanent PD monitoring for generators: Smart alarm management. In: *PES Innovative Smart Grid Technologies*, 1–6. Perth, WA, Australia: IEEE.
68 Wen, J.Y.W. (2021). Neural network based identification and localisation of partial discharge in high voltage cable. Doctoral thesis. Singapore University of Technology and Design.
69 Janani, H. (2016). Partial discharge source classification using pattern recognition algorithms. Doctoral thesis. University of Manitoba.
70 Joshi, A.V. (2022). *Machine Learning and Artificial Intelligence*, 1e. Switzerland: Springer.
71 Xavier, G.V.R., Silva, H.S., Costa, E.G.D. et al. (2021). Detection, classification and location of sources of partial discharges using the radiometric method: trends, challenges and open issues. *IEEE Access* 9: 110787–110810.
72 IEC 60270 (2015). *High Voltage Test Techniques - Partial Discharge Measurements*. Switzerland: International Electrotechnical Commission.
73 CIGRE TF 226 (2003). *Knowledge Rules for Partial Discharge Diagnosis in Service*. France: Conseil International des Grands Réseaux Électriques.

74 Rey, J.A.A., Luna, M.P.C., Valderrama, R.A.R. et al. (2020). Separation techniques of partial discharges and electrical noise sources: a review of recent progress. *IEEE Access* 8: 199449–199461.

75 Shim, I., Soraghan, J.J., and Siew, W.H. (2001). Detection of PD utilizing digital signal processing methods. Part 3: open-loop noise reduction. *IEEE Electrical Insulation Magazine* 17 (1): 6–13.

76 Barrios, S., Buldain, D., Comech, M.P., and Gilbert, I. (2021). Partial discharge identification in MV switchgear using scalogram representations and convolutional AutoEncoder. *IEEE Transactions on Power Delivery* 36 (6): 3448–3455.

77 Seo, J., Ma, H., and Saha, T. (2015). Probabilistic wavelet transform for partial discharge measurement of transformer. *IEEE Transactions on Dielectrics and Electrical Insulation* 22 (2): 1105–1117.

78 Zhang, H., Blackburn, T.R., Phung, B.T., and Sen, D. (2007). A novel wavelet transform technique for on-line partial discharge measurements. 1. WT de-noising algorithm. *IEEE Transactions on Dielectrics and Electrical Insulation* 14 (1): 3–14.

79 Liu, J. (2019). Partial discharge denoising for power cables. Doctoral thesis. University of Strathclyde.

80 Song, X., Zhou, C., Hepburn, D.M. et al. (2007). Second generation wavelet transform for data denoising in PD measurement. *IEEE Transactions on Dielectrics and Electrical Insulation* 14 (6): 1531–1537.

81 Iorkyase, E.T., Tachtatzis, C., Glover, I.A. et al. (2019). Improving RF-based partial discharge localization via machine learning ensemble method. *IEEE Transactions on Power Delivery* 34 (4): 1478–1489.

82 Evagorou, D., Kyprianou, A., Lewin, P.L. et al. (2010). Feature extraction of partial discharge signals using the wavelet packet transform and classification with a probabilistic neural network. *IET Science, Measurement and Technology* 4: 177–192.

83 Zhang, X., Hu, Y., Deng, J. et al. (2022). Feature engineering and artificial intelligence-supported approaches used for electric powertrain fault diagnosis: a review. *IEEE Access* 10: 29069–29088.

84 Lu, S., Chai, H., Sahoo, A., and Phung, B.T. (2020). Condition monitoring based on partial discharge diagnostics using machine learning methods: a comprehensive state-of-the-art review. *IEEE Transactions on Dielectrics and Electrical Insulation* 27 (6): 1861–1888.

85 IEC 60270 ED4. (2022). High-voltage test techniques-Charge-based measurement of partial discharges. https://standards.globalspec.com/std/9972885/IEC%2060270 (accessed 7 June 2023).

86 Küchler, A. (2017). *High Voltage Engineering*. Berlin: Springer.

87 Rizk, F.A.M. and Trinh, G.N. (2014). *High Voltage Engineering*. New York.: CRC Press.

5

Electromagnetic Methods for PD Detection

5.1 Introduction

Partial discharge (PD) generates extremely fast PD pulse currents that radiate electromagnetic (EM) waves in the high-frequency (HF) range, the very-high-frequency (VHF) range, and the ultra-high-frequency (UHF) range with frequencies from 3 MHz to 3 GHz.

In practice, EM PD measurement is generally divided into two categories: (i) HF/VHF PD measurement and (ii) UHF PD measurement. The decision as to which process to use (either or both) depends mainly on the test equipment available. Apart from the nature of the EM wave from the PD source, the results obtained from PD testing depend on various factors, such as the type of sensor(s) used, the type of instrumentation used, and the measurement techniques used. These factors are determined by the required or expected parameters for the process of measurement, analysis, and interpretation [1].

5.2 PD Measurement by HF and VHF Sensors

Common types of PD events in the range of 3–30 MHz to 30–300 MHz can be adequately measured using various types of commercially available HF electromagnetic radiation sensor, such as capacitive coupler (CC), current transformer (CT), Rogowski coil (RC), directional electromagnetic coupler, film electrode, and transient earth voltage (TEV) sensor [2].

5.2.1 PD Measurement by CC

5.2.1.1 Theory

A CC designed for PD measurement consists of a high-voltage coupling capacitor (C) for detecting the conducted PD pulse signal, and an impedance measurement resistor (R). When the CC is introduced into the electrical power system for PD measurement, it acts as a high-pass filter with lower cutoff frequency of f_{lo} [3]. These PD pulses may include frequencies of up to several hundred megahertz [1, 4]. However, at the CC location, the shape and magnitude of PD may be attenuated – especially for HF components. The wave spectra of the PD pulse

Partial Discharges (PD) - Detection, Identification, and Localization, First Edition. Norasage Pattanadech, Rainer Haller, Stefan Kornhuber, and Michael Muhr.

both at its origin (PD_o) and at the location of the CC (PD_t) obtained from Fourier transformation analysis (with upper cutoff frequencies of $f_{up}(PD_o)$ and $f_{up}(PD_t)$, respectively) is shown in Figure 5.1. [1]

$$f_{lo} = \frac{1}{2\pi RC} \tag{5.1}$$

allowing for the time constant of the CC: τ

$$\tau = RC \tag{5.2}$$

In practice, a bandpass filter is used to measure PDs in the HF range. This system tends to compromise PD measurement by determining the lower cutoff frequency and lower the upper cutoff frequency of the PD signal. Another problem for PD measurement in the lower frequency range is that, with respect to the setting of the lower cutoff

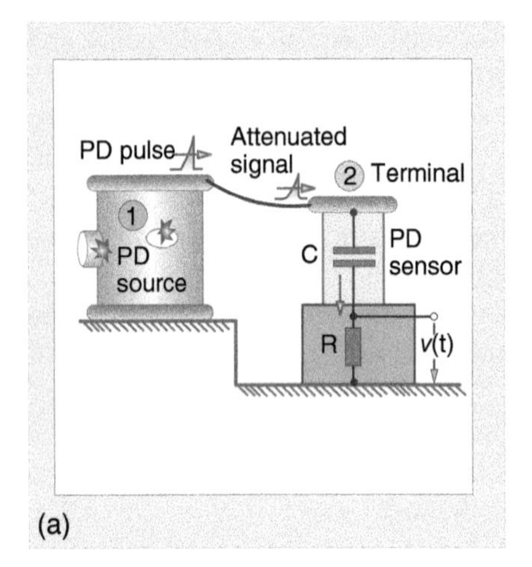

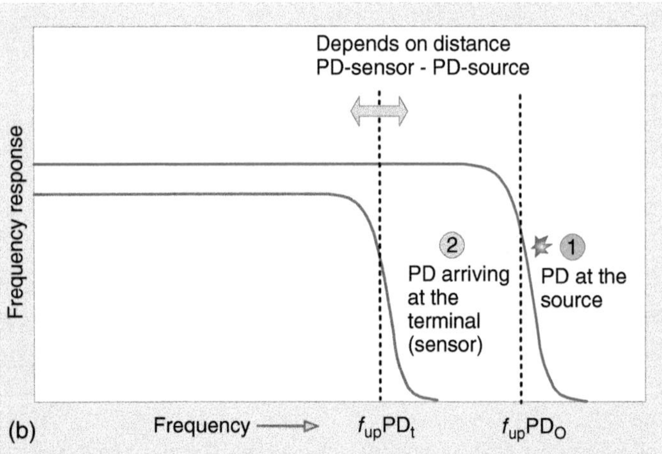

Figure 5.1 (a) PD measurement by a CC; (b) the spectrum of the PD pulses both at its origin and at the location of the CC obtained from Fourier transformation analysis.

frequency, low-frequency PD measurement is susceptible to extraneous noise signals, which may cause the problem of PD result analysis.

When setting the upper cutoff frequency of the bandpass filter, if the cutoff frequency of the bandpass filter is significantly lower than the upper cutoff frequency of the PD signal at the CC position, then the detected PD pulses will be directly proportional to the apparent charge of the PD current pulse as demonstrated in Figure 5.2. However, where the upper cutoff frequency of the bandpass filter is well above the upper cutoff frequency of the PD signal at the CC location, the detected PD pulses will not be directly proportional to the apparent charge of the PD pulses as depicted in Figure 5.3. Consequently, measurement of PD pulses in the high-frequency range is normally expressed in terms of voltage magnitude [1].

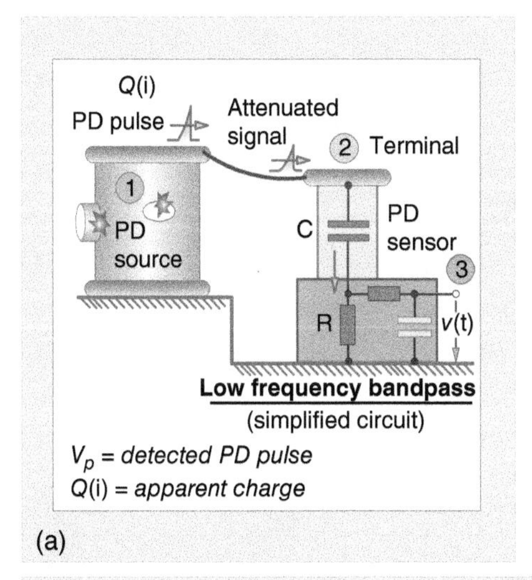

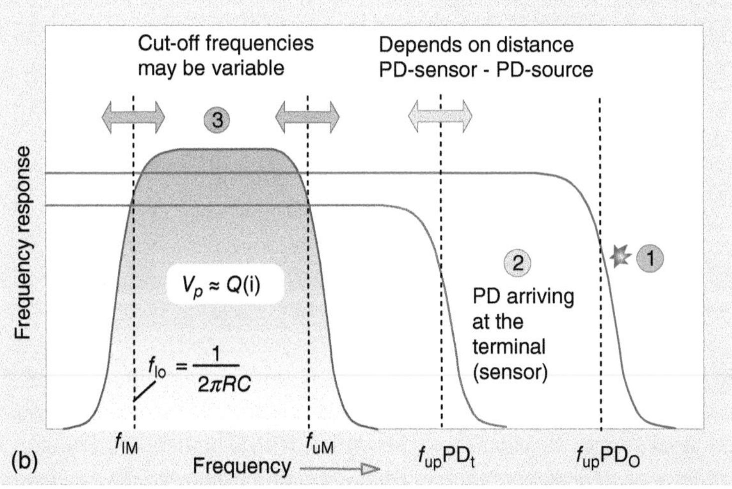

Figure 5.2 (a) PD measurement by a CC; (b) the spectrum of the PD pulses its origin and at the location of the CC obtained from Fourier transformation analysis and the low bandpass frequency response of the PD measuring system.

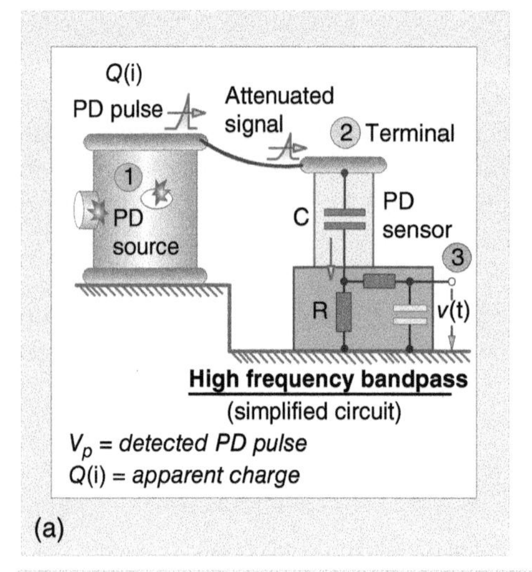

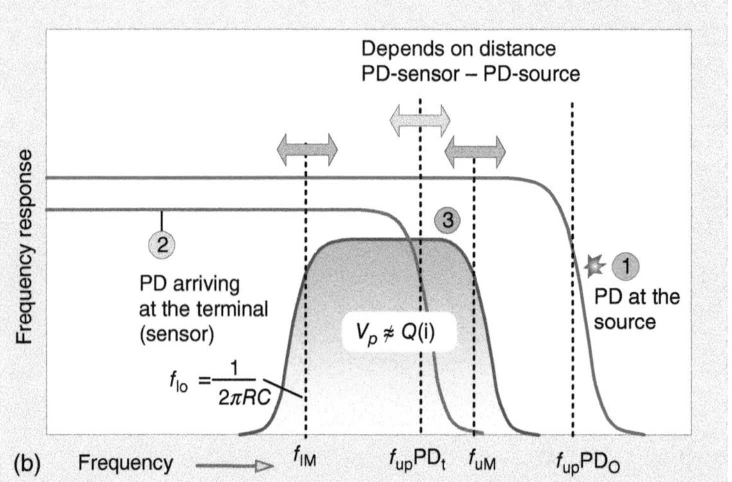

Figure 5.3 (a) PD measurement by a CC; (b) the spectrum of the PD pulses its origin and at the location of the CC obtained from Fourier transformation analysis and the high bandpass frequency response of the PD measuring system [1].

Coupling capacitors of various ratings (e.g. 80 pF to 1.5 nF) are readily available from commercial sales. Figure 5.4 shows the lower cutoff frequency of 3.18 kHz to 31 MHz for a CC at various capacitance values and constant resistance of 50 Ω. As mentioned above, PD measurement in the low-frequency range is susceptible to extraneous noise signals, which requires the use of effective noise suppression techniques or PD–noise separation. For PD measurements in the high-frequency range, the signal-to-noise ratio (SNR) is higher, and therefore the problem is reduced. However, the attenuation and distortion of the detected PD signal are much higher, different from the original PD. To increase the sensitivity of the PD measurement, appropriately higher values of resistance and/or capacitance of the CC are needed [5].

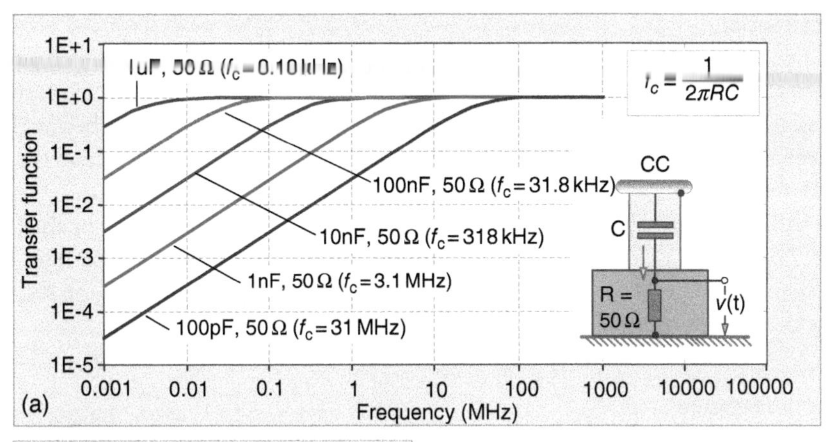

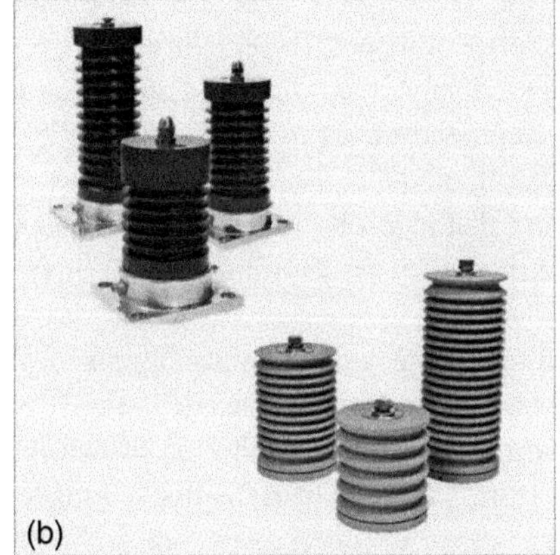

Figure 5.4 (a) The lower cutoff frequency of a CC at various capacitance values and constant resistance of 50 Ω; (b) commercial CCs.

5.2.1.2 CC Characteristics and Installation Aspect for PD Measurement

CC Characteristics

CCs are intended to be permanently installed on the high-voltage terminals of the high-voltage equipment. Accordingly, it is very important that the CCs themselves are not the cause of such equipment failure, and CCs are required to withstand voltage test standards and/or endurance voltage test standards. The following parameters must be considered:

1) Constant and low dissipation factor with a specified PD level under working condition
2) High PD inception voltage (PDIV) with a specified PD level
3) High PD extinction voltage (PDEV) with a specified PD level

Furthermore, CCs are required to function reliably under normal operating conditions with regard to temperature, vibration, and various electromechanical forces.

5.2.1.3 CC Installation

CCs are intended to be mounted on the high-voltage terminals of the high-voltage equipment so as to be in close proximity to the insulation, which is exposed to the most severe stresses from the electrical field generated during operation. However, because mounting the capacitors on the high-voltage terminals can have a detrimental effect on the designed insulation coordination and the required safety clearances, installation of the CCs is required to satisfy the appropriate insulation performance standards in order to operate reliably in service. Typically, one or two CCs are used per phase for three-phase high-voltage equipment. A single CC per phase is generally used for PD detection, with the second CC being used to determine PD location.

5.2.1.4 CC for PD Measurement

PD Measurement [1]

CCs are often used for on-line PD measurement. The capacitor detects the PD pulse current, and signals are then sent to the acquisition system unit for display, recording, and analysis, as shown in Figure 5.5. PD values and trend indications can then be used to assist in determining insulation condition or in development of an insulation model.

On-line PD monitoring is prone to interference and distortion by high levels of background noise. However, the noise problem can be minimized by the technique of PD time-domain measurement, which uses *pulse-shape analysis*.

Pulse-shape analysis is used to analyze the rise-time of detected PD pulse signals and thereby determine the noise. PD signals, especially those generated near CCs, tend to exhibit a noticeably shorter rise time compared to background noise signals, as shown in Figure 5.6. However, PD signals generated further from CCs will also display a longer rise time.

Test Procedure

In accordance with IEC 60034-27-2 [1], after the PD coupler is installed, normalization (aka sensitivity checking) of the PD measuring system should be carried out before conducting PD measurement and PD localization. IEC 60034-27-2 [1] classifies normalization methods according to measuring frequency system i.e. low-frequency normalization and high-frequency normalization.

Normalization for Low-Frequency Systems: A short-duration current pulse from a standard pulse generator in accordance with IEC 60270 [5] is applied to the high-voltage terminal of the equipment under test, in order to simulate the effect of PD pulses during PD measurement. This allows for the ratio between the recorded (mV) and the applied charge to be calculated. It is important to note that pulse generators are not able to fully replicate actual PD pulses occurring within the complex structure of the insulation system.

Normalization for High-Frequency Systems: In accordance with IEC 60034-27-2 [1], a specific rectangular voltage pulse is applied to the high-voltage terminal of the test object. It should be noted that IEC 60034-27-2 [1] only applies to on-line PD measurement of the stator winding insulation of rotating electrical machines. For conducting PD measurement

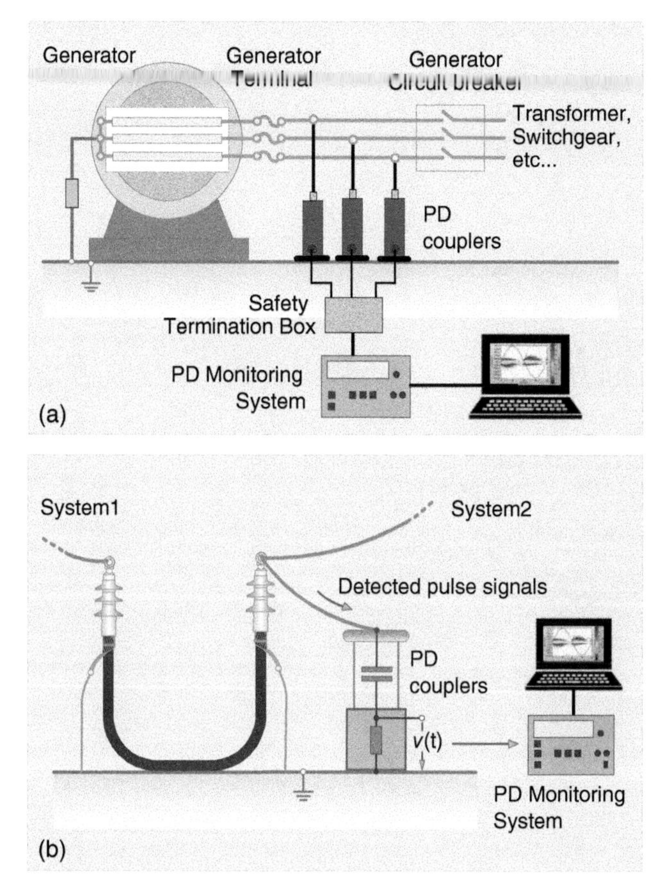

Figure 5.5 CCs for on-line PD measurement of (a) rotating machine; (b) underground cable system.

of other high-voltage equipment by CCs, it may or may not need to conduct the normalizations as mentioned.

Application of CC for PD Measurement of High-Voltage Apparatus

Application of CC for PD Measurement of Rotating Machine: PD measurement of rotating machines by use of CCs can be performed by using either a single CC or a pair of CCs per phase. Two CCs per phase are preferred for determining PD location and for PD-noise separation using the *time of pulse arrival* technique. In this process, one of the coupler units is installed at the high-voltage terminal, while the location of the second unit is away from the first coupler, determined by the capability of electronic circuit in the acquisition system, as shown in Figure 5.7. Figure 5.8 illustrates the application of CCs for on-line PD measurement of a generator rated at 11 kV 8400 kW and PD test results obtained.

Application of CCs for PD Measurement of Underground Cable: Depending on the specific application, PD measurement in underground cables can be determined by use of a single

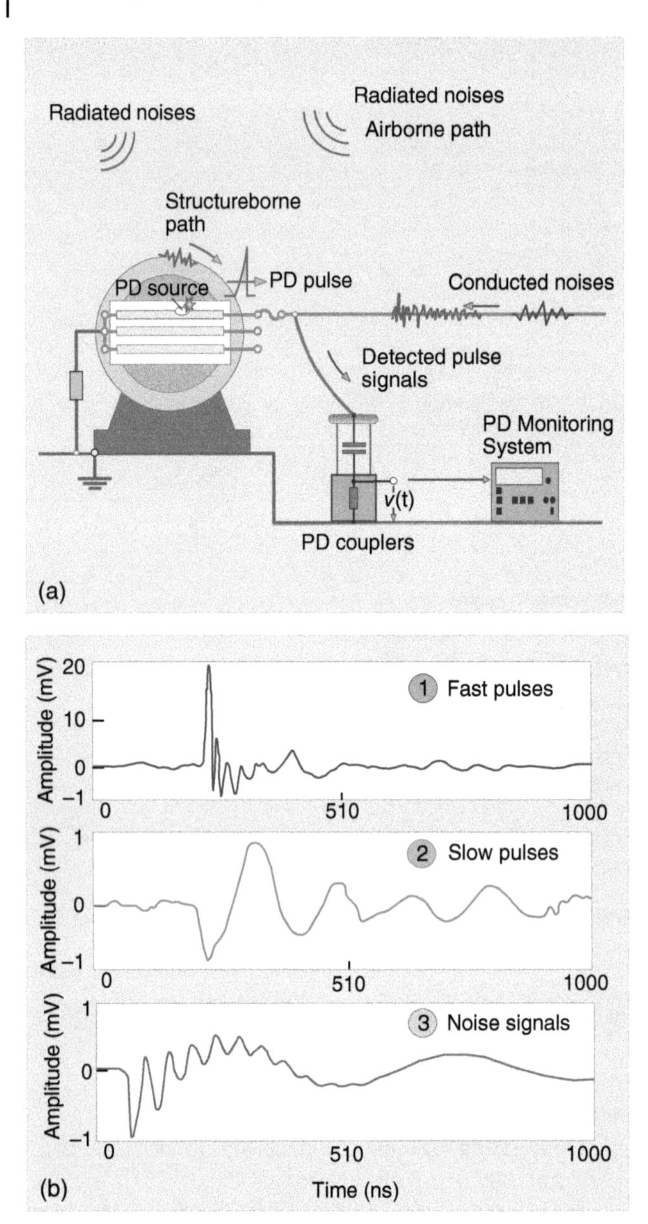

Figure 5.6 (a) PD and noise signal entering the PD measurement system; (b) pulse shape analysis.

or a double CC, as shown in Figure 5.9. Figure 5.10a shows on-site PD measurement of underground cable by a single CC and the test results obtained, both PD phase result pattern as shown in Figure 5.10 b and the estimated PD fault position presented in Figure 5.10c.

Application of CCs for PD Measurement of Other High-Voltage Equipment: Application of CCs for PD measurement for other high-voltage equipment such as dry type transformers and switchgear is available now.

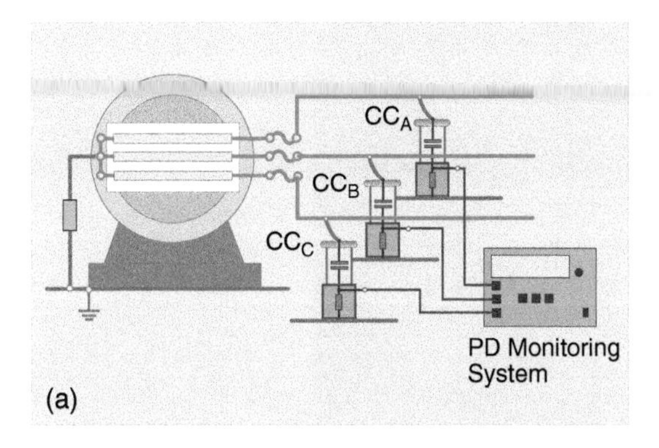

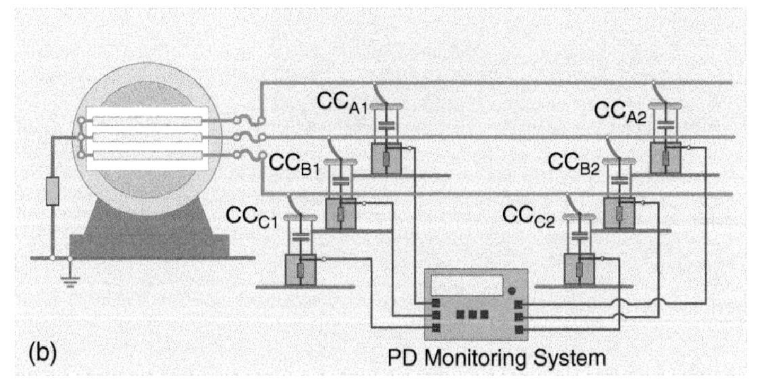

Figure 5.7 PD detection diagram: (a) one capacitive coupler (CC) per phase; (b) two capacitive couplers (CCs) per phase.

5.2.2 PD Measurement by Inductive Couplers

Inductive couplers (i.e. high frequency current transformers (HFCT) and Rogowski coils) are commonly used for on-line PD measurement of high-voltage equipment. PD current flowing through the primary conductor induces a secondary voltage in the coil of inductive couplers. The induced voltage may need to be reconstructed to the PD current measured. However, the induced signal is normally quite weak; therefore, it needs to be amplified and then transmitted to the data acquisition unit to be stored in the database for subsequent analysis and display. Typically, the inductive coupler (also called inductive converting device) functions as a high-pass filter. Therefore, only alternating or transient current can be measured [7]. Because inductive couplers do not require galvanic contact with the conductor, they do not affect the operation of the electrical power system [2].

5.2.2.1 PD Measurement by High-Frequency Current Transformers

Theory

HFCTs are widely used to detect the PD pulses (aka PD activity) generated in high-voltage equipment, such as transformers, rotating machines, switchgear, and underground cables and their accessories. The PD pulse current flowing through the primary conductor creates an induced voltage $U_t(t)$ at the secondary winding terminal of the HFCT. The induced voltage can be calculated using the equation

(a)

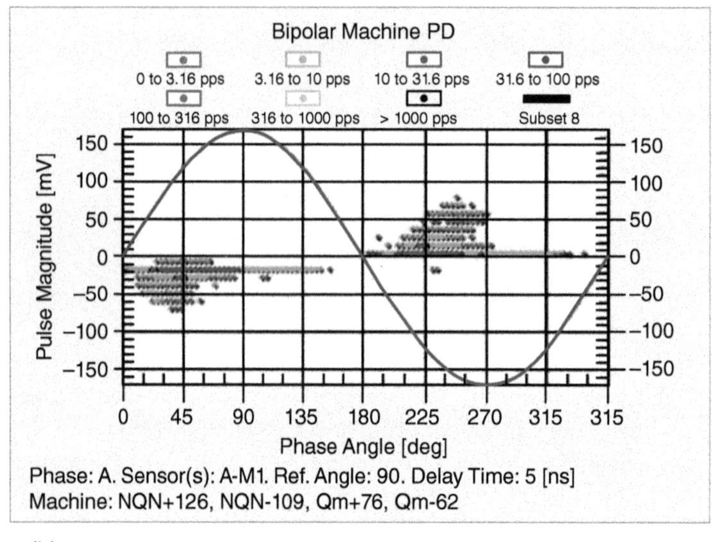

(b)

Figure 5.8 (a) Application of capacitive couplers (CCs) for on-line PD measurement of an 11 kV 8400 kW generator; (b) the PD test result obtained.

$$U_{\mathrm{t}}(\mathrm{t}) = -N_{\mathrm{s}} \frac{d\varnothing}{dt} \tag{5.3}$$

where U_t(t) is the induce voltage at the terminal of the secondary winding. N_s is the turn number of the secondary winding. ∅ is the magnetic flux passing through the magnetic core (Figure 5.11).

As already mentioned, the induced output voltage is typically quite weak. Therefore, the output voltage will be amplified, filtered, and transmitted to the data acquisition unit for storage, further analysis, and display. The magnitude of the PD pulse measured by HFCT

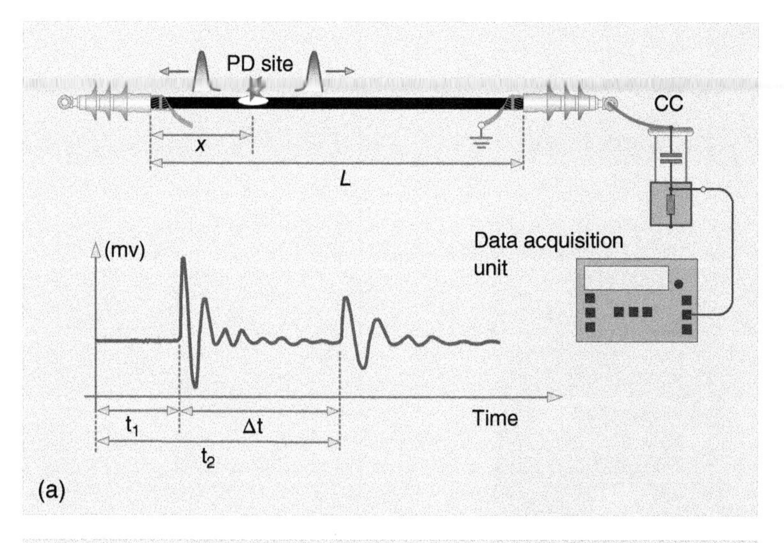

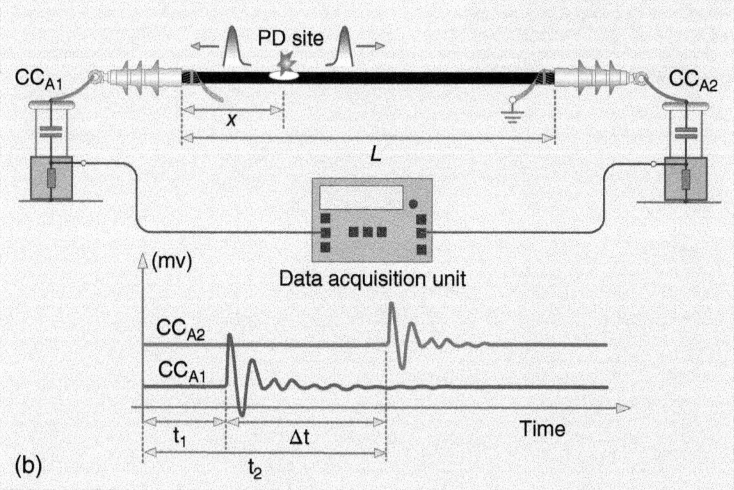

Figure 5.9 Application of capacitive couplers (CCs) for PD measurement: (a) one capacitive coupler (CC) per phase; and (b) two capacitive couplers (CCs) per phase [6].

in mV (or mA) is not comparable in term of apparent change (pC). It has been reported that the relation between mV and pC mainly depends on the calibration-pulse shapes, which are different for the real PD pulse shapes [8].

Structure

A typical HFCT consists of a toroidal core wound with insulated copper wire that terminates with a load impedance. The toroidal core material normally consists of a ferromagnetic material featuring high permeability, high resistivity, and low eddy-current loss [9]. There are two types of toroidal core in general usage: the nonsplit ferrite type and the openable-split type. Nonsplit ferrite cores provide a flat frequency response but have poor saturation characteristics, whereas openable-split ferrite cores have high saturation

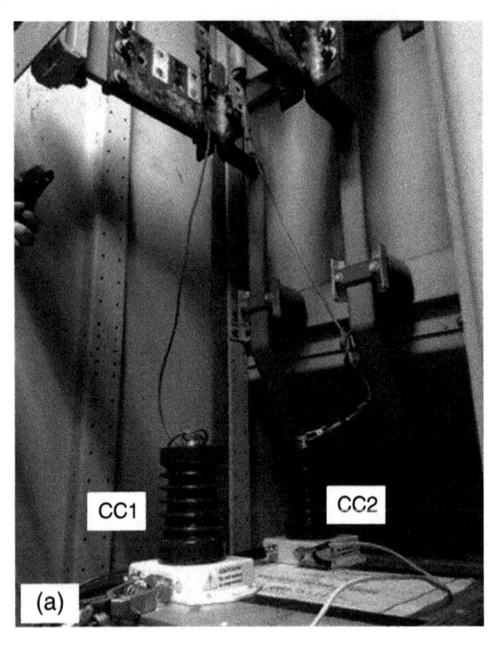

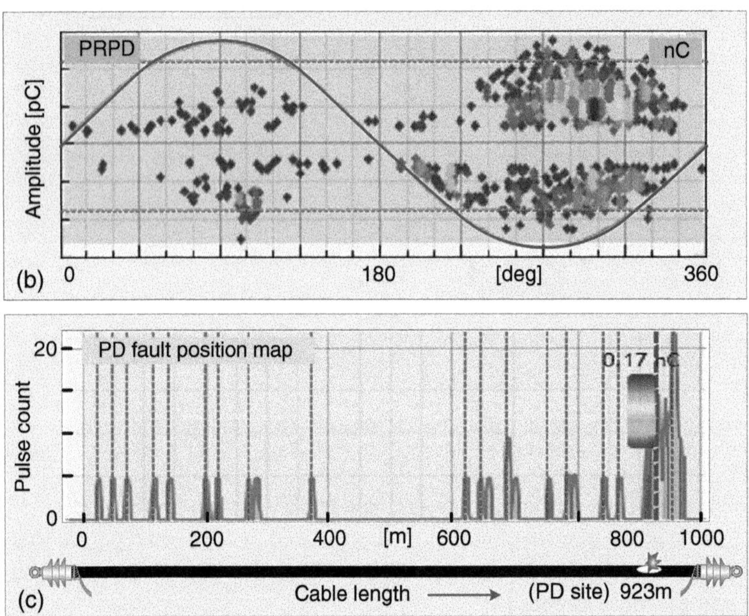

Figure 5.10 (a) On-site PD measurement of underground cable; (b) phase resolved PD pattern; (c) PD fault position.

characteristics [9]. In both types, the core winding should be wound evenly in order to reduce leakage inductance.

Characteristics

To measure PD pulse currents, at least three main characteristics of HFCT should be considered – high sensitivity, wide bandwidth, and high saturation current capacity. Sensitivity

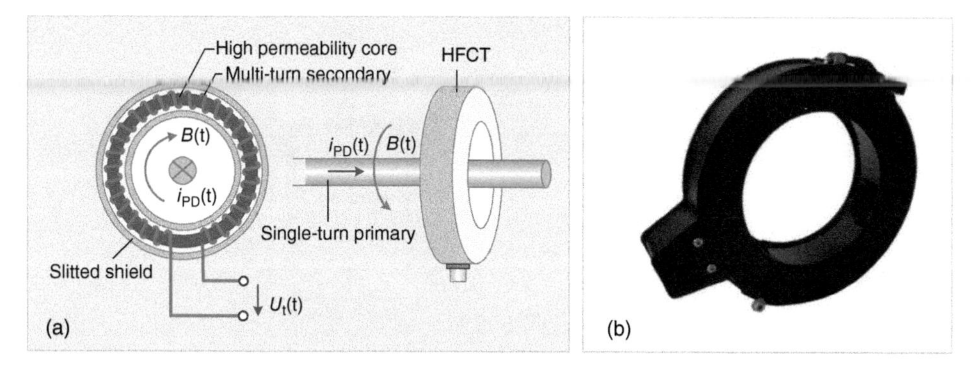

Figure 5.11 HFCT for PD measurement with a single turn primary winding and multi-turns secondary winding (a) HFCT diagram; (b) commercial HFCT.

of the HFCT is determined according to the ratio between the secondary voltage and the primary current, expressed as voltage divided by amperes, or V/A. Bandwidth of the HFCT sensor is usually determined at minus 3 dB of the transfer function. The bandwidth of commercial HFCT sensors for PD measurement is usually in the high-frequency ranges (e.g.10 kHz to 10 MHz, 30 kHz to 20 MHz). High-saturation current capacity is particularly required when the HFCT is installed in close proximity to a high-current phase conductor.

Application HFCT for PD Measurement

In order to detect PD pulse-current generated in the cable system (underground cables and their accessories), HFCTs can be clamped to the earthing strap or around the conductor by which the shield screen at the clamping point may be removed. Similarly, for other high-voltage equipment, such as transformers, rotating machines, or switchgear, the HFCT can also be clamped to the earthing strap. The diagram and the example of PD measurement by HFCT are shown in Figure 5.12 a and b respectively.

5.2.2.2 PD Measurement by Rogowski Coil

Theory

The Rogowski coils mainly used to measure high amplitude alternating current (AC) and high-amplitude impulse current. However, recently a Rogowski coil has been introduced to detect low-amplitude current with high-frequency components, such as PD pulse current [10], [11]. For PD measurement, the RC is clamped around the conductor of which PD current is to be measured. When the PD pulse current passes through the conductor, a voltage is induced at the terminals of the winding (Figure 5.13). The induced RC voltage can be calculated as

$$v_t(t) = -M\frac{d}{dt}i_{PD}(t) \tag{5.4}$$

where $v_t(t)$ is the voltage induced at the coil by the PD current i_{PD} and M is the mutual inductance or sensitivity of the Rogowski coil. The induced voltage is directly proportional to the derivative of the PD current in the conductor. However, because the waveform of the RC output signal differs from the measured PD current, the output voltage

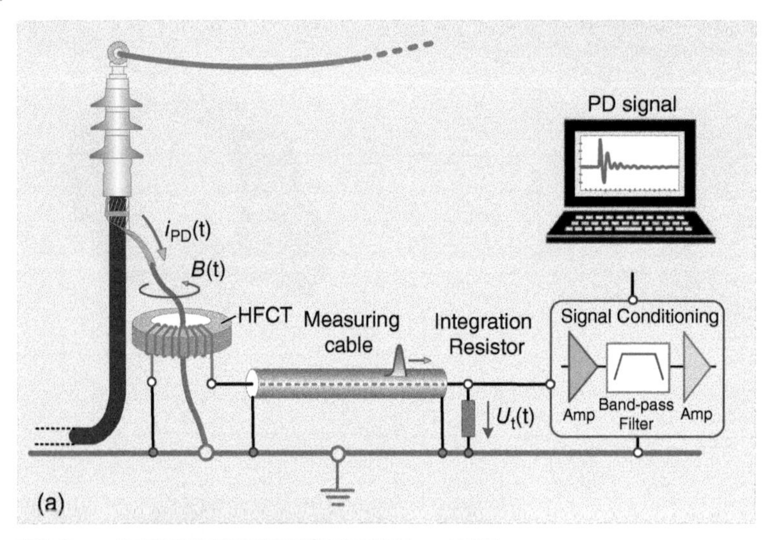

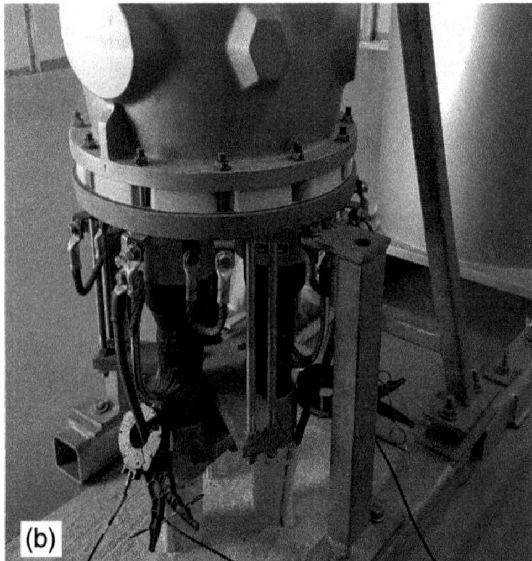

Figure 5.12 Application of HFCT for PD measurement of underground cable (a) diagram; (b) on-site underground PD measurement.

needs to be integrated to reconstructed the measured PD current as described by the following equation:

$$I_{\mathrm{PD}}(\mathrm{t}) = -\frac{1}{M}\int v_{\mathrm{t}}(\mathrm{t})dt \tag{5.5}$$

There are various types of integrators in common usage, such as the self-integrator, the resistive-capacitive integrator, the op-amp integrator, and the microprocessor integrator. Furthermore, a low-impedance resistor connected to a Rogowski coil can also be used as a self-integrator, although it provides a very low gain. Resistive-capacitive based integrators use resistors and capacitors to construct an integrator circuit capable of measuring

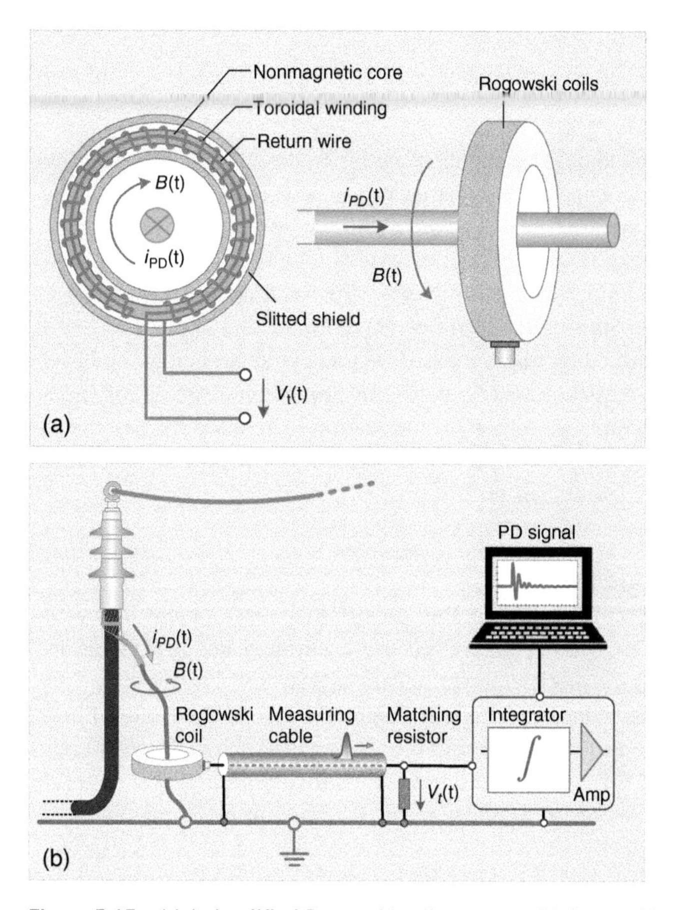

Figure 5.13 (a) A simplified Rogowski coil structure; (b) Rogowski coil with integrate circuit.

high-frequency outputs in excess of 100 MHz. By contrast, op-amp integrators are designed to measure low-frequency outputs below 100 MHz. Microprocessor integrators (aka numerical integrators) consist of a microprocessor based devices used with an analog-to-digital converter, moreover a digital filter may be added to provide a better result. Currently, a numerical technique is simplified, more effective, and easier to use [12], [13].

Structure

The design of RC devices usually includes an integrator. The Rogowski coil and its integrator applied for PD measurement are designed for high-frequency measurement from ten to hundreds of MHZ [13]. The main component of a typical Rogowski coil is the toroidal coil wound with insulated copper wire terminated by an integrator. A nonmagnetic core (aka an air core) is used for the purpose of providing the required linear characteristic. Rogowski coils come in various shapes, such as round, oval, and rectangular, with round and oval being the most common. Similarly, two types of cores – nonsplit cones and openable-split cone – are generally used. A typical Rogowski coil has two wire loops connected in electrically opposite directions in order to minimize the influence of any energized conductors in the vicinity [14].

Characteristics

In applying Rogowski coils for PD measurement, two main characteristics should be considered:

1) Sensitivity: The Rogowski coil should be highly sensitive to detect low-level PD. This sensitivity can be considered as a mutual inductance value, as it represents the ability of the Rogowski coil to produce an output voltage due to the change of the primary current (PD current) (V/A).
2) Bandwidth: Accurate measurement of PD requires a wide bandwidth measuring system. Considering the transfer impedance in frequency domain, the upper and lower cutoff frequencies are determined by the interested frequency range. Below the lower cutoff frequency, the power frequency, and its harmonic components, including interference, tend to exhibit a significant degree of attenuation. Therefore, in practice, a lower cutoff frequency in the range of 50 kHz to 1 MHz, along with a higher cutoff frequency of at least 120 MHz, are suggested [15].

Application Rogowski Coil for PD Measurement

Rogowski coils can be used for on-line, off-line, and on-site PD measurement in a wide variety of high-voltage equipment. This is largely because traditional Rogowski coils are easy to use and do not interrupt the electric power network. However, Rogowski coil are generally less sensitive than HFCT devices, although Rogowski coil do have more linear characteristics and are nonsaturable [16]. One recent innovation has been the development and adoption of a Rogowski coil for PD measurement, which incorporates a PCB (printed circuit board) [12], [16]. Examples of RC application for PD measurement are shown in Figure 5.14.

5.2.3 PD Measurement by DCS

5.2.3.1 Theory

A directional coupler sensor (DCS) is a passive device that is mainly used to monitor PD in underground cables and joints. The DCS has proven to be a reliable method of measuring both on-line and off-line PD measurement in the frequency range of 2–500 MHz [7]. The DCS operating principle is to couple energy generated by PD pulses as they propagate to various output ports (for DCS two ports are usually sufficient) in accordance with the direction of the pulse [17–19] whereby energy generated by the PD pulse as it propagates in one direction tends to generate an output signal at the one of two output ports, and energy generated by the PD pulse as it travels in an opposite direction tends to generate an output at the other output port [17].

5.2.3.2 DCS Structure and Characteristic

A standard DCS device consists of two main parts: a metallic foil that serves as an electrode and a dielectric layer above the electrode. A DCS is typically 0.1–0.3 cm in length, or approximately one-third of the cable circumference [17], [19]. DCS devices used for PD

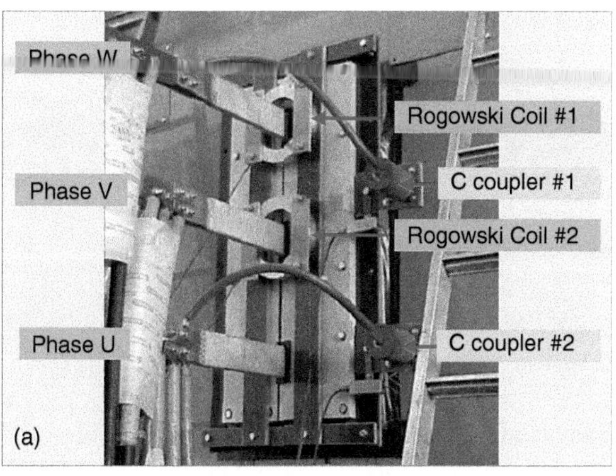

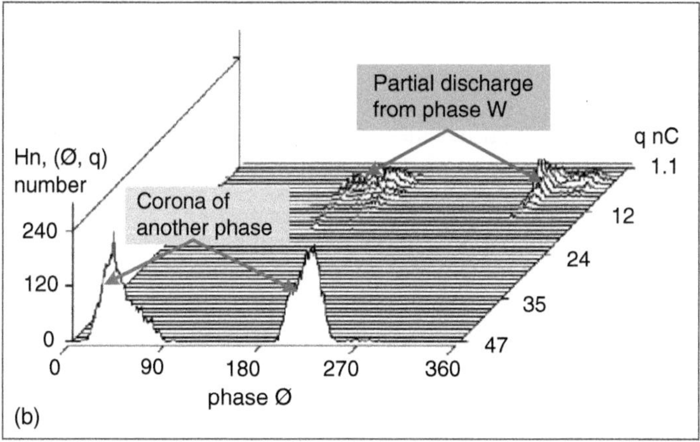

Figure 5.14 (a) Application of Rogowski coil for PD measurement of generator; (b) phase resolved PD pattern [10].

measurement of cables and joints are based on capacitive coupling and inductive coupling mechanisms.

The two most important characteristics of DCS are the *coupling factor* and *directivity*. The coupling factor (or coupling attenuation) quantifies the ability to couple the energy from PD signals [17]. Coupling factor relates to the sensitivity of the DCS, depending on the sensitivity of the DCS device itself and the components of the cable under test, especially the insulation screen (semicon) of the cable. Larger DCS devices tend to have a higher capability of coupling the energy generated from PD signals. Furthermore, the permittivity and conductivity of the cable insulation shield has a direct effect on the coupling ability of the sensors. For example, higher conductivity and/or thicker insulation tends to reduce the ability of the DCS to pick up signals generated by PD activity [17]. Directivity, on the other hand, is the capability of the DCS to accurately determine the direction of the PD pulse as it propagates by comparing the voltage amplitude recorded at each port of DCS [2].

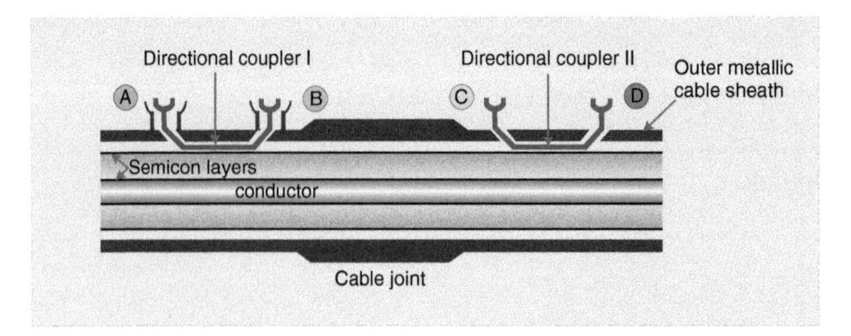

Figure 5.15 Principle setup of two DCS applied for PD measurement of cables and joints [17].

5.2.3.3 Application of DCS for Cable and Joint PD Measurement

DCS devices can be utilized to detect and measure various PD signals generated both inside and outside cable joints.

Principle of DCS Application for Cable and Joint PD Measurement

Normally, DCS devices are installed on opposite sides of the cable joint (as illustrated by Figure 5.15) where they are completely covered by the outer metallic cable sheath, which provides a high degree of protection from external noise [2], [17].

If external PD signals are generated on the left side of the cable joint, these signals are coupled to ports A and C. Similarly, if external PD signals are generated on the right of the cable joint, these signals are coupled to ports B and D. However, if PD signals are generated inside the cable joint, these signals are coupled to ports C, as shown in Figure 5.16a and summarized in Figure 5.16b.

Application of DCS for Cable and Joint PD Measurement

Following proper installation of DCS devices at the cable joint, DCS performance should be verified by the injection of a simulated PD pulse at one DCS port and measuring the output signals at the other DCS ports. For example, in Figure 5.17 simulated PD pulse is applied to port B of the DCS 1 and the output signals at ports C and D of the DCS 2 are subsequently measured [18]. Similarly, for PD measurement at a specific test voltage where the PD source was inside the cable joint, the output signals were measured at ports A–D, as shown in Figure 5.18a. In this example, it is clear that the output signal at port B is higher than at port A, and likewise that the output signal at port C is higher than at port D. In Figure 5.18b the PD source is on the left side of the cable joint, and it can be seen that the output signals recorded at ports A and C are clearly higher than those recorded at ports B and D.

5.3 PD Measurement by UHF Method

5.3.1 Theory

5.3.1.1 General Idea

As an alternative to conventional PD measurement, high-frequency EM waves generated by PD current pulses can also be measured by UHF sensors. EM waves can be detected by UHF sensor(s) inserted into or fitted to the outside of high-voltage equipment. However, as

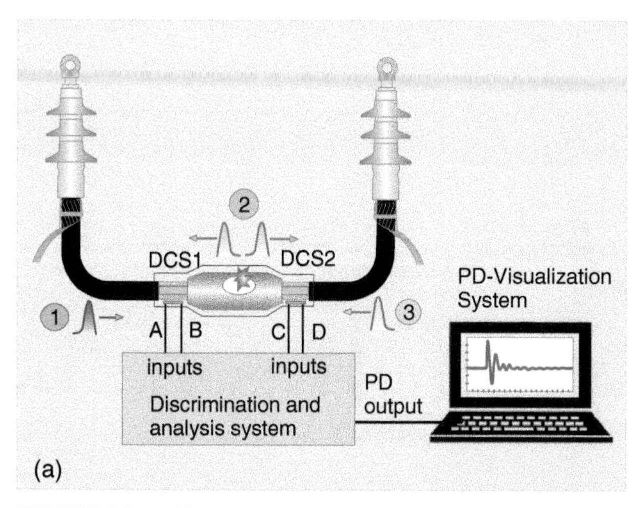

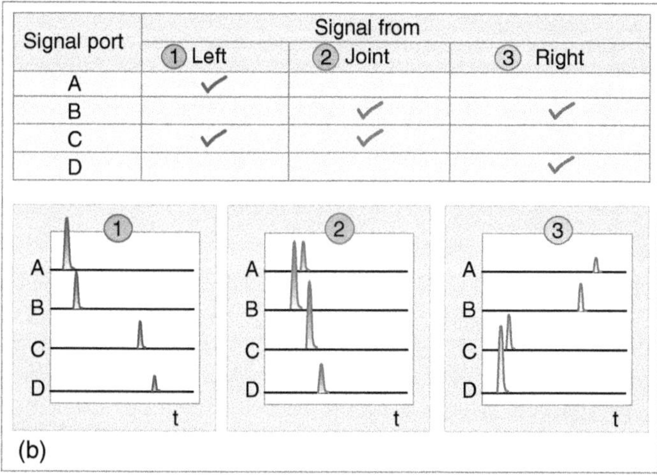

Signal port	Signal from: 1 Left	2 Joint	3 Right
A	✓		
B		✓	✓
C	✓	✓	
D			✓

Figure 5.16 (a) PD detection by DCS; (b) detected PD signals [17].

these EM waves are relatively weak, the detected signals need to be amplified and then transmitted to a suitable data processing unit for data storage, analysis, and display. The technique of high-frequency PD measurement is characterized by a good SNR, which means that it is less susceptible to noise interference from extraneous sources and is therefore well suited to both on-site and on-line PD measurement. UHF PD measurement is used for various purposes (such as PD detection, PD monitoring, and PD source localization) and on a variety of high-voltage equipment (such as high-voltage transformers, GIS, and GIL). PD measurement by detection and analysis of EM waves can be employed as a standalone technique or combined with other PD measurement techniques. However, in common with some other unconventional PD measuring techniques, calibration according to IEC60270 [5] is not possible.

5.3.1.2 Propagation and Attenuation of UHF Signal

Pulse currents produced by PDs inside high-voltage equipment generate EM waves that radiate from the PD source. The transformer tank or outer cage of gas-insulated lines

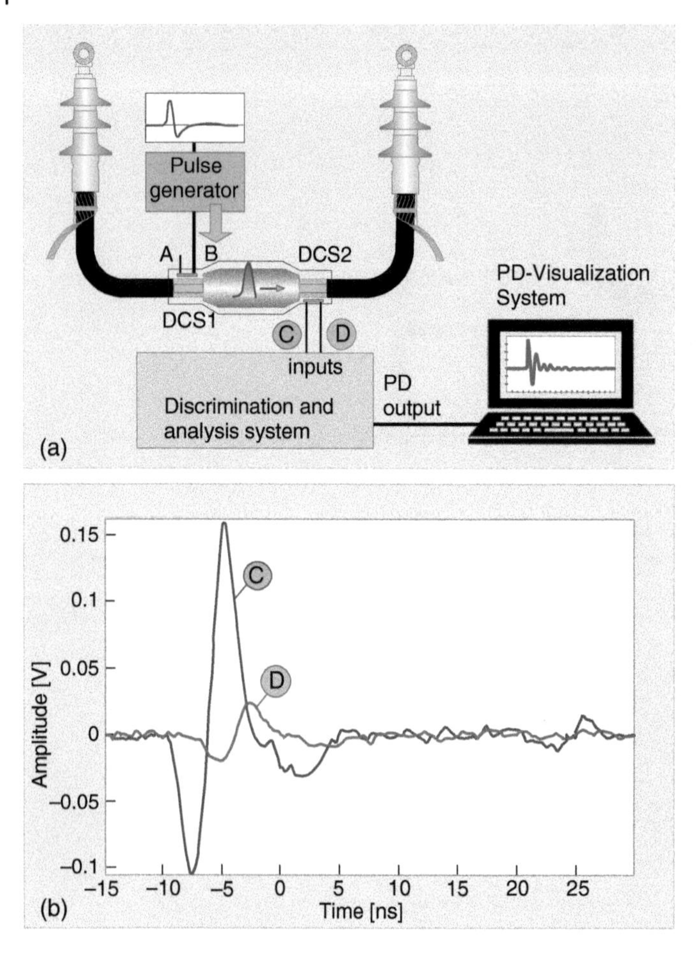

Figure 5.17 (a) DCS performance check; (b) detected PD signals at port C and port D [18].

(GILs) or gas-insulated substations (GISs) tend to act as waveguides or resonators for these EM waves, and when these waves register on a UHF coupler or on a receiving antenna, they induce an electric field in such devices. EM wave propagation in the waveguide generally consists of three modes:

1) Transverse electric (TE) mode in which there is no electric field component in the direction of propagation, only a magnetic field component
2) Transverse magnetic (TM) mode in which there is no magnetic field component in the direction of propagation, only an electric field component
3) Transverse electromagnetic (TEM) mode in which the electric and magnetic field components are transverse to the direction of propagation

The lowest frequency that EM waves can travel in a waveguide is called the cutoff frequency. In order to understand EM wave behavior inside a transformer, GIS, or GIL, the transformer tank may be regarded as a three-dimensional cavity, whereas the GIS or GIL enclosure may be described as a cylindrical waveguide [11]. Any number of modes

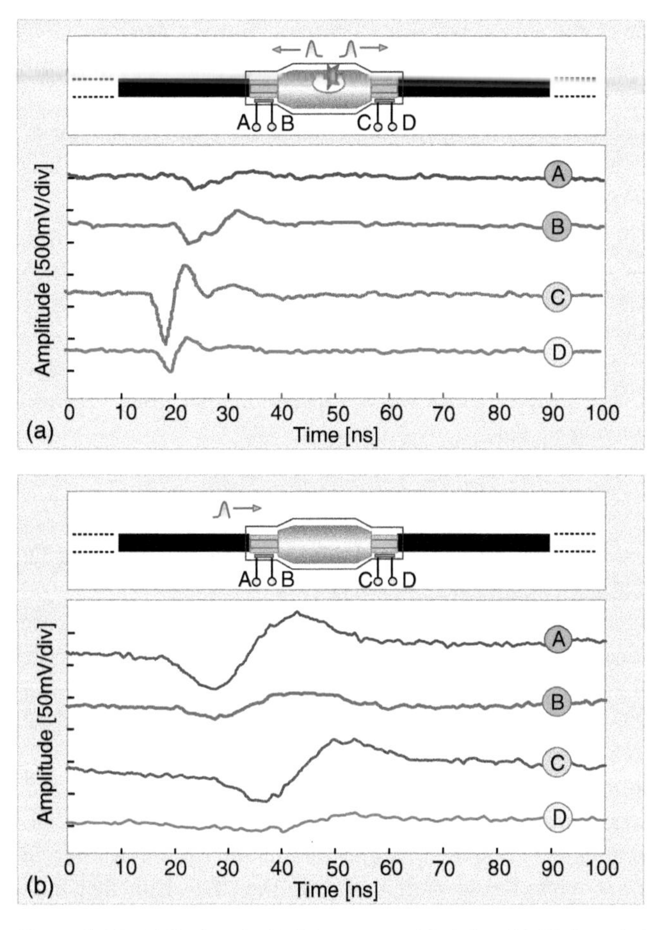

Figure 5.18 DCS signals (a) PD generated in joint; (b) PD from left of joint [18].

described above may exist in a waveguide at the same time. Therefore, in order to adequately understand the PD phenomena taking place, it is necessary to account for all modes of propagation within the measuring bandwidth in order to understand the PD signal detected by sensors used [20]. For GIS or GIL, the TEM mode is the fundamental mode of propagation in a coaxial line, whereas higher-order mode i.e. TM and TE mode needs also to be considered. The cutoff frequency ($f_c(mn)$) of TE_{mn} and TM_{mn} can be calculated from

$$f_c(mn) = \frac{k_{mn}}{2\pi\sqrt{\mu\varepsilon}} \tag{5.6}$$

where f_c (mn) is the cutoff frequency, k_{mn} is the modal eigenvalue constant, and k_{mn} is determined by evaluation of boundary condition of the EM waves in the GIS, while μ and ε are the permeability and the dielectric constants of the medium, and the symbols m and n represent the different types of wave mode.

The simulation results are reported in [21]. The examples of higher order mode TE_{mn} and TM_{mn} signals caused by the simulated PD pulse current inside the cylindrical waveguide with the diameter of 0.5 m and the distance between PD source and the detected UHF probe is 4 m are shown in Figure 5.18a. The cutoff frequency of the TE_{mn} and TM_{mn} are listed and presented in Figures 5.18b and 5.19.

In addition, the coaxial structure waveguide with the radius of the outer conductor of 0.185 m and the radius of the inner conductor of 0.072 m (as shown in Figure 5.20a.) was used to investigate the cutoff frequency of TE- and TM-mode, as listed in Figure 5.20b.

In general, cutoff frequencies are directly dependent on the structural geometry (e.g. outer casing diameter and dimensions) of the high-voltage equipment. For example, increasing the cross-section area of a GIS to allow higher operating voltage produces a lowering of the cutoff frequency, as shown in Figure 5.21.

The cutoff frequency affects the velocity of the TE and TM wave modes, especially around their cutoff frequency. The velocity of the TE mode wave propagation (aka group velocity) can be calculated using the following eq. [24].

$$V_{TE}(f) = c\sqrt{1-(f_c/f)^2} \tag{5.7}$$

where V_{TE} is the velocity of TE mode wave, c is light speed, f_c is the cutoff frequency for TE wave mode, and f is a *considering frequency* determined as $f \geq f_c$.

Examples of the propagation velocity of TE_{m1} and TEM mode in 66 and 500 kV GIS are shown in Figure 5.22 [24]. As illustrated in Figure 5.22, the TEM mode propagates with the speed of light. The lower the cutoff frequency, the closer TE mode propagation velocity approaches the TEM velocity.

5.3.1.3 UHF Signal Attenuation

Generally, there are two main reasons for UHF signal attenuation when traveling in high-voltage apparatus: attenuation caused by the *skin-effect* and attenuation caused by internal barriers and discontinuities. The *skin-effect* causes attenuation of the UHF signal by which power loss is dissipated in the metallic enclosure of the high-voltage equipment, but the magnitude of this attenuation or loss tends to be very low (theoretically around 2–5 dB/km). However, internal barriers and discontinuities tend to cause greater attenuation of the UHF signals, particularly in compactly constructed GIS where junctions, gas barriers, and bends are in closer proximity to each other, and the magnitude of UHF attenuation is around 1–2 dB/km [20], [22].

Discontinuities such as flat discs or conical shapes in GIS tends to cause only minor attenuation. Similarly, for long GIL with relatively few discontinuities (such as post insulators) tend to have very low levels of attenuation, and in such cases PD signals can sometimes be detected 100 m or more from the PD source [20]. In transformers, pressboards or disc-windings also tend to cause attenuation of UHF signal, as reported in [25].

5.3.2 UHF Sensors

UHF sensors can be categorized according to where they are installed (i.e. *internal sensors* and *external sensors*). Internal UHF sensors, such as disc sensors, cone sensors, loop sensors, and field grading sensors, are usually mounted in an internal recess where they should not compromise the integrity of the transformer insulation. This type of sensor is used to monitor the EM field – particularly the radial component. UHF PD detection by internal

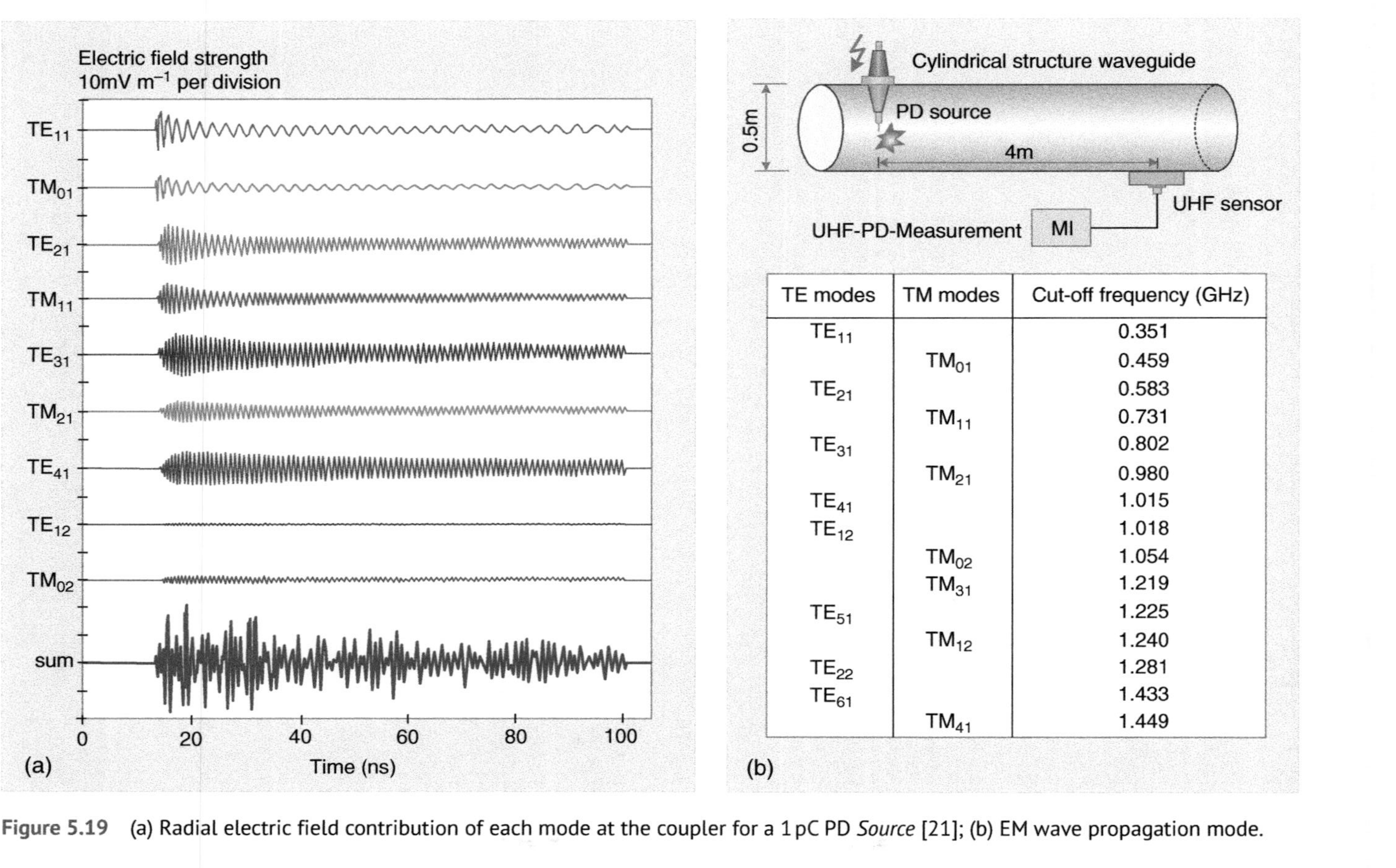

TE modes	TM modes	Cut-off frequency (GHz)
TE_{11}		0.351
	TM_{01}	0.459
TE_{21}		0.583
	TM_{11}	0.731
TE_{31}		0.802
	TM_{21}	0.980
TE_{41}		1.015
TE_{12}		1.018
	TM_{02}	1.054
	TM_{31}	1.219
TE_{51}		1.225
	TM_{12}	1.240
TE_{22}		1.281
TE_{61}		1.433
	TM_{41}	1.449

Figure 5.19 (a) Radial electric field contribution of each mode at the coupler for a 1 pC PD *Source* [21]; (b) EM wave propagation mode.

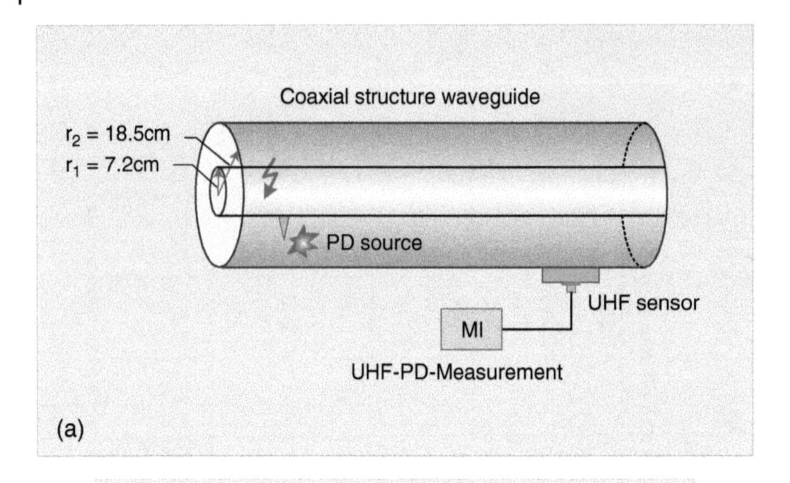

TE modes	TM modes	Cut-off frequency (GHz)
TE_{11}		0.380
TE_{21}		0.737
TE_{31}		1.063
	TM_{01}	1.312
TE_{01}	TM_{11}	1.368
TE_{12}		1.441
	TM_{21}	1.521
TE_{22}		1.645
	TM_{31}	1.742

(b)

Figure 5.20 (a) Experiment with coaxial structure; (b) cutoff frequency of TE- and TM-wave modes in a coaxial structure [22].

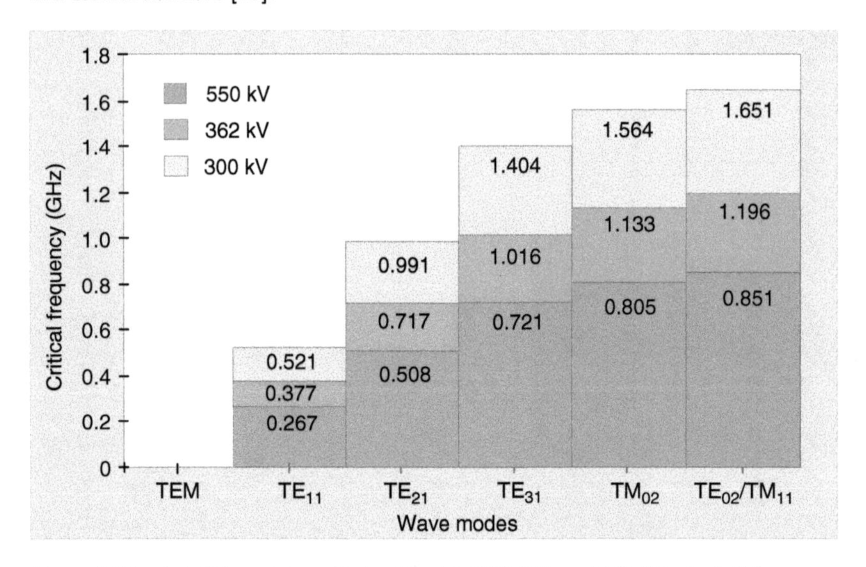

Figure 5.21 Cutoff frequency with in GIS with 300, 362, and 550 kV rated [23].

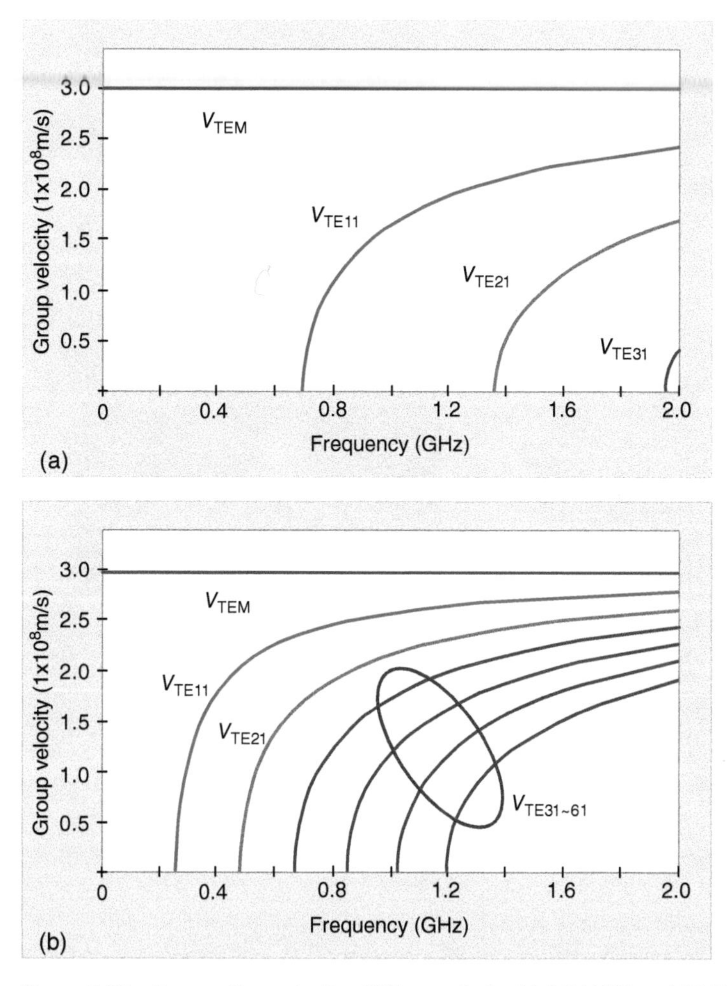

Figure 5.22 Propagation velocity of TE_{m1} mode for (a) 66 kV GIS and (b) 500 kV GIS [24].

sensors tend to have a high SNR as they are not sensitive to ambient or background noise. These sensors may either be installed inside GIS, GIL or transformers during the manufacturing process or be retrofitted during maintenance [11], [26].

External UHF sensors, such as window sensors, pocket sensors, and barrier sensors, are usually mounted on exterior casing apertures of GIS or transformers (e.g. on the inspection windows of GIS). The UHF barrier sensors are normally on the barrier between GIS compartment. For PD detection in transformers, sensors may be mounted on the transformer external casing or on the oil-filled value. Moreover, external sensors are movable, easy to retrofit, and suitable for periodical PD measurement [11], [26]. However, discontinuities at aperture material or barrier insulation where the sensors are installed tend to cause localized attenuation of the EM field and signal distortion. Moreover, external UHF sensors are susceptible to ambient noise, which tends to decrease sensitivity. Figures 5.23–5.24 show examples of UHF sensors utilized for PD measurement in GISs, GILs and other transformers.

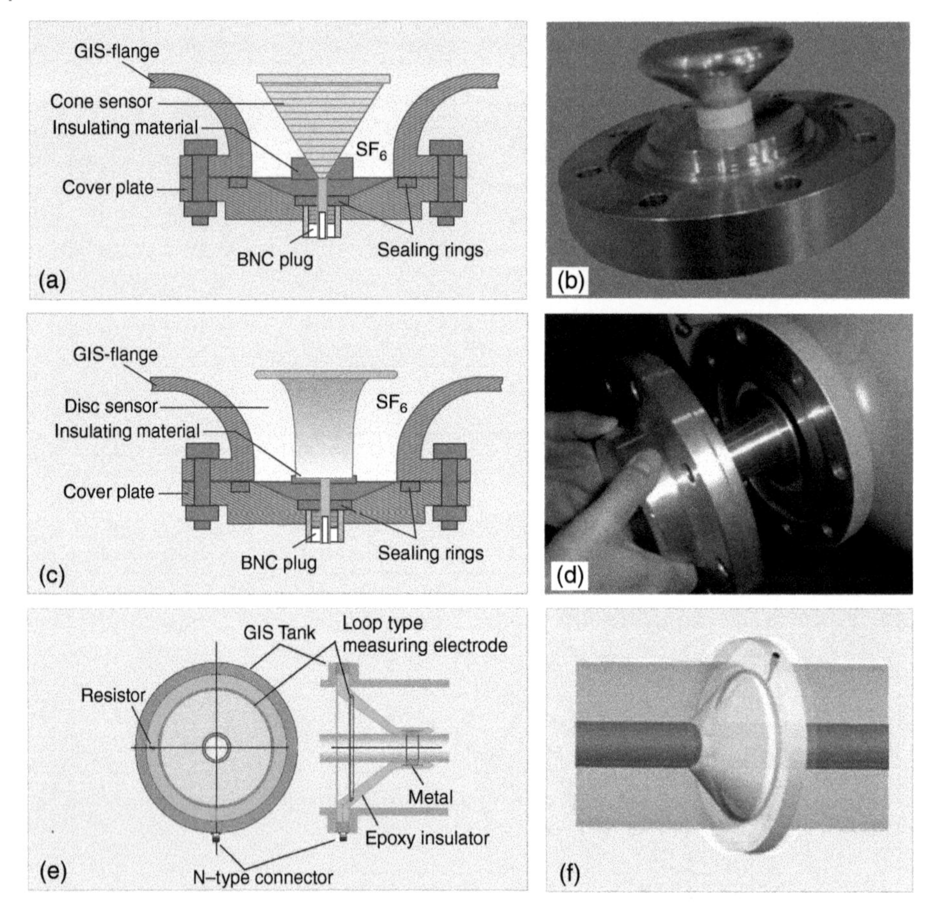

Figure 5.23 Internal sensors PD sensor (a, b) section view and mounting of UHF PD cone sensor [11]; (c, d) section view and mounting of UHF PD disc sensor [11]; (e, f) section view and mounting of UHF PD loop-shaped sensor [11], [27].

5.3.3 UHF PD Measurement System

Bandwidth and sensitivity of the UHF PD measuring system are two very important parameters to be considered before deciding on which UHF PD measurement technique to apply.

a) Bandwidth
UHF signals obtained from UHF sensors are processed in the time domain or in the frequency domain. In the time domain, the wave-shape of a single PD pulse is detected and analyzed. Similarly, in the frequency domain, spectra representing the amplitudes of various frequency resonances stimulated by the PD pulse are measured and analyzed. The signal processing technique applied for the narrow-band frequency domain tends to provide better noise suppression capability and hence higher PD sensitivity, thus making it suitable for both on-site and on-line PD measurement. When UHF PD measurement is conducted with narrow-band technique, the UHF signal is analyzed for only a part of the frequency range approximately lower than 1% bandwidth

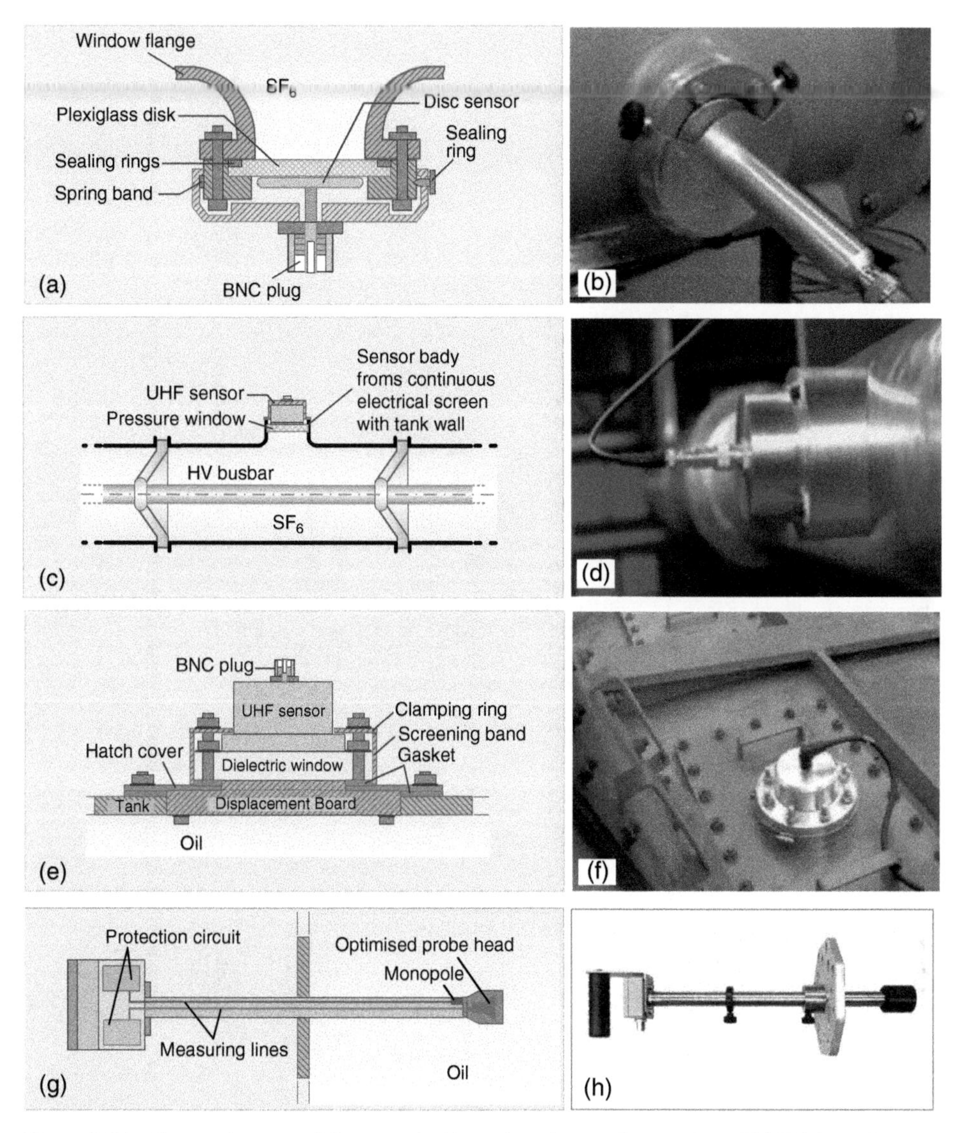

Figure 5.24 External sensors PD sensor (a, b) section view and mounting of UHF PD window coupler [11], (c, d) section view and mounting of external UHF PD sensor for GIS PD measurement [11], (e, f) section view and mounting of external UHF PD sensor for Transformer PD measurement [11], (g, h) UHF PD sensor with section view for Transformer PD measurement for installation on the oil valve [28].

(<<1% BW e.g. 3–10 MHz). Whereas the UHF PD signal over wide frequency range (e.g. 200–2000 MHz) is analyzed if UHF PD measurement is conducted with wide-band technique, as shown in Figure 5.25 [6], [26].

b) Sensitivity

The sensitivity of UHF PD measurement is mainly determined by the configuration of high-voltage equipment, the specification of the UHF sensor, the positioning and

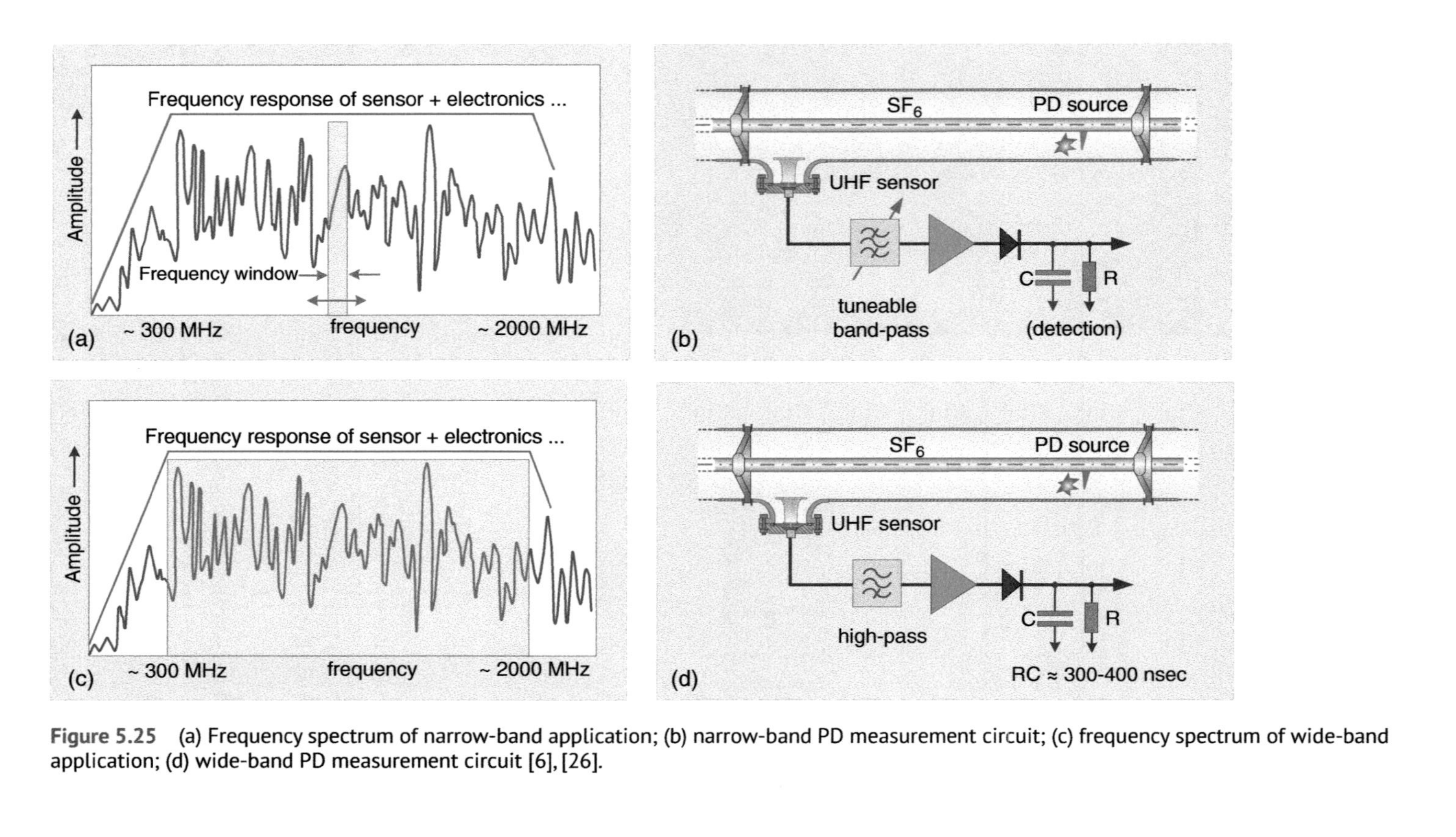

Figure 5.25 (a) Frequency spectrum of narrow-band application; (b) narrow-band PD measurement circuit; (c) frequency spectrum of wide-band application; (d) wide-band PD measurement circuit [6], [26].

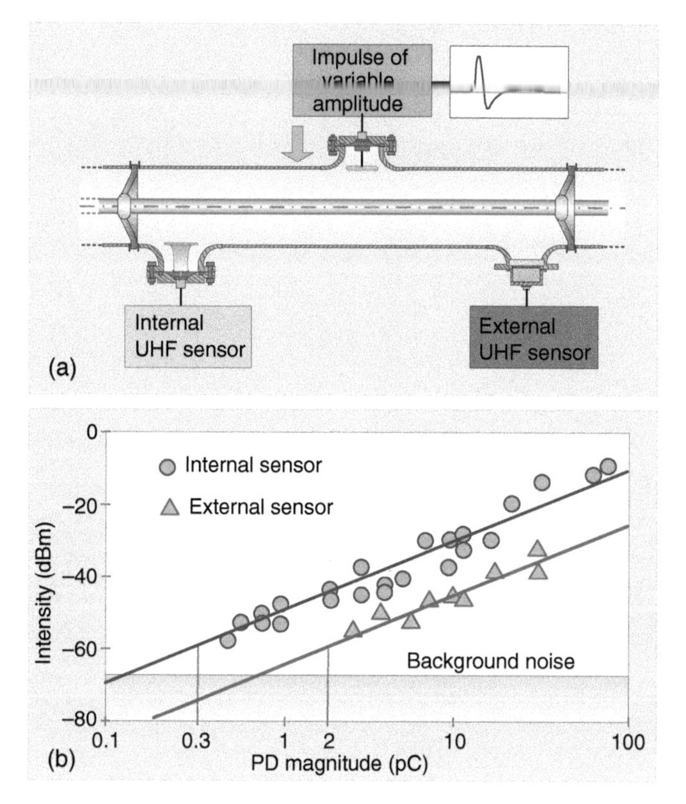

Figure 5.26 (a) Sensitivity PD measurement; (b) sensitivity comparison of UHF internal sensor and UHF external sensor [11].

installation of the sensor, the condition of the measuring cable and amplifier circuit, and the overall integrity of the measuring system and the defect type. Figures 5.26–5.27 provide examples of sensitivity characteristics in relation to sensor types and sensors position [6], [11].

5.3.3.1 Sensitivity Verification for GIS PD Measurement

Before performing PD measurements, verification of sensitivity to the UHF PD detection should be performed in order to confirm that the PD measuring system is functioning properly and able to detect PD signals to the degree of sensitivity required, for example the ability to detect UHF PD signals corresponding to a 5 pC apparent charge measured according to IEC 60270 [5]. Standard sensitivity verification usually consists of a two-step procedure, as described next.

Step 1 Testing Is Performed in the Laboratory: Its purpose is to determine the amplitude of an artificial PD pulse equivalent to 5 pC apparent charge, as measured in accordance with IEC 60270 [5]. In the test procedure, a real defect is situated close to UHF sensor A1, as shown in Figure 5.28a and the GIS conductor is energized. When the apparent charge generated by the defect reaches the threshold of 5 pC, the corresponding UHF

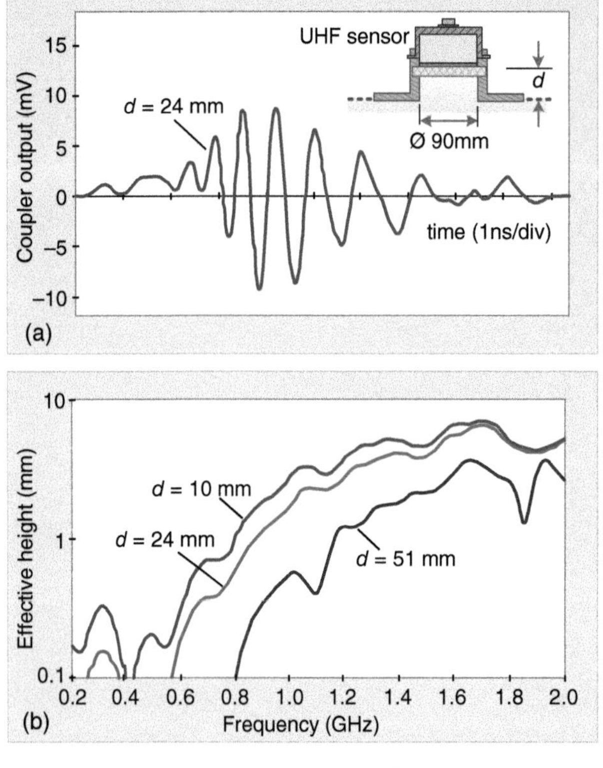

Figure 5.27 Sensitivity of a window UHF coupler at various window depths: (a) mounting configuration of the coupler and step response coupler output voltage when d = 24 mm; and (b) frequency response at three depths: d = 10, 24 and 51 mm [11].

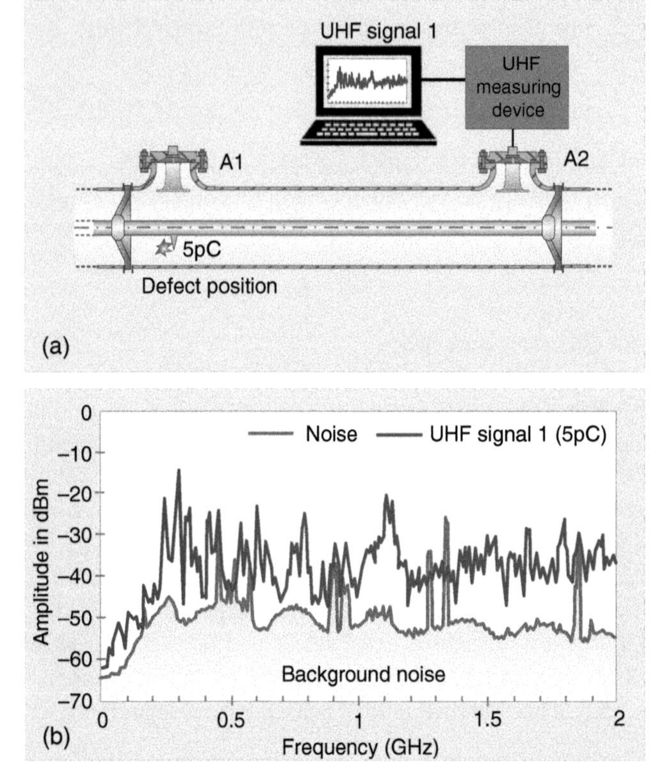

Figure 5.28 Sensitivity verification step 1 (a) laboratory setup; (b) UHF signal 1 and background noise [26].

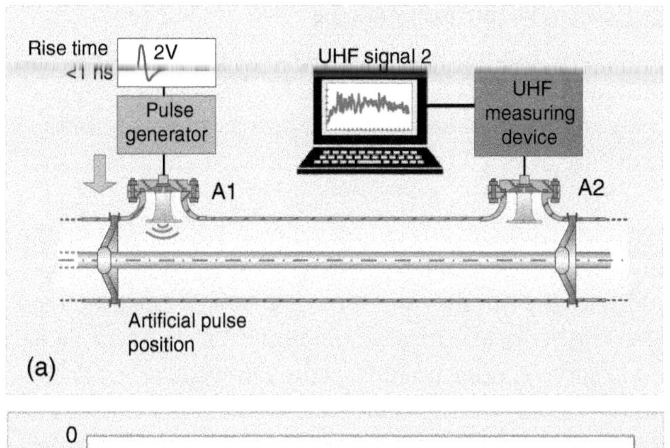

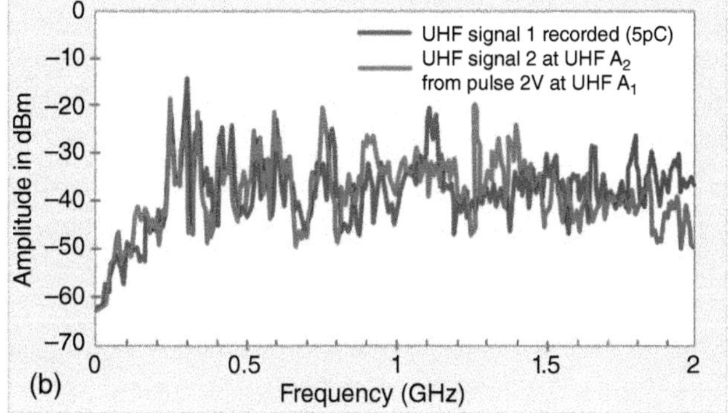

Figure 5.29 Sensitivity verification step 2: (a) laboratory test setup with an artificial pulse; (b) UHF signal 1 and 2 comparison [26].

signal is picked up by UHF sensor A2 and recorded as UHF Signal 1, as depicted in Figure 5.28b. Following this, an artificial pulse from a pulse generator are fed into UHF sensor A1, as shown in Figure 5.29a, and the amplitude of the artificial pulse is varied until the amplitude spectra of the UHF signal 2 is equivalent to the amplitude spectra of the UHF signal 1 (within a tolerance of ±20%), as illustrated in Figure 5.29b. Statistical analysis may then be used to verify the similarity of amplitude spectra produced by UHF signals 1 and 2.

Step 2 Testing Is Performed On-Site: Its purpose is to verify that the UHF PD measuring system, including sensors used, is sufficiently sensitive to detect PD signals equivalent to those from a specific PD defect. In the test procedure, an artificial pulse of specified amplitude (as determined in Step 1) is fed sequentially into all UHF sensors. The UHF PD signals thereby generated should be detected by at least the two UHF sensors adjacent to the injected PD signal sensor. Furthermore, the UHF PD measuring system and the pulse generator used for Step 2 verification must be the same (specification) as used in Step 1. Note that the number and location of the UHF sensors used affect the sensitivity, and therefore the validity of the UHF PD measurement used in the experiment.

5.3.3.2 Determination of PD Measurement by UHF PD Technique

In general, the UHF PD measurement technique mostly applies to GISs, GILs, and transformers. However, the technique can be used to measure PD in rotating machines as well. The process of UHF PD measurement requires proper preparation as follows.

Internal or external UHF sensors are selected according to various parameters, such as the type and structure of the high-voltage equipment to be tested, and the anticipated duration of PD testing (e.g. periodic measurement or long-term monitoring). In modern high-voltage apparatus especially GIS or GIL, these internal UHF sensors are incorporated during manufacture. In addition, the number and type of UHF sensors used tend to have a significant effect on the sensitivity of PD measurement. Generally, for older-type high-voltage equipment still in service, external sensors are more suitable, particularly for making periodic PD measurements. Furthermore, the sensitivity of the sensors needs to be verified, particularly for UHF PD measurement of GIS and GIL and transformers. Finally, the UHF background noise needs to be recorded before conducting UHF PD measurement at the specified test voltage. Figure 5.30 shows typical examples of background noise and PD spectra.

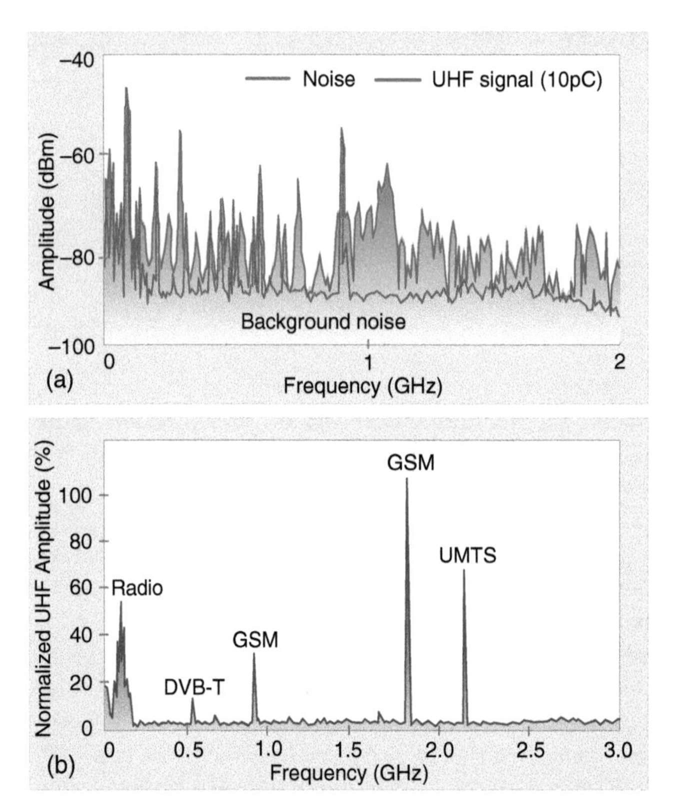

Figure 5.30 UHF signal spectrum: (a) spectrum of background noise and a PD corresponding to 10 pC measured in accordant with IEC 60270 [29]; (b) example of narrow band disturbances during performing PD testing [28].

5.3.4 Application of UHF PD Measurement

5.3.4.1 PD Detection by UHF in GIS and GIL

In practice, UHF sensors are the preferred technique for both on-site and on-line PD measurement in GIS and GIL. Normally, PD defects in GIS and GIL are categorized as either stationary defects or moving particles, as shown in Figure 5.31. Stationary defects include interior protrusions, voids inside the supporting insulators, and particles that become attached to the inner or outer surface of the supporting insulators.

The EM waves emitted from a PD event can be detected by UHF sensors, and by analyzing the waves detected, the location of the PD event can be determined. Figure 5.32 shows an example of on-site PD detection using UHF sensors in order to discover defects inside a GIS.

5.3.4.2 UHF PD Detection in Transformers

PD measurement in transformers, both on-site and on-line, by applying UHF sensor detection is a fairly recent technical innovation that can be used to determine PD location and monitor transformer condition. The UHF sensors are normally used with acoustic sensors for transformer PD measurement, although UHF sensors alone are often sufficient. Figure 5.33 shows various aspects of on-site PD detection by UHF sensors.

5.3.4.3 Application of UHF PD Detection to Other High-Voltage Equipment

UHF PD measurement has application not only to PD detection in GIS, GIL, and transformers, but also to the detection of PD in the slot component of generators [30]. A type of strip-line antenna called *a stator slot coupler* (SSC) has been developed recently for PD measurement in rotating machines [31]. The SSC is installed under the wedges at the end of the core slots at the top of the stator winding, as shown in Figure 5.34. This type of antenna is able to detect electrical signals in the frequency range of 10 to 1000 MHz. Thus, the SSC is sensitive enough to detect surface PD at the stator end-winding,

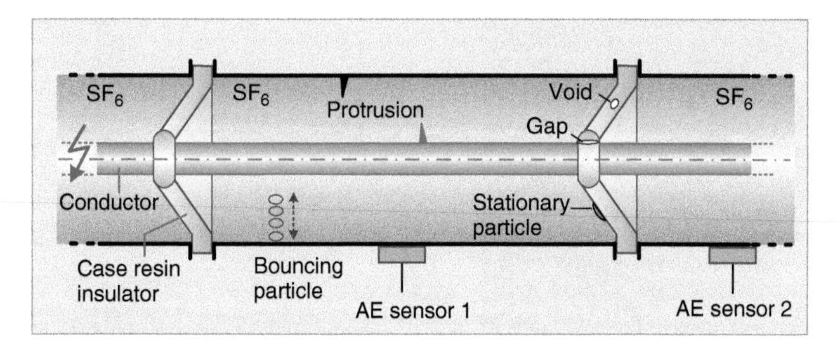

Figure 5.31 PD sources in GIS or GIL.

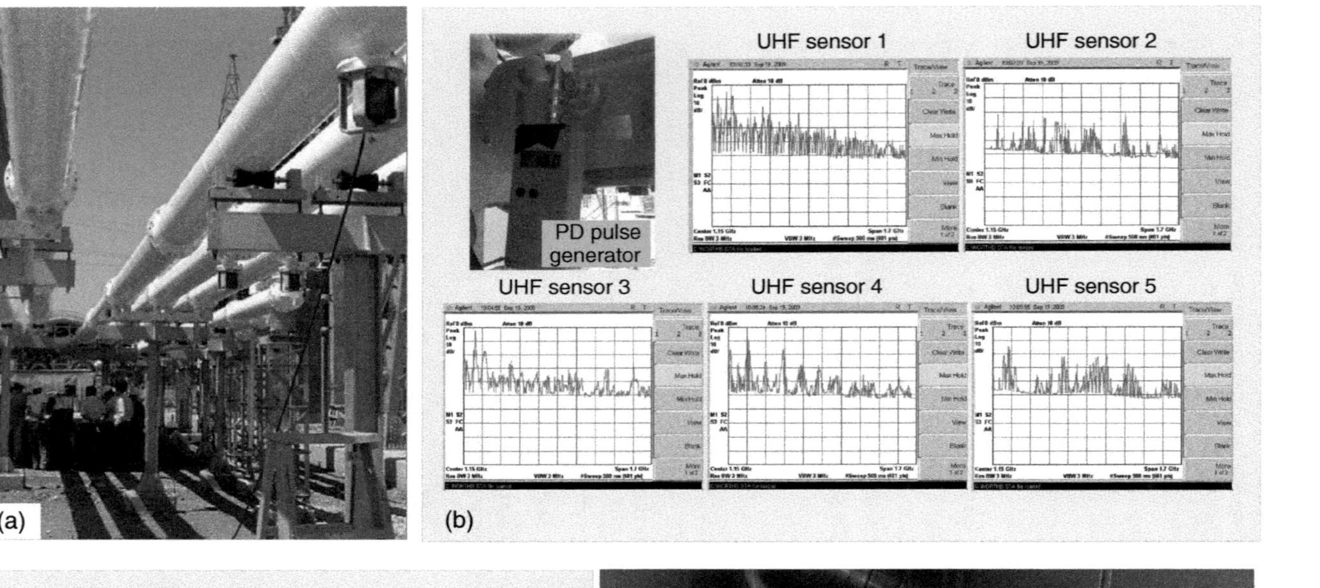

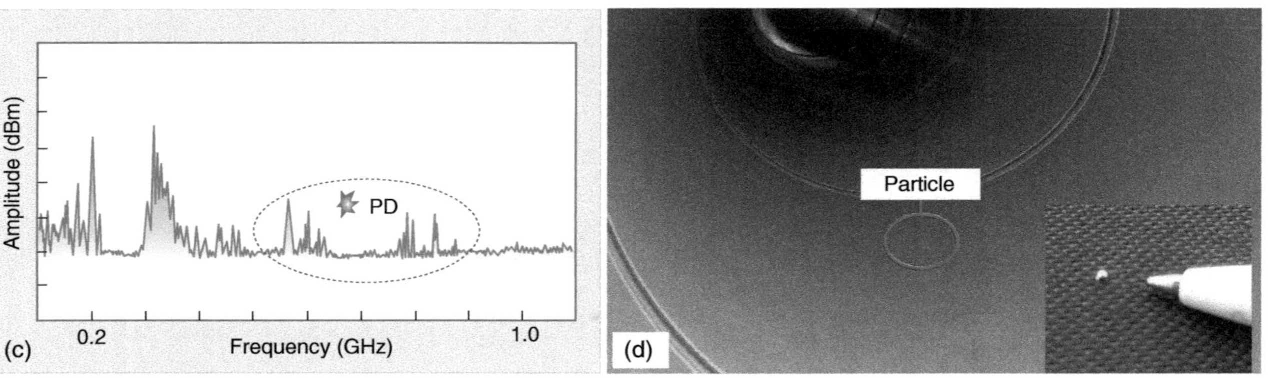

Figure 5.32 UHF PD measurement in a GIL system: (a) Overview of on-site PD measurement by UHF technique; (b) PD pulse generator and UHF signals obtained from sensitivity check process; (c) UHF PD signal with noise; (d) particle found inside GIS system tested.

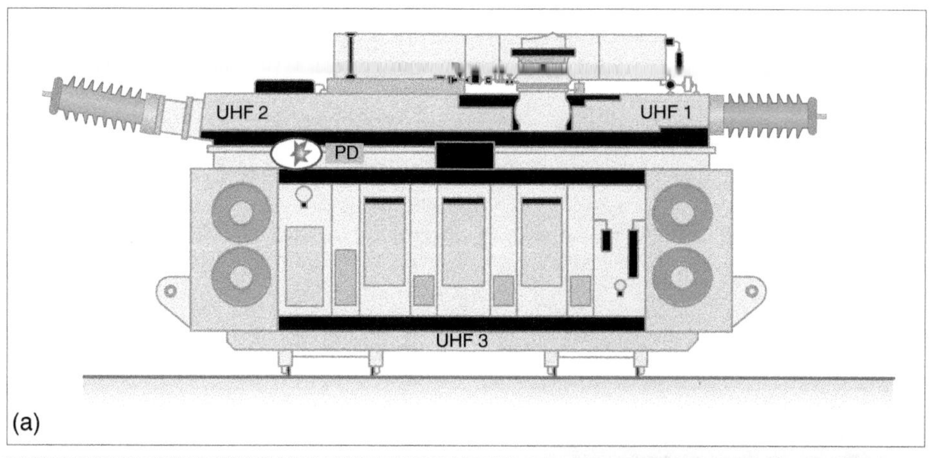

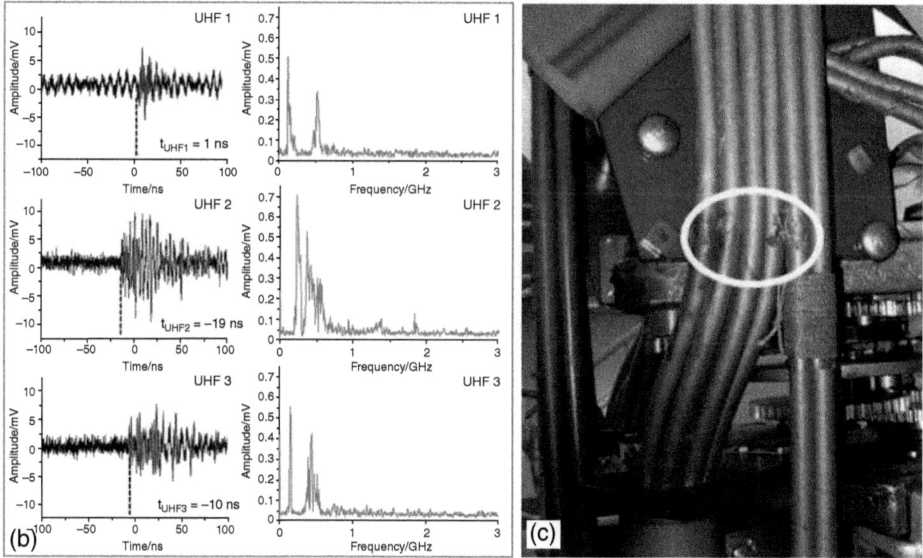

Figure 5.33 UHF PD measurement for a power transformer PD detection: (a) UHF sensor positions of the investigated power transformer rated 333 MVA 400/220/30 kV; (b) UHF PD signals obtained from the sensors; (c) insulation problem found [28].

providing it occurs close to the tip of the antenna. Moreover, sensitivity of PD detection using the SSC method depends on several variables, such as antenna location, distance between antenna and PD source, and bandwidth of UHF sensors and PD measuring system. It should be noted that the emitted EM wave tends to be attenuated by the screening effects of various generator components, such as the iron core and conductive coatings [1], [30].

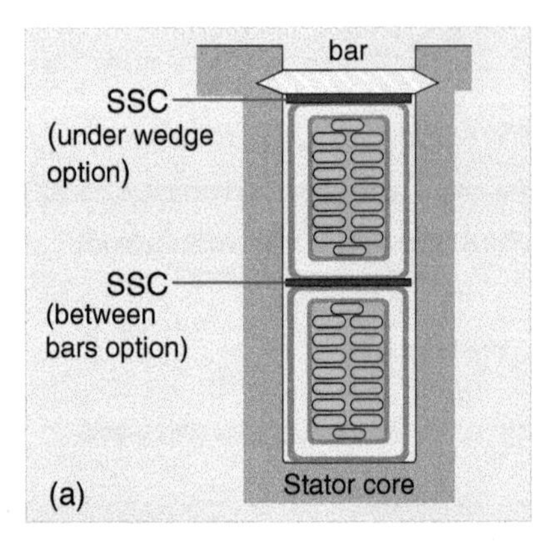

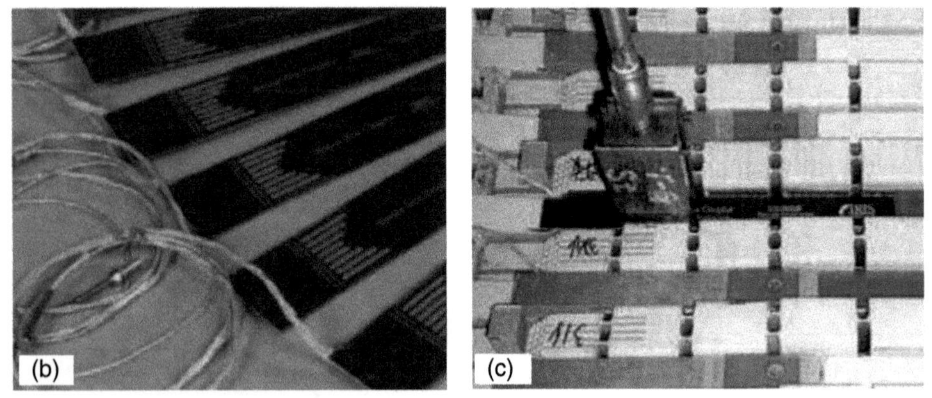

Figure 5.34 SCC antenna installation for PD measurement: (a) SCC position; (b) examples of SCC antenna; (c) SCC installation [32].

References

1 International Electrotechnical Commission. (2012). *IEC 60034: rotating electrical machines – Part 27–2: on-line partial discharge measurements on the stator winding insulation of rotating electrical machines*. Geneva, Switzerland: IEC.

2 International Electrotechnical Commission. (2016). *IEC 62478: high voltage test techniques – measurement of partial discharges by electromagnetic and acoustic methods*. Geneva, Switzerland: IEC.

3 International Electrotechnical Commission. (2017). *IEC 60034: rotating electrical machines – Part 27–1: off-line partial discharge measurements on the winding insulation*. Geneva, Switzerland: IEC.

4 Konig, D. and Rao, Y.N. (ed.) (1993). *Partial Discharges in Electrical Power Apparatus*. Berlin: VDE-verlag gmbh.

5 International Electrotechnical Commission (2015) *IEC 60270:20+AMD: high-voltage test techniques - partial discharge measurements*. Geneva, Switzerland: IEC.

6 Conseil International des Grands Réseaux Électriques (2016). Guideline for partial discharge detection using conventional (IEC 60270) and unconventional methods. *Technical Brochure. Ref. 662*.

7 Hauschild, W. and Lemke, E. (2014). *High-Voltage Test and Measuring Techniques*. Berlin Heidelberg: Springer-Verlag.
8 Rodrigo, A., Llovera, P., Fuster, V. et al. (2011). Influence of high frequency current transformers bandwidth on charge evaluation in partial discharge measurements. *IEEE Transactions on Dielectrics and Electrical Insulation* 18 (5): 1798–1802.
9 Siddiqui, B.A., Pakonen, P., and Verho, P. (2017). Novel inductive sensor solutions for on-line partial discharge and power quality monitoring. *IEEE Transactions on Dielectrics and Electrical Insulation* 24 (1): 209–216.
10 Conseil International des Grands Réseaux Électriques (2003). Knowledge rules for partial discharge diagnosis in service. *Technical Brochure Ref. 226*.
11 Conseil International des Grands Réseaux Électriques (2010). Guidelines for unconventional partial discharge measurements. *Technical Brochure Ref. 444*.
12 Samimi, M.H., Mahari, A., Farahnakian, M.A. et al. (2015). The Rogowski coil principles and applications: a review. *IEEE Sensors Journal* 15 (2): 651–658.
13 Shafiq, M., Kutt, L., Lehtonen, M. et al. (2013). Parameters identification and modeling of high-frequency current transducer for partial discharge measurements. *IEEE Sensors Journal* 13 (3): 1081–1091.
14 Institute of Electrical and Electronics Engineers (2008). *IEEE C37.235–2007: Guide for the Application of Rogowski Coils Used for Protective Relaying Purposes*. IEEE.
15 Zhang, Z.S., Xiao, D.M., and Li, Y. (2009). Rogowski air coil sensor technique for on-line partial discharge measurement of power cables. *IET Science, Measurement and Technology* 3 (3): 187–196.
16 Hemmati, E. and Shahrtash, S.M. (2013). Digital compensation of Rogowski Coil's output voltage. *IEEE Transactions on Instrumentation and Measurement* 62 (1): 71–82.
17 Pommerenke, D., Strehl, T., Heinrich, R. et al. (1999). *Discrimination between Internal PD and Other Pulses Using Directional Coupling Sensors on HV Cable Systems*, 814. Institute of Electrical and Electronics Engineers.
18 Craatz, P., Plath, R., Plath, R., and Kalkner, W. (ed.) (1999). Sensitive on-site PD measurement and location using directional coupler sensors in 110 kV prefabricated joints. In: *Proceedings of Eleventh International Symposium on High Voltage Engineering August 27–23, 1999*. London, UK: The Institution of Electrical Engineering.
19 Conseil International des Grands Réseaux Électriques (2001). Partial discharge detection in installed HV extruded cable systems. *Technical Brochure. Ref. 182*.
20 Gaddad, A. and Warne, D. (ed.) (2004). *Advance in High Voltage Engineering*. London: The Institution of Electrical Engineering.
21 Judd, M.D., Hampton, B.F., and Farish, O. (1996). *Modelling Partial Discharge Excitation of UHF Signals in Waveguide Structures Using Green's Functions*, 66. The Institution of Engineering and Technology.
22 Kurrer, R. and Feser, K. (1998). *The Application of Ultra-High-Frequency Partial Discharge Measurements to Gas-Insulated Substations*, 777. Institute of Electrical and Electronics Engineers (Jul).
23 Hoek, S.M., Koch, M. and Heindl, M. (eds.) (2010). Propagation mechanisms of PD pulse for UHF and traditional electrical measurements. *Proceedings of 2010 International Conference on Condition Monitoring and Diagnosis* (September 6–11, 2010), Tokyo, Institution of Electrical Engineers of Japan, Tokyo.

24 Hikita, M., Ohtsuka, S., Teshima, T. et al. (2007). *Electromagnetic (EM) Wave Characteristics in GIS and Measuring the EM Wave Leakage at the Spacer Aperture for Partial Discharge Diagnosis*, 453. Institute of Electrical and Electronics Engineers.

25 Conseil International des Grands Réseaux Électriques (2017). Partial discharges in transformers. *Technical Brochure. Ref. 676.*

26 Conseil International des Grands Réseaux Électriques. (2016). UHF partial discharge detection system for GIS: application guide for sensitivity verification. *Technical Brochure. Ref. 654.*

27 Conseil International des Grands Réseaux Électriques (2017). Benefits of PD diagnosis on GIS condition assessment. *Technical Brochure. Ref. 674.*

28 Coenen, S. (2012) Measurement of partial discharges in power transformer using electromagnetic signals. Dissertations. University Stuttgart.

29 Kurrer, R., Feser, K., and Herbst, I. (ed.) (1994). Calculation of resonant frequencies in GIS for UHF partial discharge detection. In: *Proceedings of the Seventh International Symposium on Gaseous Dielectrics April 24–28, 1994, Knoxville, TN.* New York: Springer Science+Business Media, LLC.

30 Conseil International des Grands Réseaux Électriques (2004). Application of on-line partial discharge tests to rotating machines. *Technical Brochure. Ref. 258.*

31 Campbell, S.R., Stone, G.C., Sedding, H.G. et al. (1994). *Practical On-Line Partial Discharge Tests for Turbine Generators and Motors*, 281. Institute of Electrical and Electronics Engineers.

32 Qualitrol (2016). Iris power stator slot couplers™. https://irispower.com/wp-content/uploads/2016/11/Iris-Power-Stator-Slot-Couplers-Brochure.pdf (accessed December 2, 2022).

6

Non-electrical Methods for PD Measurement

6.1 Introduction

Insulation systems undergoing partial discharge (PD) events generate various types of signals (e.g. light waves, acoustic waves, electrical pulses, RF signals, and chemical byproducts), which can be detected and measured. The IEEE and IEC (Institute of Electrical and Electronics Engineers and International Electrotechnical Commission) are the two main international standard organizations responsible for preparing and publishing the international standards for electrical, electronic, and related advanced global technologies. After years of debate, they published new standards for nonconventional PD measurement: IEEE std. C57.127 (2018) [1] for acoustic PD measurement of oil-immersed power transformers and reactors, and IEC 62478 (2016) [2] for electromagnetic (EM) and acoustic PD measurement. In addition, CIGRE (International Council for Large Electric Systems, as translated), an international nonprofit association, prepares and publishes CIGRE documents related to planning, design, construction, operation, maintenance, and disposal of high-voltage equipment and power and distribution systems. Figure 6.1 shows the signals generated by PD phenomena and the relevant PD measurement standards.

6.2 Optical PD Measurement

6.2.1 Theory

PD events occurring in gas media, most noticeably corona discharges, tend to be produced at the point electrodes, the sharp edge electrodes, and the surface roughness on the conductors due to gas ionization, which releases energy in the form of ultraviolet light (UV_A–UV_C – see Figure 6.2) as well as audio signals. The intense corona discharges may be observed readily in the dark after the vision of the observer has adjusted. The light spectrum emitted by corona discharge phenomena is determined by both the surrounding medium and the discharge energy, with the optical spectrum thus generated ranging from ultraviolet to infrared. Optical spectrum of corona discharges in air is dominated by the spectrum of nitrogen, where hydrogen found in mineral oil used as insulating liquid in oil-immersed transformers provides the main optical spectrum in the visible light with a lesser

Partial Discharges (PD) - Detection, Identification, and Localization, First Edition. Norasage Pattanadech, Rainer Haller, Stefan Kornhuber, and Michael Muhr.

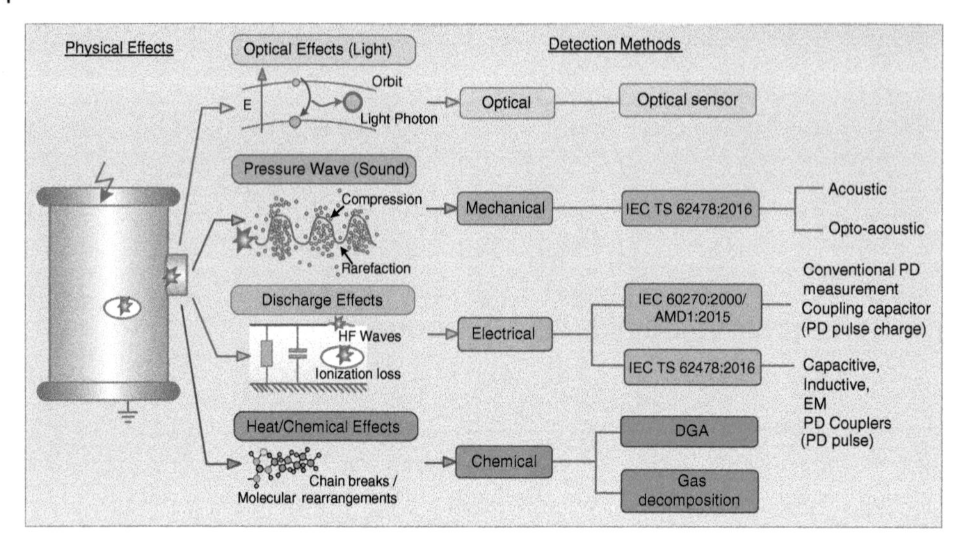

Figure 6.1 PD phenomena and detection techniques.

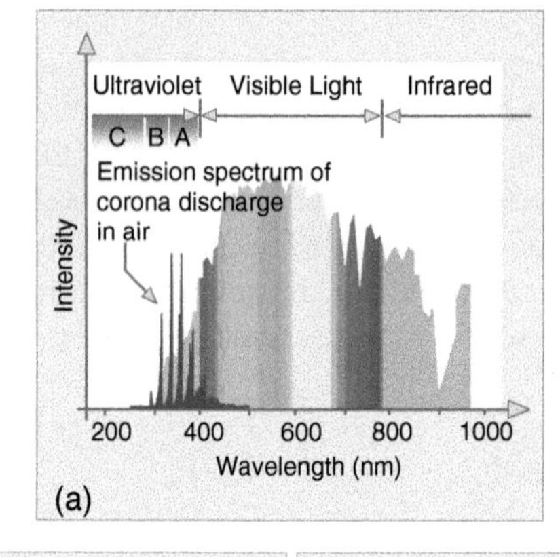

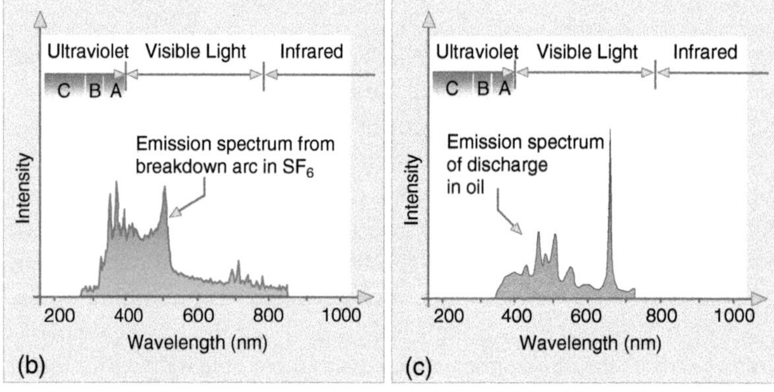

Figure 6.2 Spectrum wavelength (a) corona in air; (b) arc in SF_6; (c) discharge in oil [3].

portion in the infrared range. As Figure 6.2a illustrates, the light spectrum generated by PDs in air and/or the surface discharges surrounding the high-voltage apparatus primarily falls within the ultraviolet range of 300–400 nm. This can be detected by a corona camera, which is very useful for on-site inspection of high-voltage equipment. However, this is not possible where corona discharge originates inside high-voltage equipment, such as transformers and gas-insulated substations, where embedded light sensors or fiber optics may be required. By contrast, the emission spectrum of discharges and arcing in SF_6 gas insulation lies in the ultraviolet and blue-green region of the visible light spectrum, as shown in Figure 6.2b. However, the situation is more complex in surface discharges, wherein the light spectrum produced along the solid dielectric, due to the influence of factors such as solid material and surface conditions, including the composition of gases surrounding the solid material [3]. As for liquid insulation, such as mineral oil, the emission spectrum is predominantly formed by hydrogen and hydrocarbon compounds, which produce spectral emissions in the range of 350–700 nm as shown in Figure 6.2c.

6.2.2 Principle for Optical PD Measurement Technique

Optical PD detection is one of the most challenging PD measurement techniques, which may reveal the long time-unknown complex mechanism of discharge. The optical PD detection is based on the detection of the light generated due to ionization, excitation, and recombination process during the existence of discharge. Different approaches and optical sensors that can apply for optical PD detection are summarized in Figure 6.3. In practice, corona cameras (as shown in Figure 6.4) can be used for the detection of corona and/or surface discharge, which occurs surrounding the conductor and/or high-voltage equipment. Corona cameras are generally available in two types: nonsolar blind and solar blind. Nonsolar blind cameras (intended for both indoor and outdoor use at night) are sensitive to UV_a and UV_b, which are abundant in sunlight. Therefore, they cannot detect light emissions generated from corona or surface charges during the day. The construction diagram of the nonsolar blind camera is illustrated in Figure 6.4b. By contrast, solar blind cameras have special filters that block most sunlight frequencies in the visible spectrum except UV_c. The solar blind corona camera, therefore, can detect the corona discharges or surface discharges in daytime.

The corona cameras and their construction diagrams are demonstrated in Figures 6.4 and 6.5. In order for corona cameras to meet technical requirements, they must comply with specifications related to four parameters: (i) minimum discharge detection in pC or in

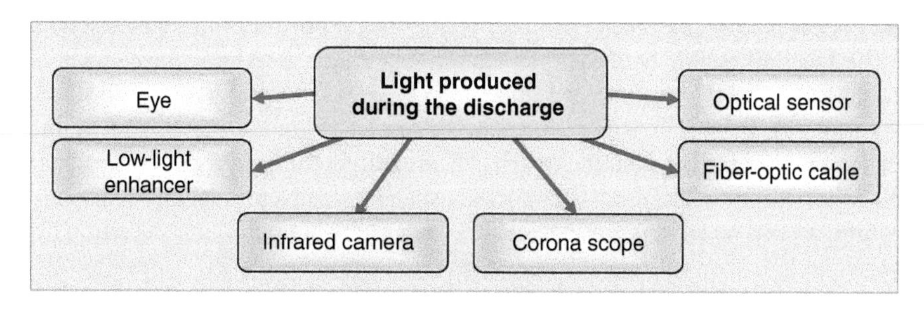

Figure 6.3 Approaches and devices for optical PD measurement [4].

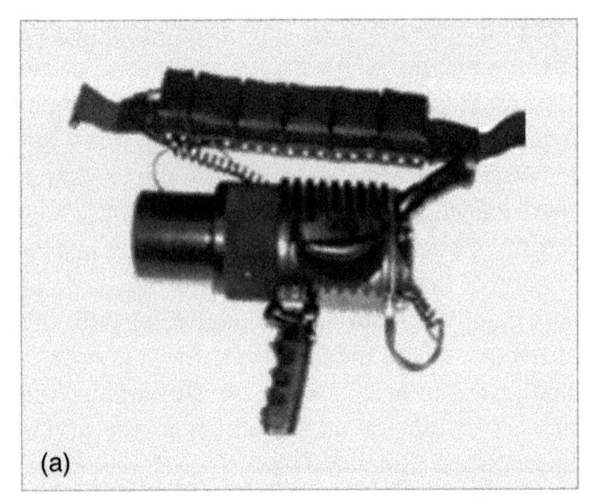

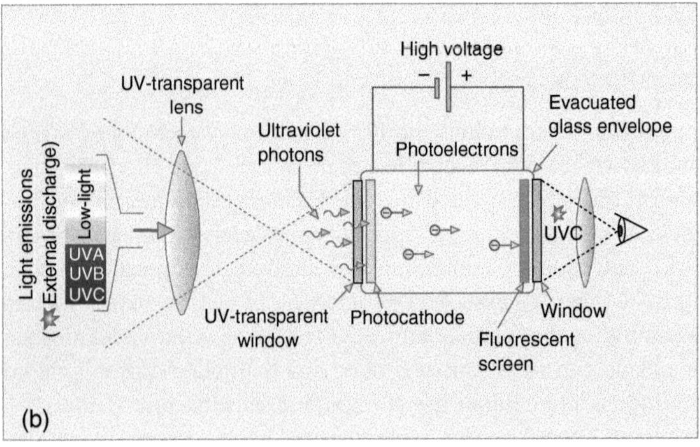

Figure 6.4 (a) A nonsolar blind camera; (b) construction diagram of the nonsolar blind camera [5].

other accepted units, (ii) minimum UV sensitivity in Watt/cm^2, (iii) spectral range in nm, and (iv) focus distance. Other parameters – such as power consumption, display characteristics, data processing, and data storage characteristics – should also be considered.

In recent years, there has been interest in optical PD measurement applied for corona and/or surface discharge detection inside high-voltage equipment, and it has been developed in the laboratory scale. Optical sensors – such as photomultipliers (PMTs), photo diodes, and optical fibers – have been utilized. Figure 6.6a,b provides PMT construction diagrams. The spectral sensitivity of such optical sensors, which determine the characteristics of the optical PD detection, is shown in Figure 6.6c. Figure 6.7 shows the application of optic fibers for corona discharge detection occurring in air or liquid insulating medium.

Generally, in laboratory research, the optical PD measurement is performed and compared with other PD measurement techniques, especially conventional PD measurement, according to IEC 60270. Figures 6.8 and 6.9 present the laboratory PD detection by optical technique and conventional PD measurement. The detected PD shapes by the optical PD detection as shown in Figures 6.8 and 6.9, have a rise time of approximately

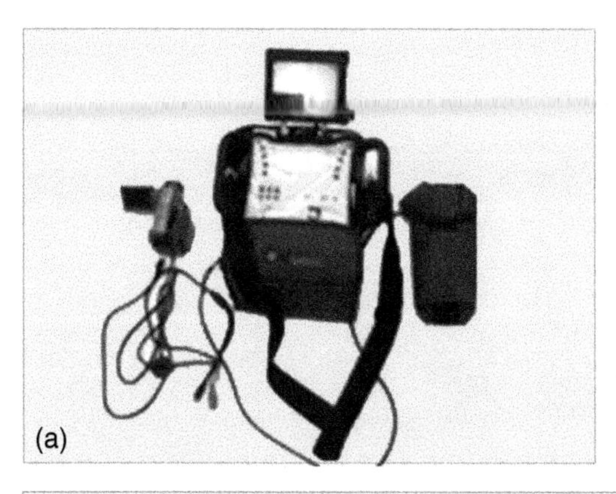

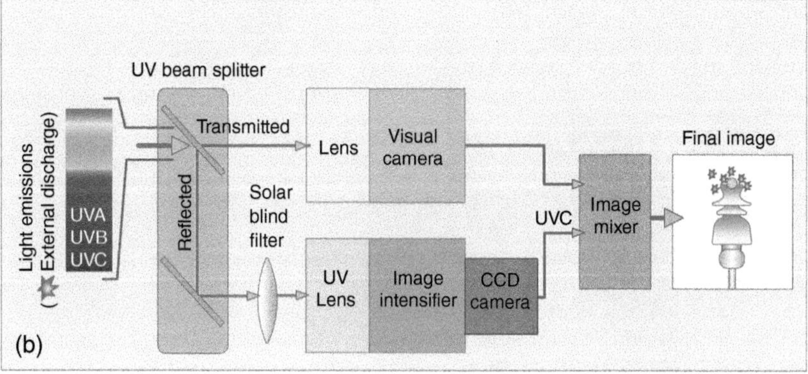

Figure 6.5 (a) A solar blind camera; (b) construction diagram of the solar blind camera [5].

5 ns and a pulse duration of about 20 ns. Figure 6.10a illustrates the advantages of the optical PD detection technique by which the complexity of PD mechanism in oil is revealed. The optical detected PD signals are clearly scattered in amplitude and shape. The positive streamers show a superposition of fast pulses, and the negative streamer comprises a growing intensity of burst pulses. It is clear that the conventional PD measurement with the limited bandwidth cannot detect the fast-impulse PDs generated in the liquid insulation system. In addition, the optical PD measurement shows the superior characteristic to detect PD signals in air excited by the impulse test voltage, as represented in Figure 6.10b.

6.2.3 Application of Optical PD Measurement

6.2.3.1 Insulators, Transformer Bushings, Surge Arrestors, Transmission Lines, and Fittings

Corona cameras are used to inspect corona discharges and/or surface discharges surrounding the high-voltage parts of insulators, transformer bushings, surge arrestors, and transmission lines. Although the corona may sometimes be visible in high-voltage stress areas, this is a very rare occurrence.

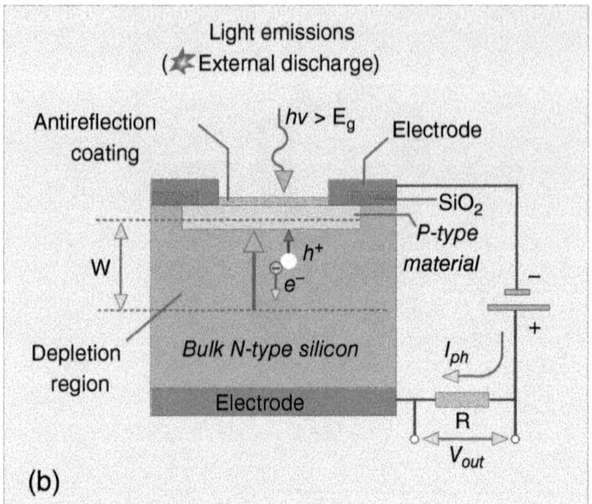

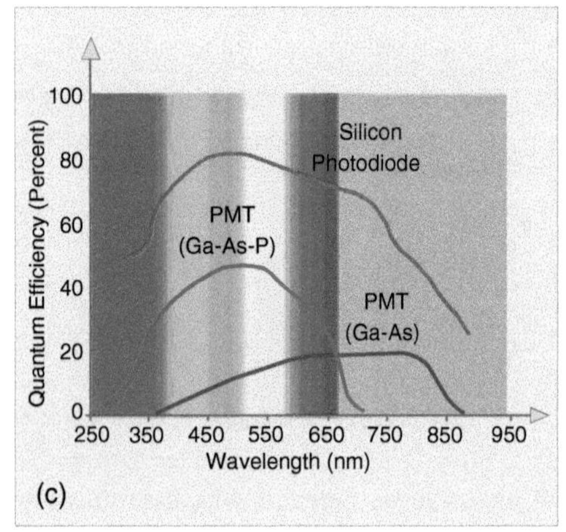

Figure 6.6 Optical sensors (a) photomultipliers and a silicon photo diode (PMT); (b) construction of PMT; (c) optical sensor spectral sensitivity [6–8].

6.2.3.2 Rotating Machines

Corona and surface discharges from the stator part of a rotating machine can be detected using a corona camera during hi-pot testing of the maintenance period. Figure 6.11 shows a solar blind camera and corona activities at the insulator connections and a stress-grading ring in a transmission line system and at the stator slot of the generator.

6.3 Acoustic Emission PD Measurement

6.3.1 Theory

PD activity generates EM waves, acoustic waves, current PD pulses, and byproducts. Transient acoustic PD signals are produced in three main ways:

1) Super-heated gas bubbles or gas channels are created by PD pulse current [2].
2) Energy is rapidly released from defective insulation material due to PD activity [1].

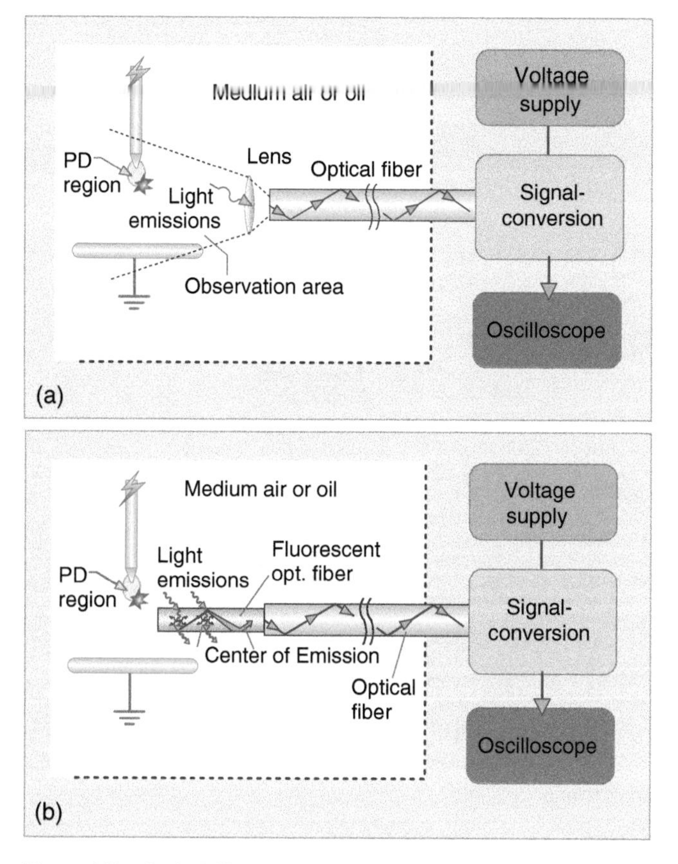

Figure 6.7 Optical fiber applications for optical PD measurement: (a) optical fiber with lens; (b) fluorescent optical fiber [9].

3) The "bouncing particle" phenomenon occurs inside high-voltage apparatus while in operation – especially GIS or GIL.

Acoustic Emission (AE) signals are usually classified as either a "burst" (pulse-like) or "continuous" (random noise-like) signal [12], as shown in Figure 6.12. The specific AE PD signal characteristics detected depend on three main variables: PD source, the propagation medium, and the operational characteristics of the AE sensor with its instrument. Generally, there are five main parameters used to analyze AE signals: amplitude, rise time, frequency, pulse duration, and the number of AE PD pulses. Figure 6.13 illustrates the parameters of a typical AE signal.

Common sources of acoustic waves usually generate a wide-band AE signal of around 1 MHz [1]. Similarly, AE waves generated by PD (or so-called AE PD signals) while traversing the mineral oil inside transformers are generally in the ultrasonic range of 20 kHz up to 1 MHz [2], meaning that these AE PD signals are able to penetrate the surrounding insulation material. Furthermore, AE PD signals may also be generated from PD sources such as any sharp point on a high-voltage conductor or from a defective outdoor insulator (of the type commonly used in transmission or distribution line systems as well as in an

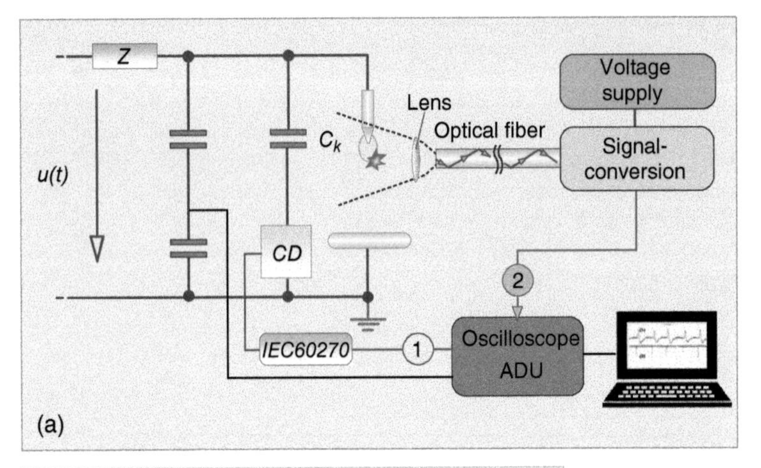

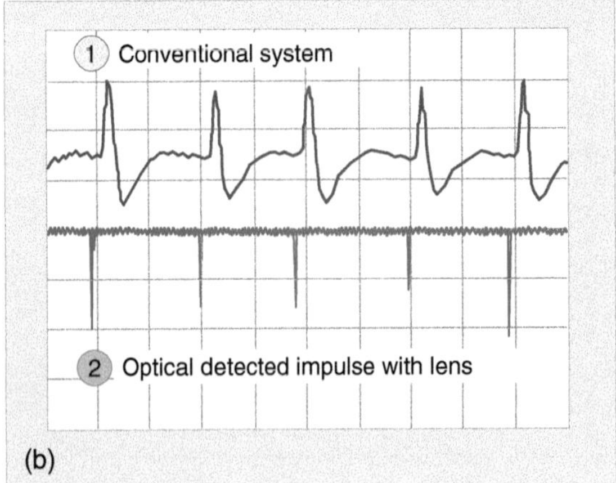

Figure 6.8 Laboratory PD detection by optical fiber with lens and conventional PD measurement: (a) test circuit diagram; (b) detected PD signals [9].

air-insulated substation [AIS]) and these signals are able to propagate through the surrounding air directly to an acoustic monitoring device. For a PD emitted from a source inside high-voltage apparatus, AE PD signals are able to traverse to the AE PD sensor (usually installed on the exterior surface wall of the apparatus) by different means, (i) direct propagation, (ii) through internal barriers, and (iii) direct propagation or penetration through the internal barriers to the surface wall of the apparatus and then traveling into the solid media of the surface wall (structure borne). However, during the propagation process, reflection and refraction of the AE signal tend to occur when the signal reaches the boundaries of each medium, which results in attenuation, absorption, and scattering of the AE signal. The propagation modes of AE PD signal are depicted in Figure 6.14.

Considering modes of wave propagation, pressure waves (*longitudinal waves*) are able to propagate in solid material as well as gas and liquid medium whereas shear waves (*transversal waves*) are able to propagate only in solid material. Further the transversal waves generally travel at around half the velocity of longitudinal waves [13] with the speed of the AE signal being determined in part by the temperature of the transmission media. Table 6.1 lists

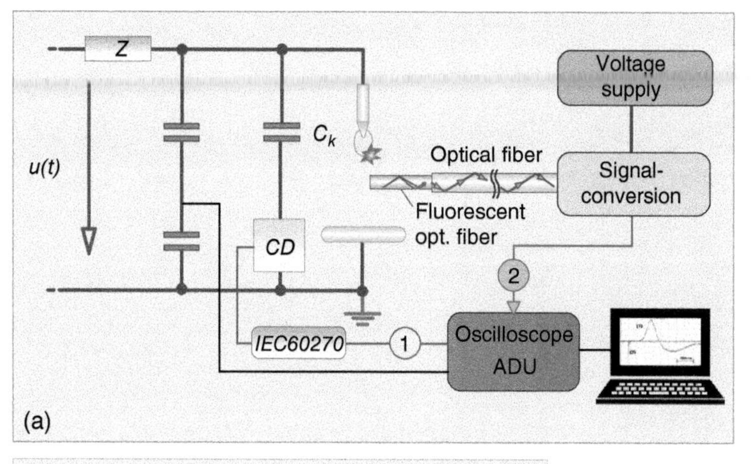

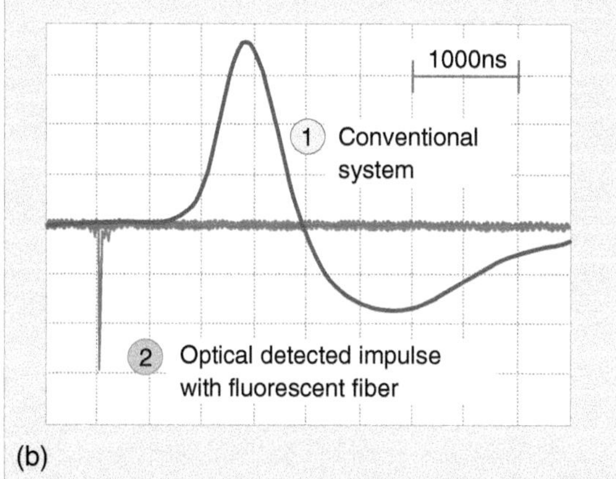

Figure 6.9 Laboratory PD detection by fluorescent optical fiber and conventional PD measurement: (a) test circuit diagram; (b) detected PD signals [9].

the longitudinal velocity of various materials commonly used as components in high-voltage apparatus. Table 6.2 illustrates the effect of mineral oil temperature on AE velocity.

6.3.2 Acoustic Receivers and Acoustic Sensors

There are two techniques used for the detection of AE PD signals: (i) by hand-held AE PD receiver, and (ii) by instrument-based AE detection.

6.3.2.1 Hand-Held AE PD Receivers

Hand-held AE PD receivers are generally classified by types as either airborne ultra-probe or ultra-probe with a pointed-tip extension as shown in Figure 6.15.

Airborne Ultrasonic Probes

Various types of airborne ultrasonic probe (particularly the ultrasonic directional microphone type) are used to detect the ultrasonic waves emitted by corona discharge from the

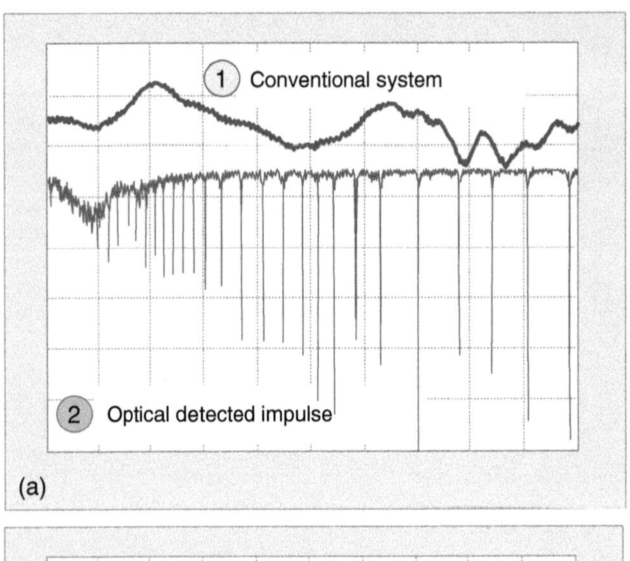

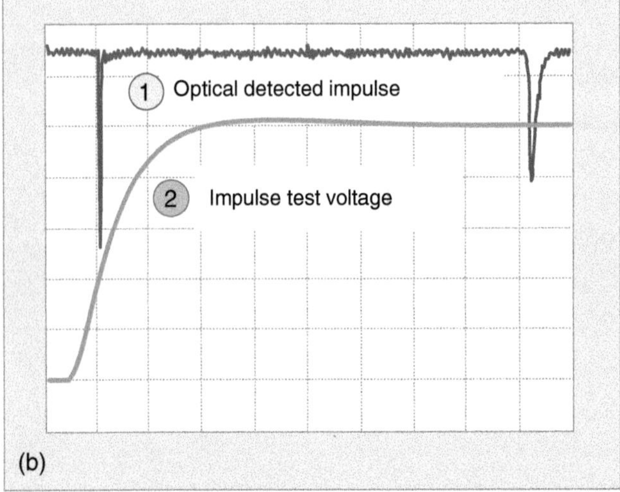

Figure 6.10 Application of optical PD measurement technique: (a) detected PD signals in mineral oil; (b) detected PD signals under impulse test voltage in air [9].

points or edges of high-voltage conductors in power substations and transmission lines. In addition, corona discharges and surface discharges from the outdoor insulators can also be detected using this type of equipment. The use of directivity microphones is generally preferred in field tests because they are able to detect AE PD signals from specific AE PD sources while also suppressing interferences from other sound sources in the operating environment. The airborne ultrasonic probes are able to detect pressure wave signals in the range of 20 to 100 kHz (with a center frequency of about 40 kHz). Then the frequencies of such pressure wave signals are down to the audio range of 100 Hz to 3 kHz for analysis by a skilled person. Furthermore, an ultrasonic wave concentrator (UWC) can be employed with the airborne ultrasonic probe, in order to enhance sensitivity where the detection of ultrasonic waves at a long distance is desired.

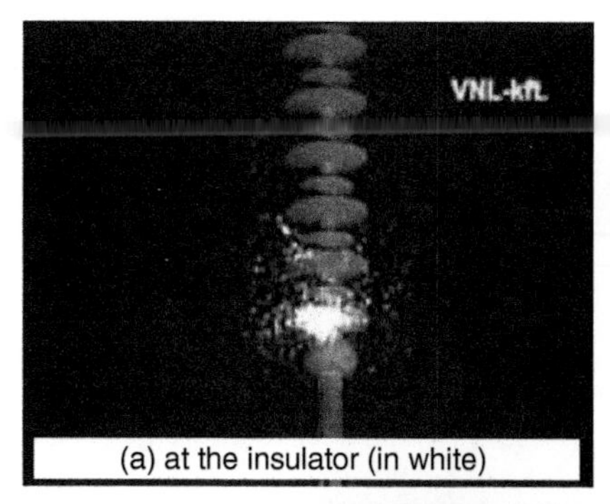

(a) at the insulator (in white)

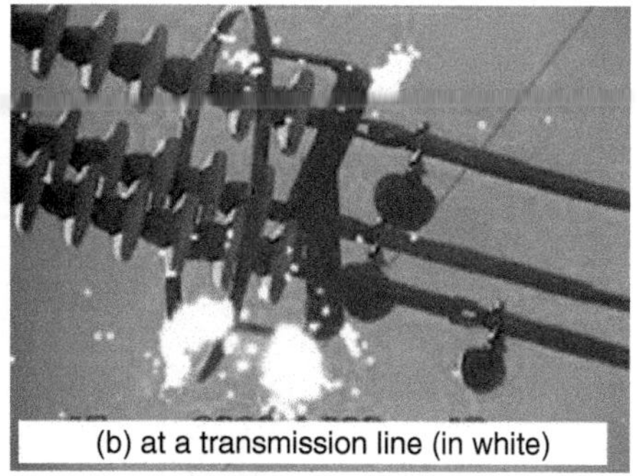

(b) at a transmission line (in white)

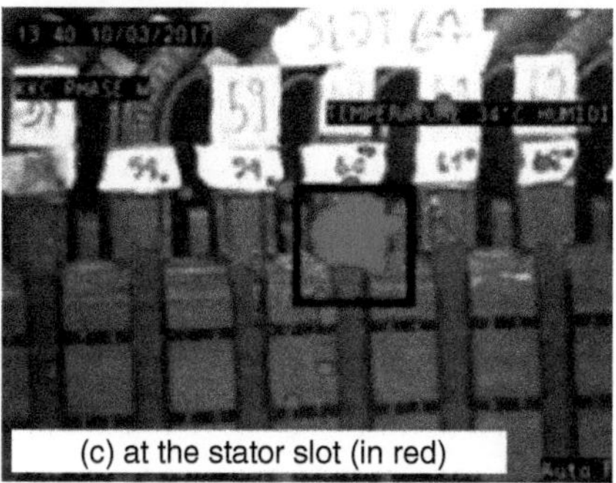

(c) at the stator slot (in red)

Figure 6.11 Application of corona discharge and surface discharge measurement by optical PD detection (a) at the insulator (in white); (b) at a transmission line (in white); (c) at the stator slot (in red) [10, 11].

Ultrasonic Probe with a Pointed-Tip Extension

Practically, the technique for determining the presence of moving particles or loose components inside high-voltage enclosures – especially gas-insulated substations (GISs) and gas-insulated transmission lines GILs – is by AE PD measurement using an ultrasonic probe with pointed-tip extension. During this process, the pointed tip is pressed on the exterior surface of the GIS unit in order to detect any AE PD signals, which are then amplified, filtered, and modulated to become audio signals transmitted to the headphone unit of the operator for expert interpretation. The more modern ultrasonic probes usually provide a digital display of the AE PD signal output from the ultrasonic probe, and may also feed data into the PRPD displaying system so that the new data from other PD tests by different techniques may be analyzed and compared on-screen.

6.3.2.2 Instrument-Based AE PD Detection

The technique of instrument-based AE PD detection usually requires that AE PD sensors be securely attached directly to the outside surface of high-voltage component to be tested, such as transformers, GIS, GIL, and underground cables. Furthermore, AE PD detection is mostly used in combination with other PD detection techniques, both to confirm the PD signal and to assist in locating the PD source(s). For example, one or more AE PD sensors may be used in combination with one or more UHF sensors to detect and analyze PD signals generated in GIS, GIL, or transformers. Similarly, AE PD sensors can also be combined

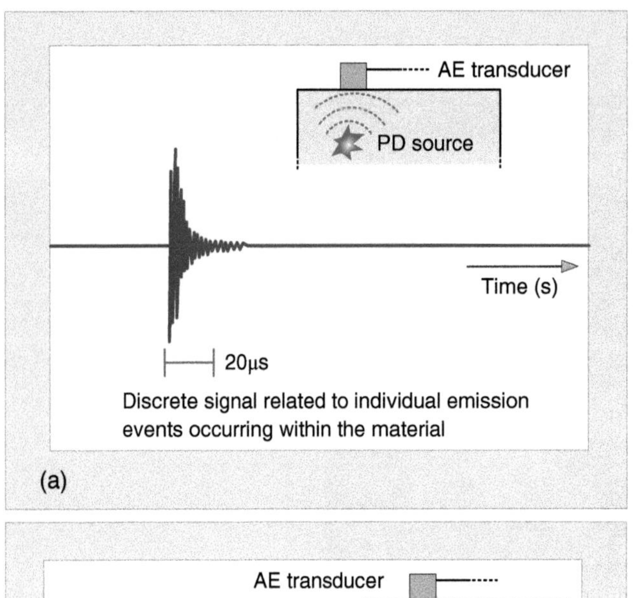

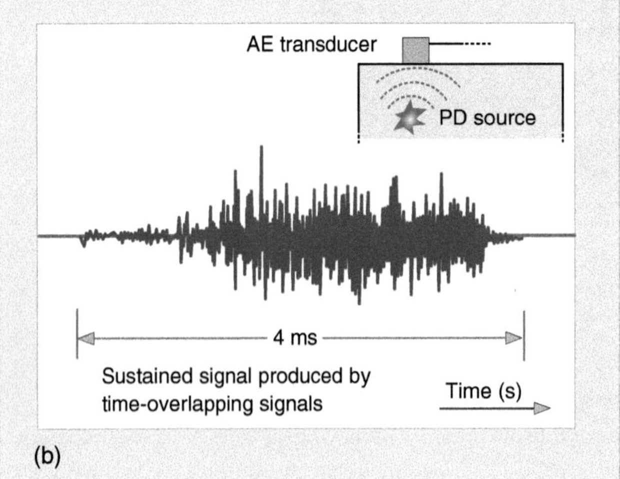

Figure 6.12 Typical AE signals: (a) burst type; (b) continuous type.

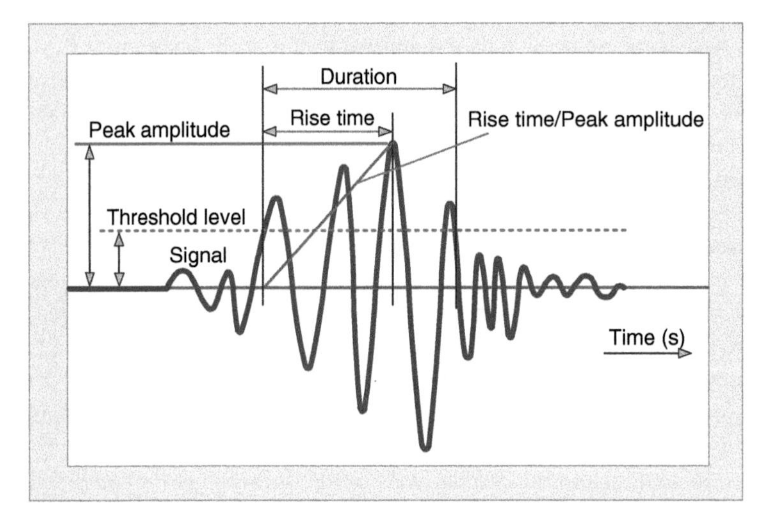

Figure 6.13 Characteristics of a typical AE signal.

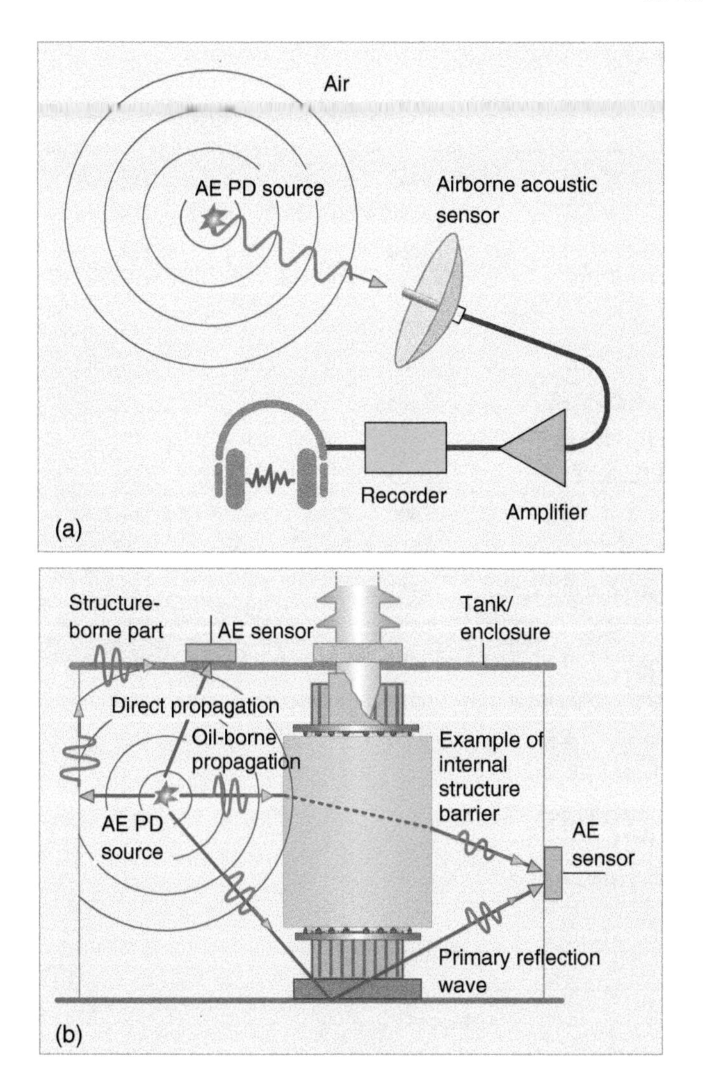

Figure 6.14 AE PD signal propagation in various media: (a) AE PD signal propagation through the surrounding air; (b) AE PD signal propagation inside a transformer tank.

with high-frequency current transformers (HFCT) and/or dissolved gas analysis (DGA) to monitor PD signals and to located PD sources in transformers.

Acoustic PD Detection System

A typical AE PD detection system consists of three main components: (i) an AE PD sensor, (ii) a data-processing unit, and (iii) a data acquisition unit. The typical PD detection system used to monitor high-voltage equipment is illustrated in Figure 6.16. The AE PD signals detected by AE PD sensors generally need to be amplified and filtered by a data processing unit. After this, the signals are transmitted digitally to the data acquisition unit to be stored in its database (data logger) for subsequent analysis and display. A run history file from the data logger is used for trending analysis of the insulation condition. Data analysis and diagnostic test results as well as trending analysis recommend the operator for action needed.

Table 6.1 Longitudinal velocity of common material used in high-voltage apparatus.

Material		Longitudinal velocity (m/s)	Reference
Metal	Aluminum 20 °C	6400	[13]
	Brass 20 °C	4280–4700	[13]
	Copper 20 °C	4700	[13]
	Iron 20 °C	5900	[14]
	Cast iron 20 °C	3500–5800	[14]
Solid insulation	XLPE 20 °C	2000	[15]
	EPR M150 25 °C 2 MHz	1563	[16]
	Wood (density 750 kgm^{-3})	4000	[17]
	Epoxy, unfilled 20 °C	2700	[14]
	Epoxy, filled 20 °C	2600	[14]
Liquid insulation	Mineral oil 25 °C	1415	[18]
	Natural ester (olive oil) 25 °C	1431	[19]
	Silicone oil 30 °C	873–929	[19]
Gas insulation	Air 0 °C 1 atm	331	[13]
	Nitrogen 20 °C	333	[13]
	Hydrogen 20 °C	1286	[13]
	SF_6, 25 °C, 1 atm	140	[13]
Mixed insulation	Pressboard impregnated with mineral oil 20 °C	2000	[13]
	Impregnated crepe paper 20 °C	1500–1600	[14]

Table 6.2 Wave velocity in mineral oil at various temperatures [1].

Temperature of mineral oil	Velocity
50	1300
80	1200
110	1100

Figure 6.15 Hand-held AE PD receivers: (a) ultra-probe with its accessories and ultrasonic wave concentrator; (b) ultra-probe with a pointed tip.

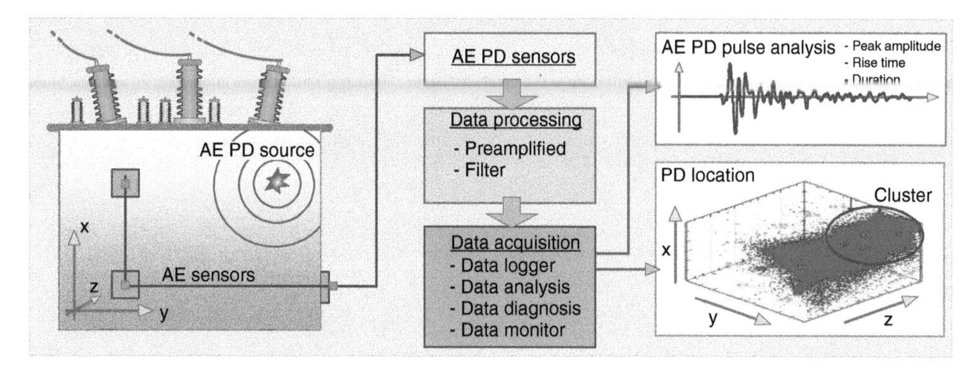

Figure 6.16 Acoustic PD detection system for the instrument-based AE PD detection.

External Acoustic PD Sensor

Due to the application of new "disruptive technologies," the industry has adopted various innovative fabrication techniques in the production of more advanced types of acoustic sensors used for AE PD detection. The most common of these is the piezoelectric sensor, which is manufactured from ferroelectric ceramic materials such as lead zirconate titanate (PZT). Piezoelectric sensors operate by converting the mechanical force exerted on the piezo material into an electrical output signal. PZT is usually the preferred material due to its high electromechanical coupling factor, and also that it can be produced in various customized shapes and sizes at a reasonable price.

External AE PD sensors can be classified into two types – the traditional single AE PD sensor and the more recently introduced three-sensor system [1]. Traditional single external AE PD sensors are usually classified as either accelerometers or AE sensors. Accelerometers function by generating an electrical signal that is proportional to the acceleration of the surface on which the sensor is located. Similarly, AE sensors function by generating an electrical signal that is proportional to the velocity of the surface attached with the sensor.

Accelerometers generally have a useful operating range of up to 50 kHz, whereas AE sensors commonly operate in a range of 30 kHz to 1 MHz [20]. Modern external AE PD sensors may feature other innovative technologies, such as using a built-in pre-amplifier in order to amplify the detected signals. Recently, optical fiber-acoustic sensors and surface acoustic wave sensors have been introduced for AE PD measurement [21]. Three-sensor AE PD detection systems are a more recent development, intended to improve (both) the sensitivity of AE PD detection system and increasing noise rejection efficiency. The three-sensor system employs three or more acoustic sensors mounted relatively close together in a rigid yet movable frame. In transformer testing, an additional electrical PD sensor is sometimes used in combination with the AE sensors in order to determine the time of origin for the PD event, which helps to locate the PD site in a transformer.

Internal Acoustic PD Sensor

Significant research on internal AE sensors with a main focus on the "waveguide" type was conducted by R.T. Harrold, who observed that (both) Pyrex glass rods and epoxy fiberglass rods as well as fiberglass optical guides when immersed in mineral oil were able to transmit AE PD signals very efficiently to an external ultrasonic sensor [22]. It seems that the most

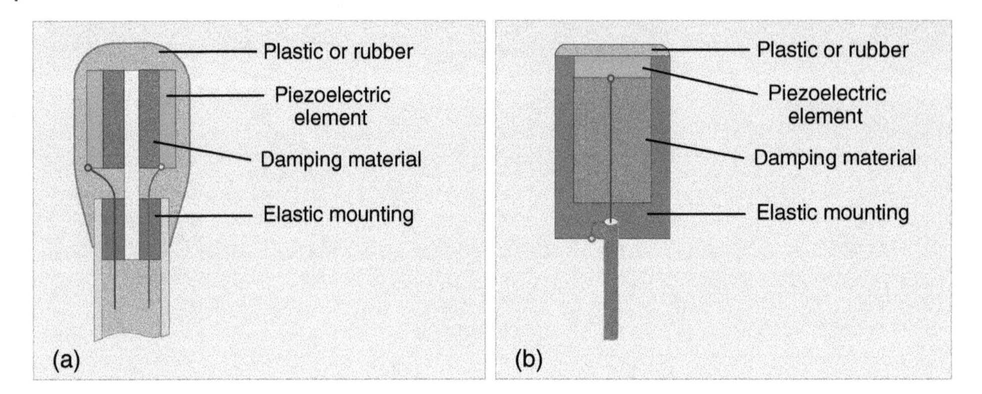

Figure 6.17 Design principles for a probe hydrophone and a membrane hydrophone: (a) a cylinder shape of piezoelectric hydrophone; (b) a disk shape of piezoelectric hydrophone [23, 24].

practical application of this research would be for AE PD detection in power transformers, whereby a solid fiberglass rod would be immersed in the insulating liquid to couple acoustic energy, which the rod would convey to an externally mounted sensor [1]. Hydrophone is one internal acoustic PD sensor. Figure 6.17 shows the principle design for hydrophones.

Acoustic PD Sensor Characteristics

The application of AE sensors for PD measurement must satisfy at least two main parameters: (i) sensitivity; and (ii) frequency response.

Sensitivity: The sensitivity of an AE PD sensor may be defined as the ability of the sensor to detect an AE PD signal from a given PD source. AE PD signal levels are measured in mV or μV whereas a known discharge level is measured in pC. Piezoelectric sensors are the most commonly used type for AE PD measurement. These devices convert sound pressure generated by a PD source to an electrical signal. The sensitivity of AE PD piezoelectric sensors therefore may be defined in mV/Pa. According to [25], the relationship between discharge level and acoustic media pressure was reported as a 1 pC discharge generating the rms sound pressure 0.2 Pa at 10 cm from the PD source in oil.

Frequency Response: As shown in Figure 6.18, AE sensors used for PD measurement can be divided into two types according to frequency response: the wide-band type (broadband) and the narrow-band type. Wide-band AE sensors are generally preferred for monitoring AE PD signals cover the flat region frequency of interest by which resonance may occur at a frequency on such region. For narrow-band AE sensors, the highest gain occurs at resonance frequency where the response curve reaches a sharp peak and then decreases. Wide-band AE PD sensors are usually capable of providing all the PD measurements required, although they may exhibit lower sensitivity to some types of discharges. Thus, where higher sensitivity is required, the use of narrow-band sensors is preferable; however, they might be suitable only for monitoring a particular type of discharge due to their limited bandwidth [26, 27].

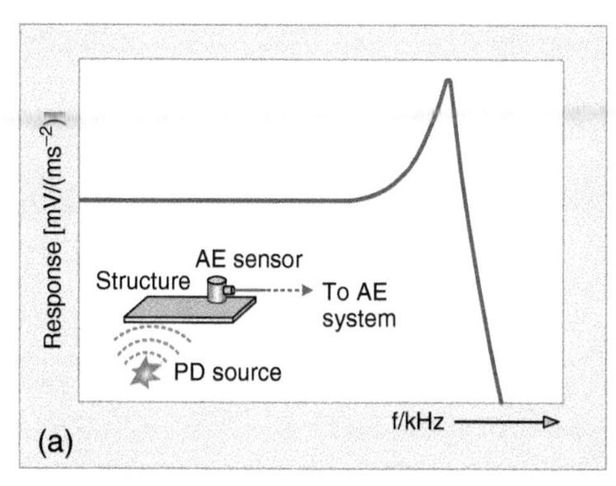

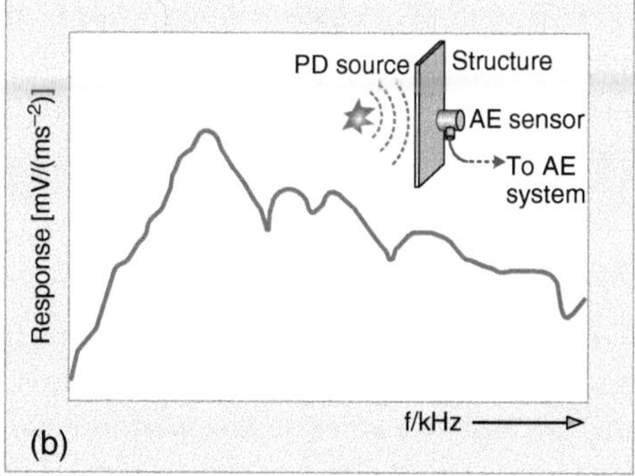

Figure 6.18 Examples of frequency response: (a) wide-band sensor type; (b) narrow-band sensor type.

6.3.3 Acoustic Noises in AE PD Measurement

Various types of acoustic noise occur during AE PD measurement, mainly low-frequency background noise. Most acoustic noise sources are associated with high-voltage equipment and their accessories. The influence of environmental noise sources from rain, snow, wind, and so on is highly dependent on where high-voltage equipment is installed. Table 6.3 lists

Table 6.3 Examples of acoustic noise sources and their characteristics, which may occur during performing AE PD measurement of a transformer [1].

Noise source	Main characteristic
1) Core magneto striction noise (Bark-Hausen effect)	1) The main frequency at twice of the power frequency 2) The same noise amplitudes in both half cycles 3) Considerable noises with harmonic frequency up to 30 to 40 kHz or above caused by over fluxing
2) Pump liquid noise (occurring in oil-circulating pump)	1) No correlation with power frequency waveform
3) Loose nameplate, fan noise, etc.	1) The repetition rate may close to PD signal 2) Longer noise signal waveform normally under 150 μs
4) Loose shielding connection	1) Large PD indication may be found but not be harmful transformer operation
5) Not properly shield of sensor wiring	1) Spurious PD
6) Switching and load tap change movement	1) Random signals
7) Thermal fault	1) Random signals
8) Environmental noises	1) Random signals with longer waveform >1 ms

the main sources of acoustic noise sources and the characteristics of each, as commonly occurs during AE PD transformer testing. In some cases, particularly field testing, reduction of (acoustic) noise as well as a PD-noise classification technique is required in order to perform PD tests properly.

6.3.4 General Idea for AE PD Measurement

6.3.4.1 Sensitivity Check for AE PD Measurement

Sensitivity check for AE PD measurement is recommended in IEC 62478, and the procedures are the same whether it be performed on-site or in a laboratory.

The sensitivity check in the laboratory is to establish the relationship between the acoustic signal level and a known apparent change in pC generated from a real PD source. In the experiment, the frequency spectrum of the detected AE signal is also recorded. In addition, an artificial AE signal emitter capable of generating AE signals, which are similar in both intensity and frequency spectrum to the AE signal obtained from the actual PD source, can be used as a reference for sensitivity check. The on-site sensitivity check should be done with the tested high-voltage apparatus in the same way as performing in the laboratory by using the same measuring system.

6.3.4.2 AE PD Measurement

External PD phenomena such as the corona discharge and surface discharge, which occur in transmission and distribution lines as well as in substations, propagate from the PD source through the surrounding air, where the PD signals can be detected by AE PD receiver, as mentioned in Section 6.3.2.1. Whereas PDs in high-voltage apparatus can be detected by AE sensors mounted in or on the exterior surface of such equipment. The detected AE PD signal is then transmitted to an AE measurement device for data display, storage, and analysis. AE PD measurement is preferred for several main reasons: (i) its nondestructive nature, (ii) low susceptibility to external EM interference, (iii) ease and speed of setup and operation, and (iv) ability to locate PD sources. However, it also has a big disadvantage in that further analysis of the detected AE PD signal is problematical because there is no correlation between the AE PD signal and the electrical PD level in pC – especially for AE PD measurement made in field tests. Therefore, acoustic PD measurement is usually conducted in conjunction with other PD detection methods in order to increase PD detection sensitivity and confidence in the accuracy of detected PD signals.

6.3.5 Application of Acoustic PD Measurement for High-Voltage Apparatus

An AE PD sensor is generally applied for PD measurement of high-voltage apparatus, especially for on-site PD test.

6.3.5.1 Detection of Corona and Surface Discharge from Outdoor Insulators or High-Voltage Conductors

Corona discharges and surface discharges occur at the defect outdoor insulators or at a sharp point as well as the edge of high-voltage conductors in an AIS, a transmission line system or a distribution line system can be detected by using an airborne ultrasonic probe

Figure 6.19 The airborne ultrasonic probe used for AE PD measurement and PD source location of: (a) voltage transformer; (b) disconnecting switch; and (c) air-to-SF_6 gas insulated bushing.

or a directional microphone as shown in Figure 6.19. The detected ultrasonic wave is transformed to audible wave before interpreted by the expert.

6.3.5.2 PD Detection in Transformers

The use of AE PD sensors for PD measurement and PD localization in transformers is a nondestructive test technique recommended in [1, 2, 28]. There are two types of AE PD sensors for the purposes above – internal AE PD sensors and external AE PD sensors. Internal AE PD sensors installed inside the transformer can pick up higher-intensity AE signals than external AE sensors attached to the outside. Utilizing only an external AE PD sensor(s) is relatively simple and sufficient for most purposes, as depicted in Figure 6.20.

6.3.5.3 PD Detection by AE PD Measurement Technique in GIS and GIL

AE PD sensors are commonly used for PD measurement and PD location in GIS and GIL. Generally, PD signals in GIS or GIL are generated from two main sources – stationary defects and moving particles, as shown in Figure 6.21. Many types of stationary defects can be found in GIS and GIL, for example: interior protrusions, particles that become attached to the outer or inner surface of supporting insulators (post insulators or conical spacers), and voids inside the supporting insulator.

PD signals caused by a stationary defect in GIS and GIL generate a pressure wave in the gas medium, which then propagates throughout the interior. External AE PD sensors are used to detect the signals generated by stationary defect. These sensors are placed at different positions on the outer casing of the GIS or GIL, as shown in Figure 6.21. A detected AE PD signal is converted to an electrical signal, which is generally needed to amplify, filter, and transmit the signal to data acquisition unit for further analysis.

Some AE PD signals may be quite weak and therefore difficult to detect, such as AE PD signals generated from a fixed particle located at the surface inside a conical spacer or generated from a void inside the insulating supporters. These PD signals are weak due to high attenuation while propagating through insulation material such as cast epoxy resin. Moreover, during propagation from the stationary PD source to the sensors, both the amplitude and frequency of the AE PD signal become more attenuated because the gas acts as a low-pass filter. The attenuation rate depends mainly on the gas pressure and angle of

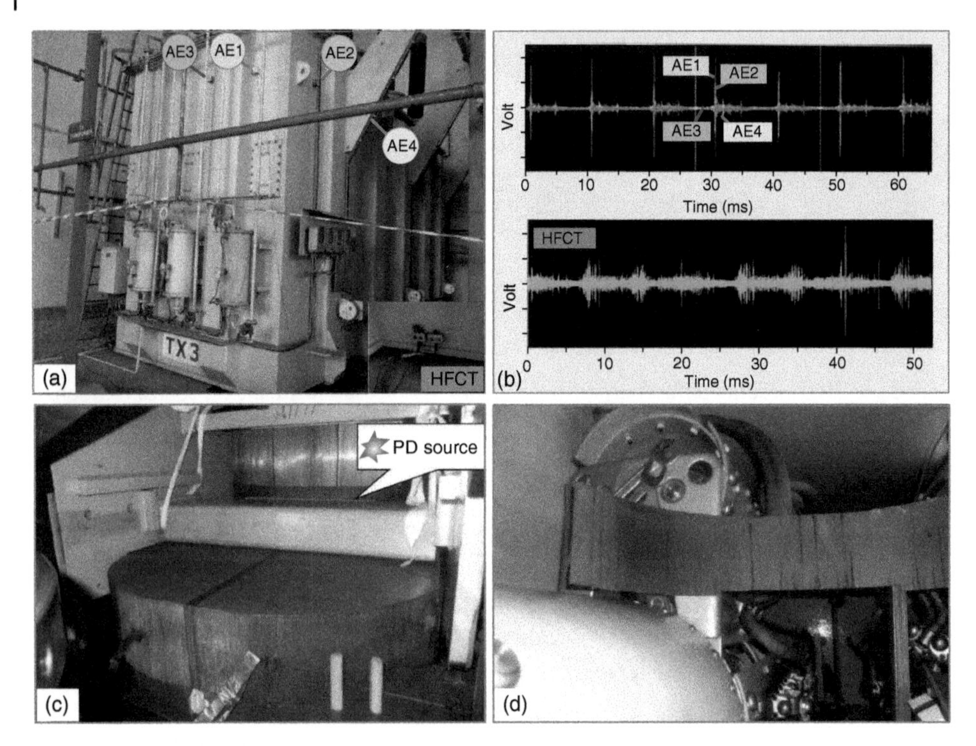

Figure 6.20 PD detection and location in a transformer rated 230 kV, 250 MVA in field test: (a) PD measurement by auscultatory method; (b) detected PD signals from acoustic PD sensors (AE1–AE4) and the HFCT; (c) PD occurring at the insulation between core laminate and flux collector; and (d) the insulation between cores laminate and flux collector.

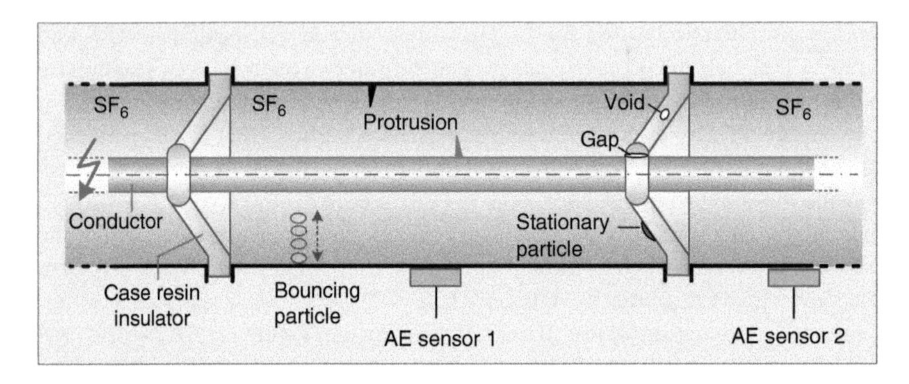

Figure 6.21 PD sources in GIS or GIL.

incidence. It should be noted that because acoustic PD detection using AE PD sensors has only a limited time for resolution, it cannot monitor every single gas discharge that occurs.

By contrast, moving particles (aka bouncing particles) generate a mechanical wave in the interior space, and their location can be determined using an ultrasonic probe with a pointed tip extension. The pointed tip of the extension is brought into contact with the GIS and/or GIL enclosure in order to pick up the AE signal(s), which are then amplified, filtered, and modulated to produce audio signals, which the operator can then analyze. Normally, a moving particle will generate a higher amplitude AE signal in the wide-band frequency range compared to the AE signal from a stationary defect. Indeed, there is a strong consensus that AE PD measurement, when compared with other commercially available diagnostic techniques, generally provides superior sensitivity in the detection of moving particles [2].

6.3.5.4 PD Detection in Rotating Machine by AE PD Measurement Technique

Generally, AE sensors, such as piezoelectric sensors and fiberoptic accelerometers, are used to detect mechanical problems in rotating machines [26, 27]. Moreover, AE sensors are sometimes also used on-site to determine PD measurement and PD location in stator winding insulation [2, 29]. However, conducting on-site PD measurement on rotating machines by AE PD measurement is sometimes made very difficult due to high noise levels in the operating environment. In such cases, highly sensitive AE PD sensors and AE measuring equipment with high noise-filtering capability and/or noise reduction capability are required. Consequently, in many countries this is not a popular method for detecting PD issues in rotating machines. Figure 6.22 shows the on-site AE PD measurement of a generator rated 6.9 kV, 13.25 MW.

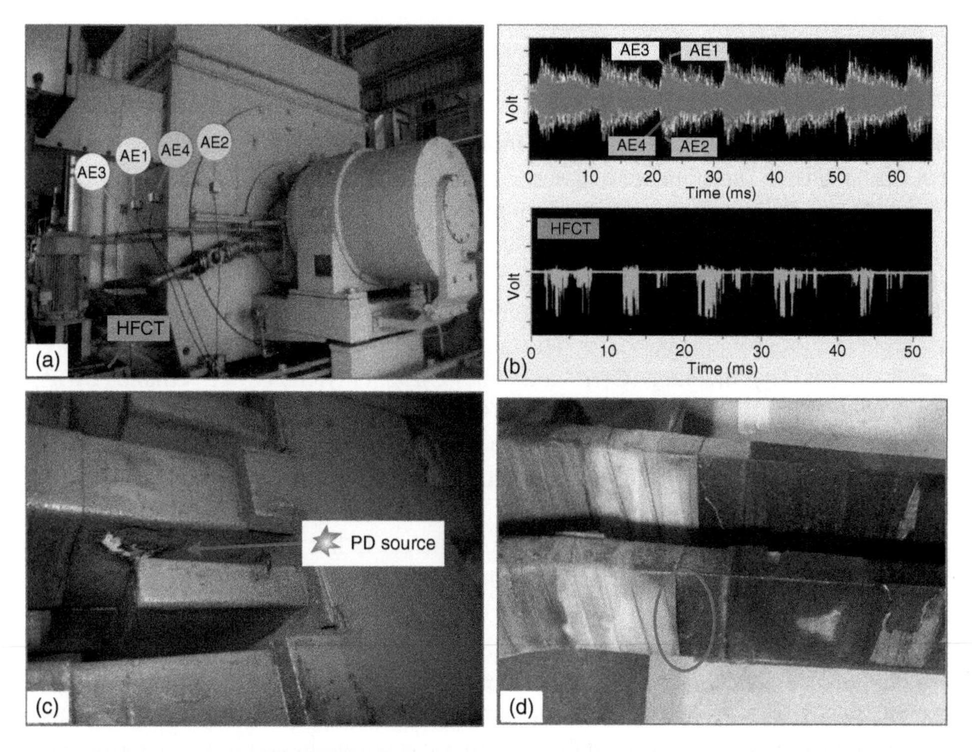

Figure 6.22 On-site AE PD measurement of a generator rated 6.9 kV, 13.25 MW (a) AE sensor positions for PD detection of generator rated 6.9 kV; (b) detected AE signals and HFCT signals; (c) PD problem happening at the stator bar before removing from the generator; (d) discharge at stator bar.

6.3.5.5 PD Detection for Other High-Voltage Equipment

AE PD measurement and PD location also have application for determining the operational condition of various other types of high-voltage apparatus, such as instrument transformers, transformer bushings, and underground cables. The same basic technique used for AE PD measurement of power and distribution transformers (as described in Section 6.3.4.2) can also be applied in the case of instrument transformers. Similarly, the general concept of AE PD measurement for GIL can also be applied to underground cable AE PD measurement.

6.4 Chemical Byproducts

6.4.1 Theory

PD activity generates various types of measurable signals, including the chemical byproducts. These byproducts have a strong tendency to dissolve in the liquid insulation present or react with a catalyst to form a corrosive gas. Therefore, detection and identification of these chemical byproducts may be used to discover PD problems occurring in the insulation system of high-voltage equipment. The two most commonly employed techniques in such analysis are dissolved gas analysis and decomposition gas analysis.

6.4.2 Dissolved Gas Analysis for Liquid Insulation

6.4.2.1 Dissolved Gas Generation in Liquid Insulation

DGA has long been the standard diagnostic technique for the reliable detection of faults such as PD, arc discharge, or overheating in high-voltage apparatus that use dielectric oils and oil-impregnated papers (OIPs) as insulation. During the degradation of liquid insulation, gaseous byproducts are formed and then dissolved in the oil. Quantitative analysis is able to determine the type and concentration of dissolved gases, which facilitates fault prediction and preventative measures. The most common of the gases produced are as follows: hydrogen (H_2), methane (CH_4), acetylene (C_2H_2), ethylene (C_2H_4), ethane (C_2H_6), carbon monoxide (CO), and carbon dioxide (CO_2) [29]. These gases are mainly used for DGA. Besides, other generated gases such as propylene (C_3H_6) and propane (C_3H_8) may be required. The amount and type of gas generated depends on the temperature of the oil, as illustrated in Figure 6.23.

To perform DGA techniques, first the liquid dielectric has to be sampling from the high-voltage apparatus. Then, the gas extraction from the liquid sample has to be carried out. Three gas-extraction methods – vacuum extraction, stripper column extraction, and head space sampling – are described in [30]. The vacuum gas-extraction techniques can be classified as a partial degassing technique (a single-cycle vacuum extraction) and a Toepler technique (a multicycle vacuum extraction) details in [31]. Then the specified value of the extracted gases is introduced into a gas chromatograph by which the gas components are separating on a column and then detecting individually. In the final process, a DGA interpretation scheme such as Duval Triangle, IEC ratios, Dörnenburg ratios, Roger ratios, or Key gas method will be selected to interpret the tested data.

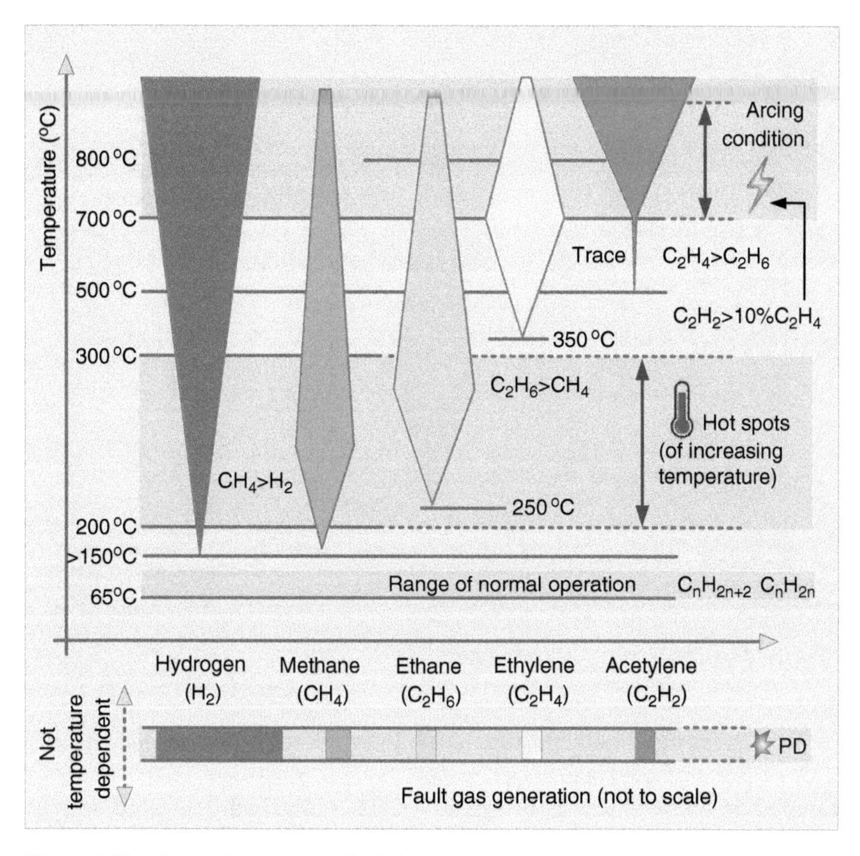

Figure 6.23 Approximate mineral oil decomposition versus temperature above 150 °C [29].

Syringes, sampling tubes, or sampling bottles can be used to contain the oil samples; each type of liquid sampling containers has different advantages and disadvantages that the users need to consider [32].

6.4.2.2 Application of DGA for PD Analysis

Application of DGA for PD Analysis of Transformers

Offline DGA for PD Detection in Liquid Insulating Immersed Transformers and Reactors

Transformers use dielectric liquids for electrical insulation and cooling purposes, and various solid insulation types for electrical insulation and mechanical purposes. Examples of insulation used in liquid-immersed transformers are mineral oil, silicone oil, synthetic ester, and natural ester, whereas examples of solid insulation are paper, pressboard, and solid wood, which are all cellulose-based materials. Internal arcing, overheating, or partial electrical discharge (PD) result in the production of various gases – hydrogen (H_2), methane (CH_4), acetylene (C_2H_2), ethylene (C_2H_4), and ethane (C_2H_6), which subsequently tend to dissolve in the insulating liquid, as shown in Figure 6.24. This allows quantity analysis of gases extracted from samples of the insulating liquid, and DGA interpretation, using various interpretative methodologies in order to determine the likelihood of the fault occurring, in the transformers.

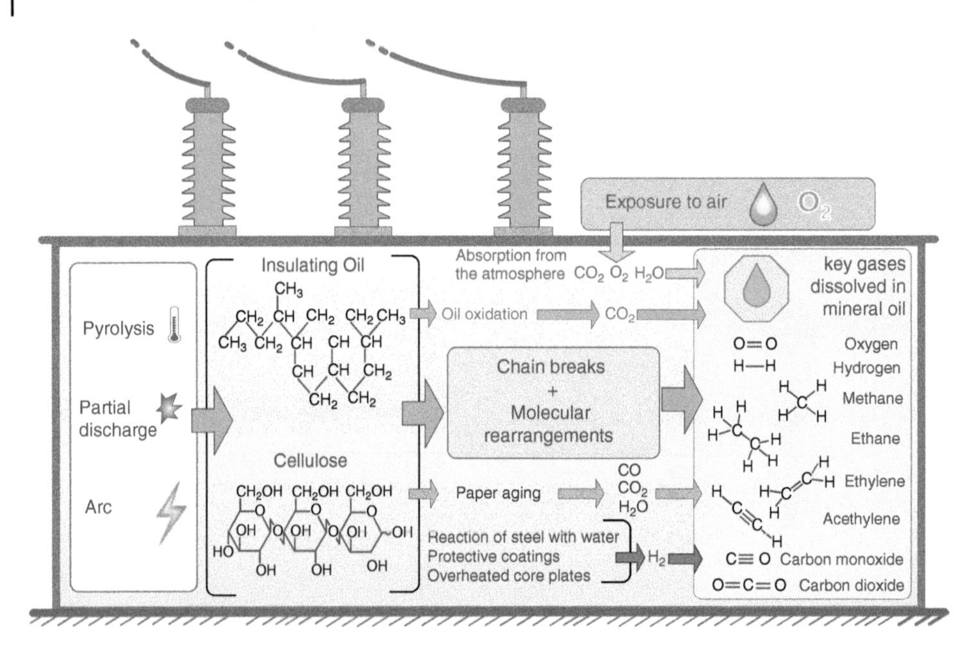

Figure 6.24 Dissolved gases occurring in transformers.

The five recommended methodologies for DGA interpretation associated with PD phenomena [33, 34] are discussed below.

Duval Triangle: Only three hydrocarbon gases – C_2H_2, C_2H_4, and CH_4 – are used to classify fault types in Duval Triangle 1 (the classical Duval Triangle for transformers). The concentration of C_2H_2, C_2H_4, and CH_4 are calculated in percentages using equations (6.1) to (6.3).

$$\%C_2H_2 = \frac{100x}{x+y+z} \tag{6.1}$$

$$\%C_2H_4 = \frac{100y}{x+y+z} \tag{6.2}$$

$$\%CH_4 = \frac{100z}{x+y+z} \tag{6.3}$$

where x is the amount of C_2H_2, y is the amount of C_2H_4, and z is the amount of CH_4. Consequently, all ratios will be located in the triangle map to indicate the types of faults.

For PD case discussed in this chapter, PD zone locates in a small area at the tip of the triangle as shown in Figure 6.25. Details of other fault types are described in IEC [33]. Michel Duval also develops the classical Duval Triangle 3 for alternative insulating liquids [35]. Equations (6.1) to (6.3) are also applied for alternative insulating liquid triangles. Moreover, additional Duval Triangles 4 and 5 have been developed for low temperature faults in transformers, which can only apply for fault identification by the Duval Triangle 1 to help users in case of difficulty for distinguishing the expected problem PD, T1, or T2 [35].

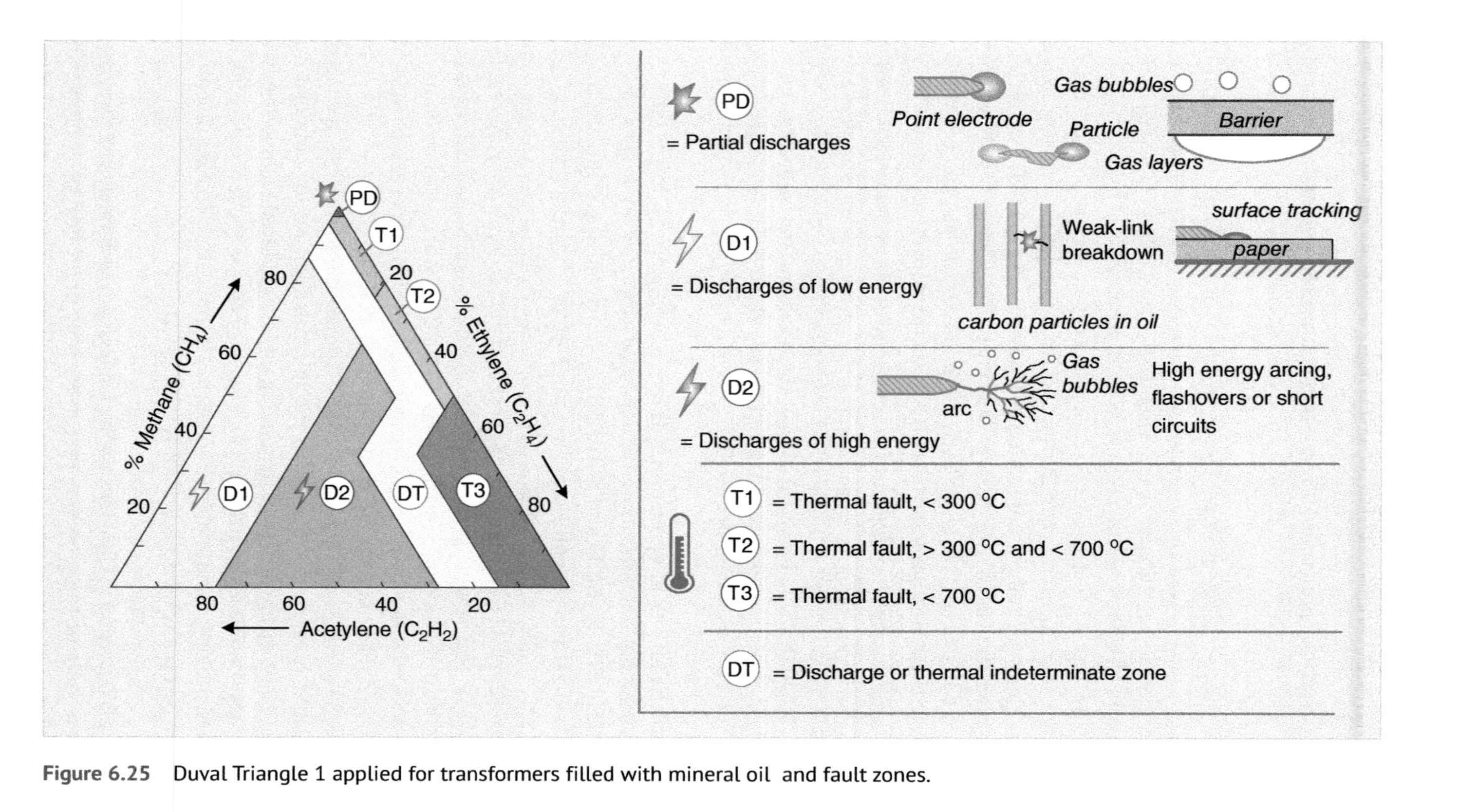

Figure 6.25 Duval Triangle 1 applied for transformers filled with mineral oil and fault zones.

Nowadays, Duval Pentagons have been introduced for interpreting the test results of DGA applied for mineral oil [36] and alternative liquids [37]. Duval Pentagon needs different calculation formula to get the position located in the pentagon described in [36, 37]. In Duval Pentagons, stray gas is included; therefore, all gases dissolved in the insulating liquid are considered. Consequently, the capacity for gas interpretation in practice of Duval Pentagon seems higher than that for Duval Triangle 1. However, considerable data is needed to develop the accuracy of Duval Pentagon, and that could take more than 10 years.

IEC Ratios: IEC ratio is one of DGA interpretation techniques recommended in IEC [33]. IEC ratio concentrates the ratios of these gases C_2H_2/C_2H_4, CH_4/H_2, C_2H_4/C_2H_6 and it shows the symptom of PD inside the oil-immersed transformers.

Dörnenburg Ratios: At first, Dörnenburg proposed two ratios of gases, CH_4/H_2 and C_2H_2/C_2H_4, to diagnose PD, arcing, and thermal deterioration [38]. Currently, Dörnenburg ratios have been improved and recommended in IEEE std. C57.104 [34] by which four gas ratios are introduced, such as CH_4/H_2, C_2H_2/C_2H_4, C_2H_2/CH_4, C_2H_6/C_2H_2. If in the mineral oil $CH_4/H_2 < 0.1$, $C_2H_2/CH_4 < 0.3$, $C_2H_6/C_2H_2 > 0.4$, and in the gas space $CH_4/H_2 < 0.01$, $C_2H_2/CH_4 < 0.1$, $C_2H_6/C_2H_2 > 0.2$ where C_2H_2/C_2H_4, in the mineral oil, and in the gas space is not significant value it reveals that PD problem occurs in the oil-immersed transformers. To apply these criteria, the user needs to compare the measured gases with the limit values, if they meet the conditions, Dörnenburg ratios are suggested to use with being valid [34].

Rogers Ratios: This method had been used four gas ratios at the beginning, i.e. CH_4/H_2, C_2H_6/CH_4, C_2H_4/C_2H_6, C_2H_2/C_2H_4 [38]. Currently only three gas ratios, C_2H_2/C_2H_4, CH_4/H_2, C_2H_4/C_2H_6 are used as recommended in IEEE standard [34]. It should be noted that three gas ratios utilized in the Rogers ratio method is the same as three gas ratios for IEC ratios, but the criteria for gas interpretation is different. Applying the Rogers ratio method is similar to the procedure for Dörnenburg ratio method. Applying the Rogers ratio method if $C_2H_2/C_2H_4 < 0.1$, $CH_4/H_2 < 0.1$, and $C_2H_4/C_2H_6 < 1.0$ shows the symptom of PD (low energy density arcing) inside the oil-immersed transformers. If C_2H_2/C_2H_4 and C_2H_4/C_2H_6 continually increase to above 3, it reveals the development of discharges with high intensity.

Key Gas Method: This is a simple method based on the predominant gases explained in IEEE standard [34]. The relative proportions for the gases are considered. In case of PD (low energy electrical discharges) in oil, H_2, CH_4, C_2H_4, C_2H_6 will be found about 85%, 13%, 1%, and 1%, respectively. Moreover, other interpretation techniques such as MSS (Müller, Schlliesing, and Soldner) method [39], Nomograph method [40], or diagnostic charts [41] may be applied. Even though some DGA interpretation methods have been proposed, which are available to users, the DGA with the interpretation results obtained from the same raw data sometimes are different, depending on the employed techniques. In case of ratio methods, they sometimes do not yield an analysis because the ratios may fall outside the defined zones. Besides, some analysis results may be not correct (wrong diagnosis) [42]. Therefore, the expert is always needed to deal with DGA.

On-line DGA for PD Detection in Liquid Insulating Immersed Transformers and Reactors

In order to avoid some of the problems that may occur in off-line DGA processing (e.g. improper sampling, transportation and storage), application of on-line DGA techniques has been proposed. The main technical advantage of on-line DGA is that it enables both the rate of change of gases and trend analysis of gases to be monitored in real-time, thus providing more reliable data than traditional sample analysis, which is performed following time base maintenance schedule and more timely corrective action to avoid transformer failure. Currently, on-line DGA measurement plays an important role in industry with regard to monitoring dissolved gases in oil-filled power transformers. Various types of gases may be monitored by on-line DGA according to the type of gas detector employed, hydrogen gas being the most prominent.

Application of DGA for PD Analysis of Instrument Transformers

The application of DGA to power transformers is generally successful; therefore, it should be effectively applied for Voltage transformer (VT) as well due to the quite similarity of configuration and function. PD in VTs is generated in gas-filled cavities mainly due to poor paper impregnation, high levels of humidity in the insulation paper, or creases in the insulation paper [33]. The structure and function of CTs are different from that of power transformers – one consequence being that discharges in CTs relate switching operations carried out at a nearby substation [43]. IEC provides ranges of 90% typical concentration gas values observed in instrument transformers both current transformer (CT) and VT and recommends that if $CH_4/H_2 < 0.2$, $C_2H_4/C_2H_6 < 0.2$, and C_2H_2/C_2H_4 is insignificant, whatever the value, it shows the symptom of PD inside the oil-immersed instrument transformers [31].

Application of DGA for PD Analysis of OIP Bushing

The insulation system of oil impregnated paper (OIP) bushing is more similar oil-filled cables and CT than power transformers. Therefore, DGA may not be as sensitive to the developing faults as in the transformers. Besides, the interpretation of DGA results may not be interpreted in the same way as for the power transformers. PDs, especially corona discharge in mineral oil, generate mostly H_2 gas along with very small amount of other gases such as methane. Prolonged corona discharges tend to generate large amount of H_2 and X wax deposition. It should be noted that H_2 can also be generated from rusting, electrolysis of free water, and stray gassing in new oil [43]. IEC standard [33] proposes a criterion for PD inside bushing that if $CH_4/H_2 < 0.07$, it shows the symptom of PD. This criterion may be not applicable to modern bushings containing mixtures of mineral oil and dodecylbenzene (DDB), in unknown proportion [43].

Application of DGA for PD Analysis of Underground Cable

High-pressure fluid-filled (HPFF) cables, self-contained liquid-filled (SCLF) cables, and high-pressure gas-filled (HPGF) cable all use cellulose paper impregnated with mineral oil as insulation. Under thermal or electrical stress, these insulation systems tend to generate gases such as hydrogen, hydrocarbon gases, carbon dioxide, and carbon monoxide. DGA should be able to detect incipient faults in the cable system by analyzing the presence of these gases [44]. However, performing DGA is sometimes problematical because obtaining suitable oil samples can be very difficult. Gases may also be absorbed by DDB in modern cables containing mixtures of mineral oil and DDB [33]. Thus, proper liquid sampling and handling, suitable gas extraction technique, and correct data interpretation are critical factors in achieving valid DGA results. According to the IEC, identifying faults such as PD, discharges of low or high energy, and overheating by DGA interpretation (as for power transformers) are also valid for underground cable, as mentioned in [33]. Conversely, other experts claim that DGA interpretation as applied to power transformers cannot be directly applied to underground cables due to differences in design, materials, and operational parameters [45]. Generally, recommendation for detecting PD in oil-filled high-voltage equipment is that a high ratio with more hydrogen than methane (for example, $CH_4/H_2 < 0.1$) indicates the presence of PDs [33, 45].

6.4.3 Decomposition Gas Analysis

6.4.3.1 Decomposition SF_6 Analysis for Gas-Insulated High-Voltage Equipment

SF_6 is used in gas blast circuit breakers (GCB), gas-insulated buses (GIB), gas-insulated lines (GIL), and gas-insulated substation (GIS), due to high dielectric strength, high thermal stability, low toxicity, and chemical inertness. PD, sparking, or arcing inside gas-insulated high-voltage equipment tends to cause SF_6 gas decomposition, thereby creating new chemical compounds or byproducts.

Analysis of SF_6 decomposition byproducts can be used to monitor PD activity, and thus the condition of the internal insulation condition in gas-insulated equipment [46]. Currently, decomposition gas analysis is one of the most powerful test techniques available for conducting this type of analysis (Figure 6.26 (a) and (b)). Two further advantages of application decomposition gas analysis for PD monitoring in high-voltage apparatus is that it is a noninvasive method that requires no retrofitting of complex equipment and can be performed on-line [44]. Various types of SF_6 byproducts are generated when PD occurs in gas-insulated equipment composed of SF_6 gas, solid insulation, and solid conductor material. Examples of SF_6 byproducts, including the subsequent byproducts, are SF_4, SF_2, SOF_2, SO_2F_2, SOF_4, SO_2, H_2S, HF, F, CF_4, and CO_2. The amount and rate (or degree) of SF_6 decomposition depends on a complex set of parameters, such as PD intensity and duration, electric field stress, purity of SF_6, type of impurity, gas pressure, and the presence of solid material both conductor and insulation in discharge regions [47]. For example, prolonged PD tends to cause large amounts of decomposition products to accumulate inside equipment. Details of SF_6 byproducts reported by researchers are provided in [44, 48, 49]. It should be noted that not only O_2 and H_2O but also CF_4, CO_2, N_2 tend to be present even in highly pure SF_6 [47, 49, 50].

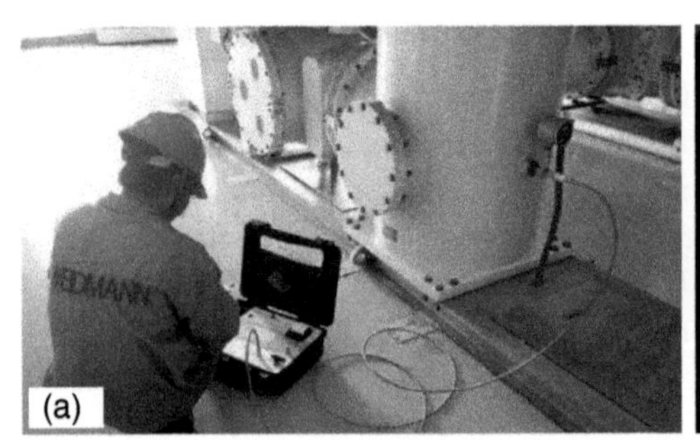

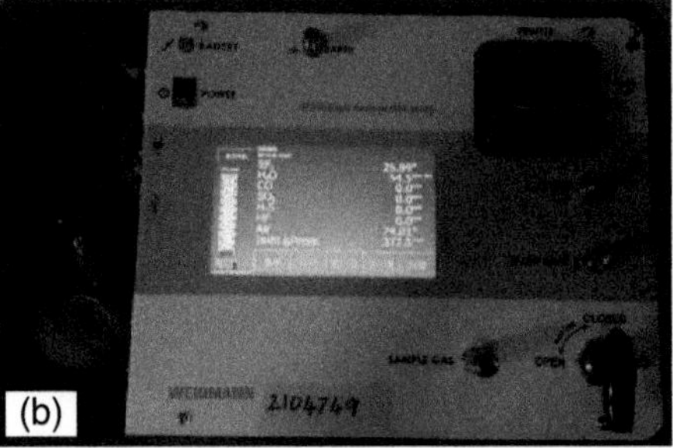

Figure 6.26 SF_6 byproducts measurement for GIS insulation monitoring: (a) SF_6 byproducts measurement; (b) SF_6 byproducts test result.

Various SF_6 byproducts are generated by PD activity in gas-insulated equipment; however, not all are suitable for use in determining the presence of PD activity. The two main issues to be considered are as follows:

- The amount of gas detected should increase in response to higher PD amplitude, higher intensity of PD activity, and increased duration of PD.
- Extremely unstable gases and highly reactive gases are unsuitable for the purpose of determining the degree of PD activity.

According to [46] the detection of SF_6 byproducts (such as H_2S, SO_2, SOF, and SO_2F_2) in SF_6 gas-insulated equipment can be used to indicate the occurrence of a PD event.

6.4.4 Ozone Measurement and Analysis for Air-Cooled Hydrogenerators

Ozone gas is one of the main chemical byproducts generated by external PD – particularly corona discharge in air or on insulation surfaces. Various types of PD, such as slot discharge, end-winding discharge, and discharge at field grading areas, can cause ionization of air on the insulation surface to produce ozone and ozone byproducts. However, ozone measurement as a means of monitoring PD activity is normally restricted to air-cooled hydrogenerators [51].

Ozone is generated not only by PD activity but also by natural phenomena (such as lightning) and operating equipment (such as paper copiers and fax machines). Ozone concentration depends on the rate of ozone generation and the rate of ozone removal by decay, leakage, ventilation, and treatment. If the detected ozone concentration is greater than the ozone ambient level of 0.1 ppm, this is considered to be a strong indication of PD activity.

References

1 Institute of Electrical and Electronics Engineers (2018). *IEEE Standard c 57.127™: Guide for the Detection and Location of Acoustic Emissions from Partial Discharges in Oil-Immersed Power Transformers and Reactors*. New Jersey: IEEE.

2 International Electrotechnical Commission (2016). *IEC TS 62478: High Voltage Test Techniques – Measurement of Partial Discharges by Electromagnetic and Acoustic Methods*. Geneva, Switzerland: IEC.

3 Schwarz, R. (2002). Optische Teilentladungsdiagnostik für Betriebsmittel der Elektrischen Energietechnik. Dissertations. Graz.

4 Muhr, M. (2007) A prospective standard for acoustic and electromagnetic partial discharge measurements. https://http://www.slideserve.com/venice/partial-discharge-pd-measurement (accessed 16 December 2018)

5 Osapublishing (2017). UVC upconversion material under sunlight excitation. https://www.http://osapublishing.org/ol/abstract.cfm?uri=ol-41-4-792 (accessed December 18, 2018).

6 Photonicsonline (2018). Silicon photomultipliers for photon detection, https://www.photonicsonline.com/doc/silicon-photomultipliers-for-photon-detection-0001 (accessed December 18, 2018).

7 Kim, K.Y. (2010). Principle of operation, structure and technology. In: *Advances in Optical and Photonic Devices*, 1e (ed. K.Y. Kim), 256. Croatia: In Tech.

8 Davidson M. W. (2018). Concepts in digital imaging technology electronic imaging detectors, http://hamamatsu.magnet.fsu.edu/articles/digitalimagingdetectors.html (accessed December 16, 2018).

9 Schwarz, R. and Muhr, M. (2007). Modern Technologies in Optical Partial Discharge Detection. *Proceeding of Annual Report Conference on Electrical Insulation and Dielectric Phenomena, 14 Proceeding of Annual Report Conference on Electrical Insulation and Dielectric Phenomena*, Vancouver, BC, Canada. (17 October 2007). IEEE.

10 Candrone (2014). Corona Camera, https://candrone.com/products/corona-camera (accessed 16 December 2018).

11 Yutcis, E. (2012). Insulation testing of HV rotating machines. https://www.ofilsystems.com/news/rotating%20machines.html (accessed 16 December 2018)

12 Rossing, T.D. (ed.) (2007). *Handbook of Acoustics*. New York: Springer.

13 Ramu, T.S. and Nagamani, H.N. (2010). *Partial Discharge Based Condition Monitoring of High Voltage Equipment*. New Delhi: New Age International Publishers.

14 Lundgaard, L.E. (1992). Partial discharge – Part XIII: Acoustic partial discharge detection – fundamental consideration. *IEEE Electrical Insulation Magazine 8* (4) (Jul-Aug), p. 25.

15 Buchholz, U.B., Jaunuch M., Stark W. et al. (2012) Acoustic data of cross linked polyethylene (XLPE) and cured liquid silicone rubber (LSR) by means of ultrasonic and low frequency DMTA. https://www.semanticscholar.org/paper/Acoustic-data-of-cross-linked-polyethylene-(XLPE)-Buchholz-Jaunich/c96daf542913ec4fa1a435f462c6f4ef99f4c00b?navId=extracted (accessed 16 December 2018)

16 Fibre Reinforced Plastic (2012). Ultrasonic inspection for fibre reinforced plastic & composite, http://www.fibre-reinforced-plastic.com/2012/01/ultrasonic-inspection-forfibre.html (accessed December 16, 2018).

17 Properties taken from Kaye & Laby Online, http://www.kayelaby.npl.co.uk (accessed December 16, 2018).

18 Schwarz, R. (2009) Messtechnik und diagnostik an elektrischen betriebsmitteln. Habilitation Dissertation. Graz.

19 Rshydro (2018). Sound speeds in water, liquid, and materials. www.rshydro.co.uk/sound-speeds/ (accessed 16 December 2018).

20 Lazarevich, A. K. (2003) Partial discharge detection and localization in high voltage transformers using an optical acoustic sensor. Dissertations. Virginia Tech.

21 Janus, P. (2012) Acoustic emission properties of partial discharges in the time-domain and their application. Dissertations. KTH Royal Institute of Technology.

22 Harrold, R.T. (1979). Acoustic waveguides for sensing and locating electrical discharge in high voltage power transformers and other apparatus. *IEEE Transactions on Power Apparatus and Systems* PAS-98 (March-April): 449.

23 Alippi, A. (1991). *Acoustic Sensing and Probing*, 4e. World Scientific Publishing Co. Pte. Ltd.

24 Kuttruff, H. (2007). *Acoustics: An Introduction*. Taylor and Francis.

25 Conseil International des Grands Réseaux Électriques. (2017). Partial Discharges in Transformers. *Technical Brochure. Ref. 676.*

26 Conseil International des Grands Réseaux Électriques. (2013) Guide for the Monitoring, Diagnosis and Prognosis of Large Motors. *Technical Brochure. Ref. 558.*

27 Conseil International des Grands Réseaux Électriques. (2010) Guide for On-Line Monitoring of Turbogenerators. *Technical Brochure. Ref. 437.*

28 Gross, D. (2016) Acquisition and location of partial discharge in transformers. Dissertation. Graz.

29 Singh, S. and Bandyopadhyay, M.N. (2010). Dissolved gas analysis technique for incipient fault diagnosis in power transformers. A bibliographic survey. *IEEE Electrical Insulation Magazine* (December), 26 (6): 41–46.

30 American Society for Testing and Materials (2017). *ASTM D3612-02: Analysis of Gases Dissolved in Electrical Insulating Oil by Gas Chromatography*. United States: ASTM.

31 International Electrotechnical Commission (2011). *IEC 60567: Oil-Filled Electrical Equipment – Sampling of Gases and Analysis of Free and Dissolved Gases – Guidance.* Switzerland: IEC.

32 Horning, M., Kelly, J., Myers, S. et al. (2004). *Transformer Maintenance Guide*, 3e. S.D. Myers Inc.

33 International Electrotechnical Commission (2015). *IEC 60599: Mineral Oil-Filled Electrical Equipment in Service – Guidance on the Interpretation of Dissolved and Free Gases Analysis.* Switzerland: IEC.

34 Institute of Electrical and Electronics Engineers (2008). *IEEE C57.104: Guide for the Interpretation of Gases Generated in Oil-Immersed Transformers*. New York: IEEE.

35 Duval, M. (2008). The Duval Triangle for load tap changers, non-mineral oils and low temperature faults in transformers. *IEEE Electrical Insulation Magazine* 24 (6): 22–29.

36 Duval, M. and Lamarre, L. (2014). The Duval Pentagon – a new complementary tool for the interpretation of dissolved gas analysis in transformers. *IEEE Electrical Insulation Magazine* 30 (6): 9–12.

37 Duval, M. and Lamarre, L. (2017). The new Duval Pentagons available for DGA diagnosis in transformers filled with mineral and ester oils. *IEEE Electrical Insulation Conference (EIC)*, Baltimore, MD (11–14 June 2017). *IEEE.*

38 Mathes, K.N. (1976). Influence of electrical discharges in oil and combinations of oil and paper. *IEEE Transactions on Electrical Insulation* 11 (4): 164–180.

39 Ivanka Atanasova – Höhlein. (2009). DGA- method in the past and for the future. *Diagnostic Conference*, Siofok, Hungary (14–16 October 2009). Siemens AG.

40 Independent Laboratory Service (2018). Dissolved gas analysis of mineral oil insulating fluids, http://www.nttworldwide.com/tech2102.htm (accessed 24 December 2018).

41 Mori, E., Tsukioka, H., Takamoto, K. et al. (1999) Latest diagnostic methods of gas-in-oil analysis for oil-filled transformer in Japan. *Proceedings of 13th International Conference on Dielectric Liquids (ICDL'99)*, Nara, Japan (20–25 July 1999). IEEE.

42 Duval, M. and Dukarm, J. (2005). Improving the reliability of transformer gas-in-oil diagnosis. *IEEE Electrical Insulation Magazine* 21 (4): 21–27.

43 Electric Power Research Institute (2006) Condition Monitoring and Diagnostics of Bushings, Current Transformers, and Voltage Transformers by Oil Analysis. *EPRITech. Rep. Ser. 1012343.*

44 Chu, F.Y. (1986). SF_6 decomposition in gas-insulated equipment. *IEEE Transactions on Electrical Insulation* EI-21 (5): 693–725.

45 Electric Power Research Institute (2000). Guidelines for the interpretation of dissolved gas analysis (DGA) for paper-insulation underground transmission cable system, final report. *EPRITech. Rep. Ser.* 1000275.

46 Zhang, X., Tang, J., Xiao, S. et al. (2017). The SF_6 decomposition mechanism: background and significance. In: *Nanomaterials Based Gas Sensors for SF6 Decomposition Components Detection* (ed. A. Farrukh), 1–11. Croatia: INTECH.

47 Heise, H., Kurte, R., Fischer, P. et al. (1997). Gas analysis by infrared spectroscopy as a tool for electrical fault diagnostics in SF_6. *Journal of Analytical Chemistry* 358 (7): 793–799.

48 Van Brunt, R.J. and Herron, J.T. (1990). Fundamental processes of SF_6 decomposition and oxidation in glow and corona discharges. *IEEE Transactions on Electrical Insulation* 25 (1): 75–94.

49 Tang, J., Liu, F., Zhang, X. et al. (2012). Partial discharge recognition through an analysis of SF_6 decomposition products Part 1: Decomposition characteristics of SF_6 under four different partial discharges. *IEEE Transactions on Dielectrics and Electrical Insulation* 19 (1): 29–36.

50 Tang, J., Liu, F., Qinghong, M. et al. (2012). Partial discharge recognition through an analysis of SF_6 decomposition products Part 2: Feature extraction and decision tree-based pattern recognition. *IEEE Transactions on Dielectrics and Electrical Insulation* 19 (1): 37–44.

51 Stone, G.C., Culbert, I., Boulter, E.A. et al. (2014). *Electrical Insulation for Rotating Machines: Design, Evaluation, Aging, Testing, and Repair*. Hoboken, New Jersey: Wiley.

7

PD Localization

7.1 Introduction

Partial discharge (PD) measurement is very important for evaluating the integrity of the electrical insulation system. Generally, PD amplitude is represented in different forms, depending on the measurement techniques. To estimate the severity of PD occurring in the insulation system, not only the PD amplitude and the rate of PD occurrence but the location of PD source is significant. To precisely locate the PD source, an engineer can prepare a suitable maintenance plan or replace the defective part, depending on the severity degree of such insulation. PD measurement might be divided into the conventional PD measurement and the nonconventional PD measurement. It seems that the nonconventional PD measurement using various types of sensors such as high-frequency current transformer (HFCT), acoustic sensor, and UHF sensor is preferable to obtain the location of PD source.

7.2 The Complexity of PD Localization

At a discharge site, a PD pulse has a very short duration as an impulse signal with up to nanosecond rise time and nano to microsecond pulse width. Each PD pulse contains a certain amount of energy that is dissipated during traveling from the PD source to the PD detection sensor. Therefore, the amplitude of the detected PD signal is generally lower than the original one. Moreover, some high-frequency components are lost during the traveling of the PD signals into the high-voltage equipment due to the attenuation, reflection, refraction characteristics, as represented in Figure 7.1. Therefore, correctly locating the PD site is very difficult due to the complex nature of the insulation system and the sophisticate of the wave propagation mechanism.

Partial Discharges (PD) - Detection, Identification, and Localization, First Edition. Norasage Pattanadech, Rainer Haller, Stefan Kornhuber, and Michael Muhr.

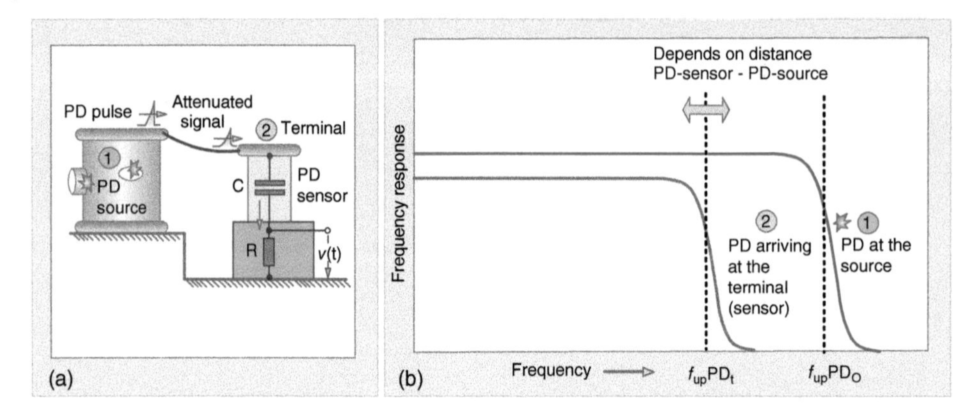

Figure 7.1 (a) PD measurement by a capacitive coupler (CC) (b) the spectrum of the PD pulses both at its origin and at the location of the CC obtained from Fourier transformation analysis [1].

7.3 Classification of PD Localization

PD localization can be divided into different ways according to the criteria used. In this chapter, the PD localization is classified into two groups, i.e. the PD localization for internal insulation and the PD localization for external insulation.

7.3.1 PD Localization for the Internal Insulation

According to IEC 60071-1 [2], the internal insulation is protected from the effects of atmospheric and external conditions. Therefore, mostly internal insulations are in high-voltage housing. All PD types, i.e. corona discharge, surface discharge, and internal discharge, may exist in the high-voltage equipment. Various PD localization methods such as pulse time arrival method, auscultatory method, and triangulation method can be applied to locate the PD sources in the internal insulation system. The mentioned techniques are most suitable for stationary defect detection. For the bouncing particles mainly occurring in the GIS and/or GIL, the ultra-probe AE sensor with a pointed tip is more efficient than other techniques.

7.3.2 PD Localization for the External Insulation

Generally, the external insulation is exposed to the environment; therefore, the atmospheric and other environmental conditions such as the contaminant strongly affect the insulation integrity. Two types of PD problems – i.e. corona discharge and surface discharge – generally occur for the external insulation. The PD localization technique for the external insulation is not that complicated; a corona camera and/or an airborne acoustic PD sensor can be utilized.

7.4 PD Localization Techniques for the Internal Insulation

PD localization techniques generally applied for the internal insulation – i.e. pulse time arrival method, auscultatory method, triangulation method, and bouncing particle localization method – are described below.

7.4.1 Pulse Time Arrival Method

7.4.1.1 Concept of Pulse Time Arrival Method

The pulse-time arrival (time of flight) method can be used to determine a PD source location (aka localization). The time difference between the time at which the direct-incident PD pulse signal is measured and the time at which the first reflected PD pulse signal is measured is used to calculate the location of the PD source. This technique is applied on the basic assumption that (i) one PD detector per phase is used and (ii) the speed of the PD signal is constant. If a pair of PD detectors per phase is used, the time difference between detection of the incident PD signal by the respective PD detector is used to calculate the location of the PD source.

7.4.1.2 Application of the Pulse Time Arrival Method for PD Localization in High Voltage Equipment

Application of the Pulse Time Arrival Method for PD Localization of a Rotating Machine

PD measurement of rotating machines by use of CCs can be performed by applying either a single CCs or a pair of CCs per phase. However, two CCs per phase are preferred for roughly determining the PD source location whether PD signal is generated from the rotating machine. One of the coupler units is installed at the high-voltage terminal, while the location of the second unit is away from the first coupler determined by the capability of the electronic circuit used in the acquisition system, as shown in Figure 7.2.

A PD signal generated inside or at the rotating machine will be first detected by the CCs installed at the near end HV terminal of the rotating machine, and then the CCs at the far end HV terminal of the rotating machine can detect such PD signal with lower amplitude and longer rise time.

Application of the Pulse Time Arrival Method for PD Localization in an Underground Cable System

PD measurement and PD localization in underground cable systems can be distinguished by using a single or double CC, as shown in Figure 7.3. When applying a single CC per phase, the time difference between the time (t_1) at which the direct-incident PD pulse signal is measured and the time (t_2) at which the first reflected PD pulse signal is measured to calculate the location of the PD source, as demonstrated in Figure 7.3a. When using a pair of CCs per phase, the time difference between detection of the incident PD signal by the respective CCs is utilized to compute the location of the PD source, as shown in Figure 7.3b.

The formula for computing the PD source location in case of using a CC per phase as a PD detector is given in Eq. (7.1):

$$X_{pd} = L - \frac{\Delta t}{v/2} \tag{7.1}$$

where X_{pd} is the distance of PD source from the near end cable terminator; L is the cable length; Δt is the time difference between the time at which the direct-incident PD pulse signal is measured (t_1) and the time at which the first reflected PD pulse signal is measured (t_2); and v is the speed of PD pulse signal in underground cable.

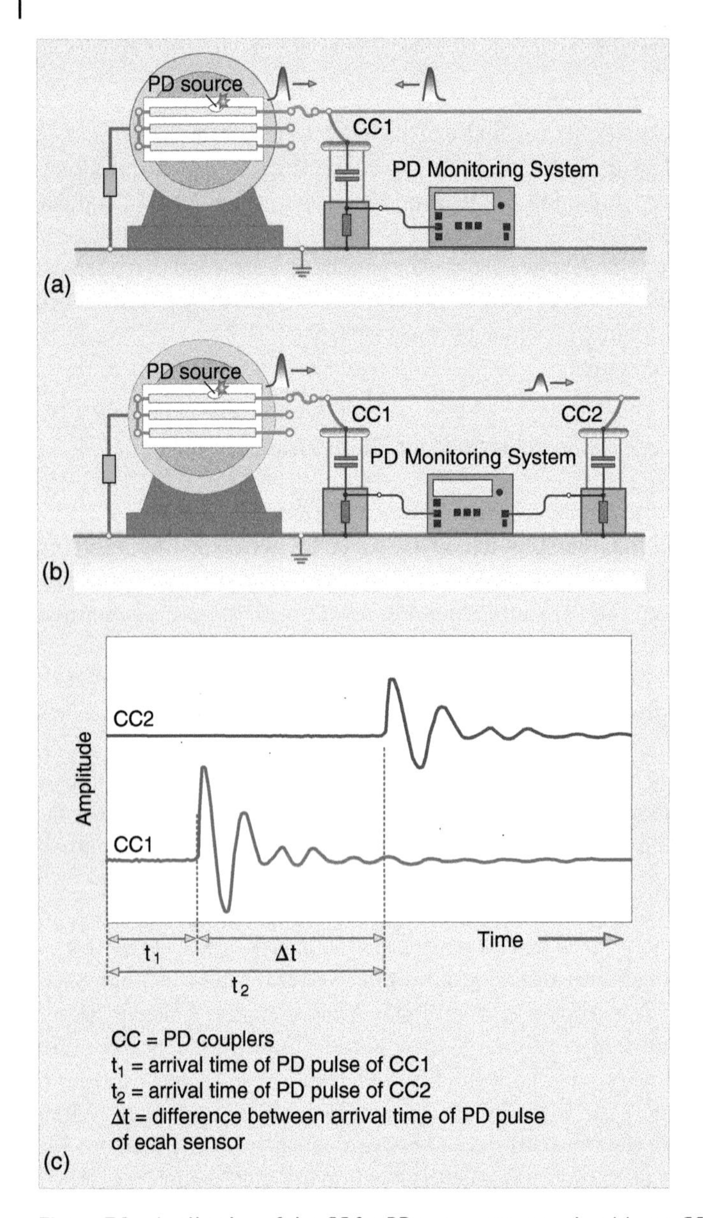

Figure 7.2 Application of the CC for PD measurement using (a) one CC per phase and (b) two CC per phase. (c) Time different measurement obtained from CC 1 and 2 used for PD source localization.

To compute the location of the PD source, in the case of using a pair of CC per phase as a PD detector, the formula used is expressed in Eq. (7.2):

$$X_{\mathrm{pd}} = \frac{L}{2} - \frac{\Delta t}{v/2} \tag{7.2}$$

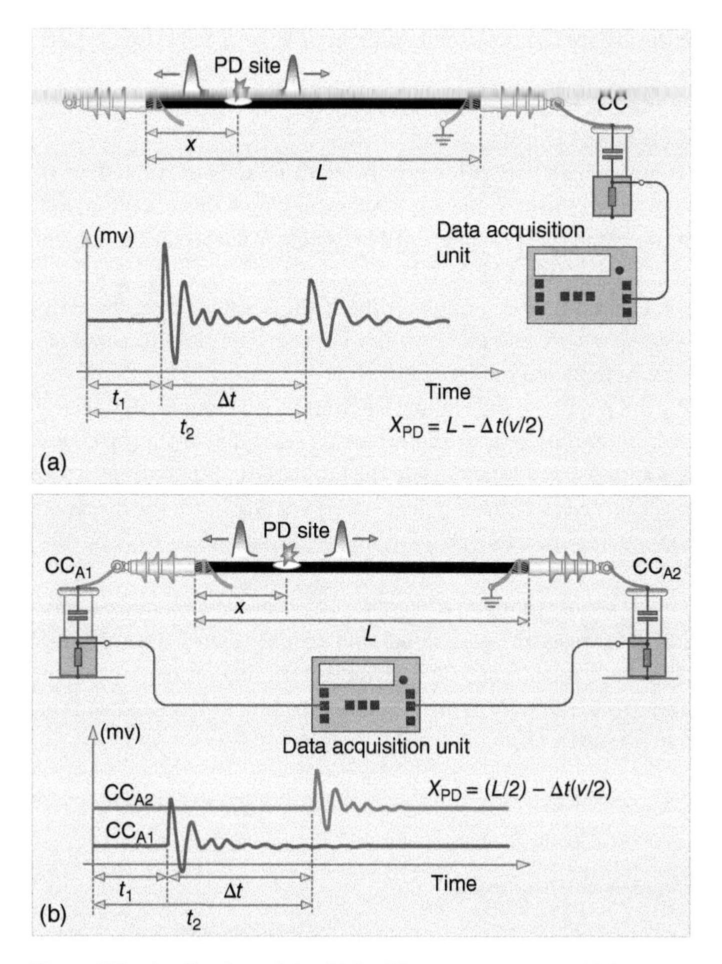

Figure 7.3 Application of the CC for PD measurement and PD localization (a) one CC per phase and (b) two CCs per phase [3].

where X_{pd} is the distance of PD source from the near end cable terminator; L is the cable length; Δt is the time difference between detection of the incident PD signal by the respective CCs; and v is the speed of PD pulse signal in underground cable.

Application of the Pulse Time Arrival Method for PD Localization in GIS and GIL

AE sensors or UHF sensors (so-called antennas) are now widely used to detect a PD signal occurring in GISs or GILs. The detected PD signal is normally not only used for evaluating the severity of PD but for PD localization as well. Analyzing such detected PD signals provides a roughly suspected area or even a specified position of a PD source. If performing the PD pulse shape analysis based on the pulse time arrival method, the roughly suspected area that the PD source locates can be identified. By applying the mathematic approach based on the pulse time arrival method, the specified position of the PD source can be determined:

1) *Pulse shape analysis for roughly suspected area identification of the PD source in GIS or GIL.* The *time-of-flight* concept is incorporated into the process of locating PD sources in GIS and GIL. The process requires a minimum of two AE or UHF PD sensors to be

installed on the surface of the GIS or GIL, as shown in Figures 7.4 and 7.5. If two AE PD sensors are used, when the PD event occurs, the AE PD signal is detected by the sensors, and the *time-of-flight* is recorded for incorporation into the analysis. There are two basic situations in which this process is performed:

Case 1: If the PD source is not located between any pairing of the sensors, then the PD source is assumed to be located closest to the AE PD sensor, which detects the signal first. This is because the detected AE PD signal will be stronger at the sensor, which is closer to the PD source, as shown in Figure 7.4.

Case 2: If the PD source is located between any pairing of the sensors, then the time required for the AE PD signal to arrive at the sensors will be shorter than the *time-of-flight* between the sensors, as depicted in Figure 7.5.

In order to properly analyze the AE PD signal emanating from the PD source, it is necessary to estimate the degree of attenuation in the detected AE PD signal as it propagates in and around the insulator support structures inside the GIS or GIL. Furthermore, the signal is also attenuated while it traverses the length of the GIS or GIL, with the decrease in amplitude of the AE signal being inversely proportional to the square root of the distance traveled.

2) *Mathematical approach for PD localization in GIS or GIL.* A PD event generates a very fast UHF electromagnetic (EM) wave or pulse signal, which propagates from the PD source to the internal UHF sensors positioned on each side of the PD source, as shown in Figure 7.6a. This signal travels to the UHF sensors at different times (or so-called time of flight) in tens of ns range (Figure 7.6b).

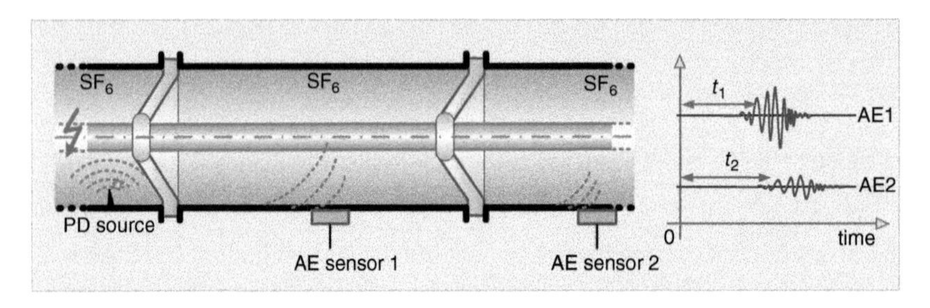

Figure 7.4 Case 1 PD source location out of the area between AE1 and AE2.

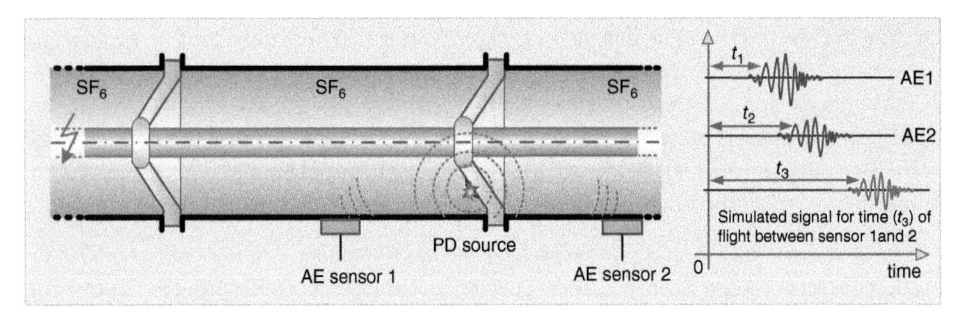

Figure 7.5 Case 2 PD source location between AE1 and AE2.

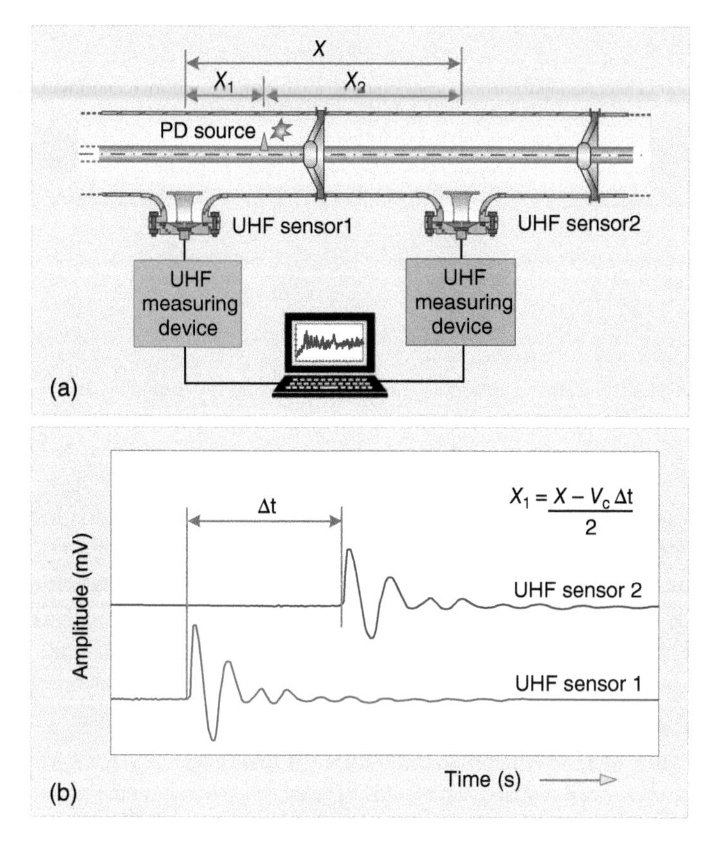

Figure 7.6 (a) On-site UHF PD measurement; (b) time of flight method for PD localization.

The distance x_1 between PD source and UHF sensor 1 can be calculated using the following equation:

$$x_1 = \frac{x - v_c \Delta t}{2} \tag{7.3}$$

where Δt is the difference between arrival times at sensor 1 and sensor 2 for the UHF signals, and v_c is the propagation velocity of the UHF signal ($\approx 3 \times 10^8$ m/s). However, it should be noted that the reliability of the above formula in determining PD location may be significantly affected by relatively high UHF signal attenuation over long distances or passage through solid objects and internal discontinuities [4].

7.4.2 Auscultatory Method

7.4.2.1 Concept of Auscultatory Method

The auscultatory method uses one or more movable AE PD sensors to detect the AE PD signals. This technique is based on the concept that the detected AE PD signal attenuates, both in magnitude and frequency, depending on the distance between the sensors and the PD source, as depicted in Figure 7.7. In other words, the longer the distance covered, the smaller the PD amplitude and lower frequency components of the PD are detected.

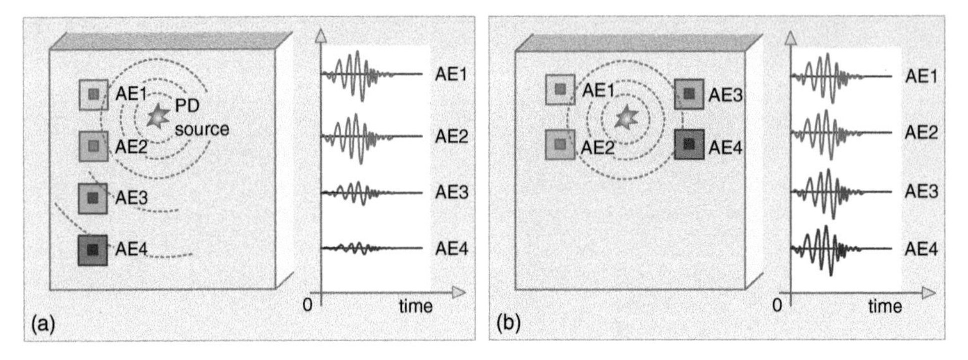

Figure 7.7 PD localization by the auscultatory method: (a) first stage to get AE PD signals with different intensity; (b) second stage after relocation PD sensors.

In addition, the shape of the detected AE PD signal also depends on PD type, as determined by the type of the PD source, the propagation media, and the specific characteristics of both the sensor and other components of the detection system. Practical application of the auscultatory method requires that when one or more sensors detect PD activity (assuming only one PD source), the other sensors (assuming multiple AE PD sensors are used) will be moved (as near as reasonably possible) to the source of the strongest AE PD signal.

7.4.2.2 Application of the Auscultatory Method for PD Localization in a Transformer

To locate the PD source, the movable AE PD sensors are used. In this case study, four AE PD sensors combined with a HFCT are utilized. These AE PD sensors are installed at various places on the exterior surface of the transformer, as shown in Figure 7.8a. When PD occurs, the AE PD signal will transmit and register at the sensors. The sensor AE1, which is at the nearest to the PD source, picks the strongest PD signal, whereas other sensors AE2–AE4 may also get the electrical PD (EE PD) signal but at lower PD intensity, as depicted in Figure 7.8b. Simultaneously the HFCT also captures the PD signal to confirm that the signals detected by the AE PD sensors are the PD signal. The PD sensors that get the lesser PD intensity will be manually moved to the vicinity where the stronger AE PD signal is generated. The place enclosed by the AE sensors is suspected to be the PD source location, as shown in Figures 7.8c, d. The PD source is verified by untanking the transformer, as demonstrated in Figure 7.8e, f.

7.4.3 Triangulation Method

7.4.3.1 Concept of Triangulation Method

The triangulation method for locating a PD source in high-voltage equipment requires a minimum of three PD sensors fixed at different positions on the exterior wall of the equipment. Starting with the consideration of the basic concept of the triangulation method, three AE PD sensors are used. When a PD event occurs, the AE PD signal generated travels from the PD source to the sensors with different traveling times (or *time of flight*). As the speed of the recorded PD signals is assumed to be constant, denoted by v, the location of the PD source can be calculated by the triangulation method by factoring in the variations in traveling time.

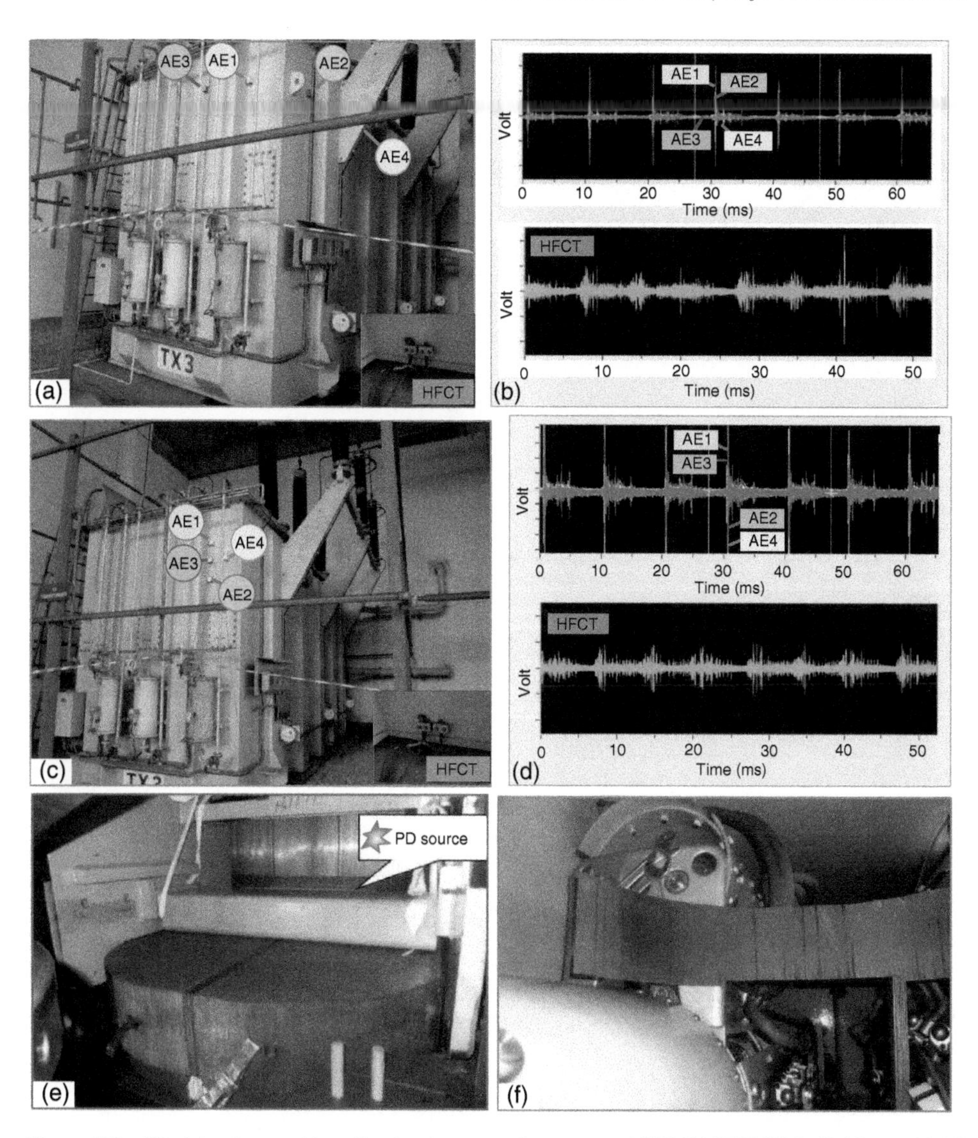

Figure 7.8 PD detection and localization in a transformer rated 230 kV, 250 MVA in field test. (a) AE sensor positions for the first step of PD localization by the auscultatory method; (b) AE PD signals and EE signal obtained from the first step; (c) AE sensor positions for the last step of PD localization by the auscultatory method; (d) AE PD signals and EE PD signal obtained from the last step; (e) PD occurring at the insulation between core laminate and flux collector; (f) the insulation between cores laminate and flux collector.

In practice, an electrical PD sensor is normally used in combination with the regular acoustic AE PD sensors in order to determine the time of origin for PD events detected by AE PD sensors. Figure 7.9a provides an illustration of a recorded AE PD signal traveling to three sensors – AE1, AE2, AE3. Figure 7.9b demonstrates the EE PD and the AE PD signals registered at the sensors. The traveling times are shown as t_1, t_2, t_3, and the distances from the PD source to the three sensors, x_1, x_2, x_3, can be calculated from Eqs. (7.4)–(7.6).

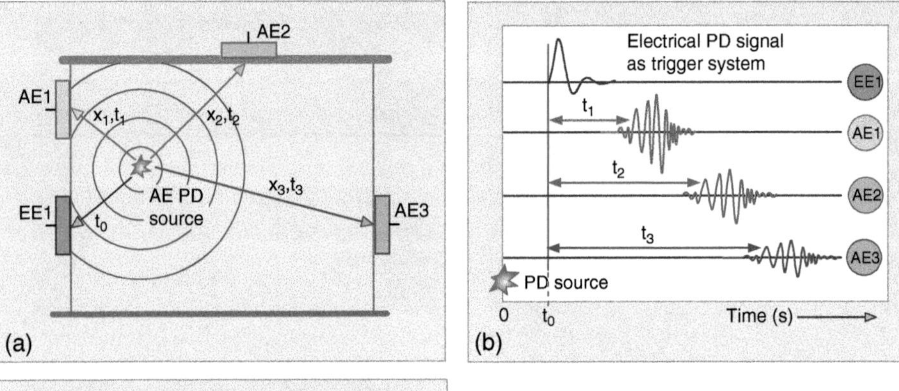

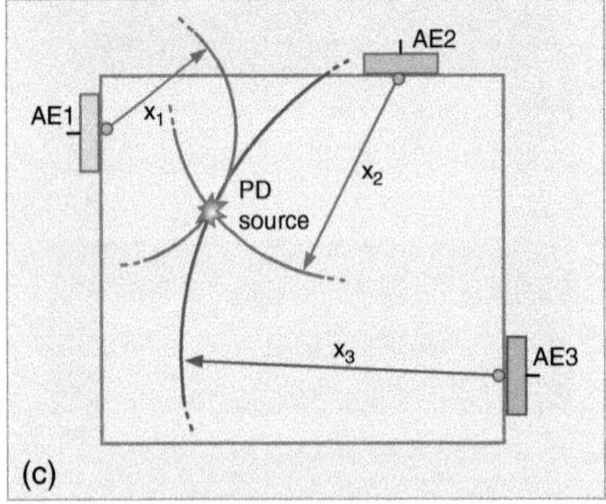

Figure 7.9 PD localization by the triangulation method: (a) AE PD sensors with an EE PD sensor for PD location; (b) EE PD and AE PD signals registered at the sensors with different times of flight; (c) PD position by triangulation method.

$$x_1 = vt_1 \tag{7.4}$$

$$x_2 = vt_2 \tag{7.5}$$

$$x_3 = vt_3 \tag{7.6}$$

The location of the PD source is obtained from the intersection of the circles with the radius of x_1, x_2, and x_3, respectively. Figure 7.9c demonstrates a visualization of the triangulation method for locating a PD source.

Locating the PD source can be performed in another way if the positions of the sensor AE1, AE2, and AE3 are represented in the Cartesian coordinate system of the given points as (x_1, y_1, z_1), (x_2, y_2, z_2), and (x_3, y_3, z_3), respectively. The speed of the recorded PD signals is assumed to be a constant expressed as v. The traveling times from the PD source to the AE1, AE2, and AE3 are designated as t_1, t_2, t_3, respectively. Finding the location of the PD source requires solving the Eqs. (7.7)–(7.9):

$$(x - x_1)^2 + (y - y_1)^2 + (z - z_1)^2 = (vt_1)^2 \tag{7.7}$$

$$\left(x-x_2\right)^2+\left(y-y_2\right)^2+\left(z-z_2\right)^2=\left(vt_2\right)^2 \tag{7.8}$$

$$\left(x-x_3\right)^2+\left(y-y_3\right)^2+\left(z-z_3\right)^2=\left(vt_3\right)^2 \tag{7.9}$$

According to CIGRE 676 recommendation, at least four UHF sensors may be used, as presented in Figure 7.10a, b. These sensors should be mounted on appropriate sites of the transformer tank, as detailed in [5]. In the case of four UHF sensors are used; by which the sensor positions are described by the space coordinates (x_1, y_1, z_1), (x_2, y_2, z_2), (x_3, y_3, z_3), and (x_4, y_4, z_4) for UHF sensor 1, UHF sensor 2, UHF sensor 3, and UHF sensor 4, respectively. As the speed of the UHF signal is assumed to be constant denoted as v_s; the traveling time to each AE sensor respected to a detected PD onset is designated t_{1i}. The location of the PD source (x, y, z) can be calculated using Eqs. (7.10)–(7.13).

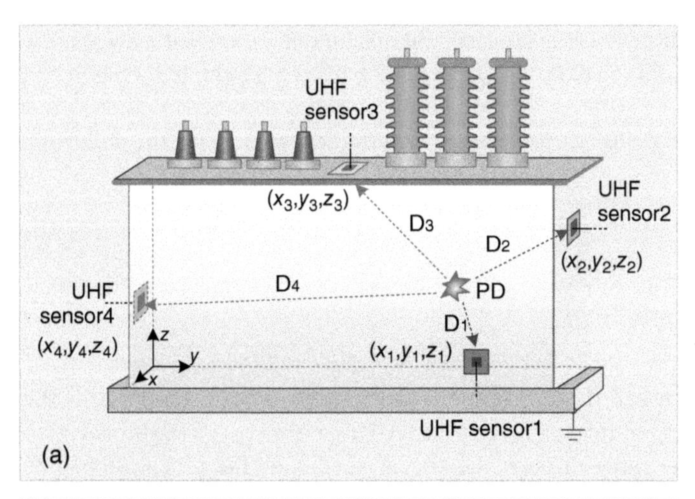

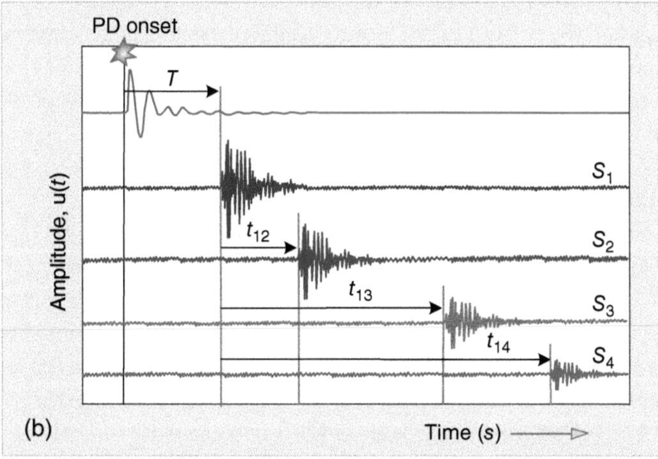

Figure 7.10 PD localization technique: (a) sensor positions with Cartesian coordination system; (b) UHF signal time difference t_{1i} with respect to detected PD onset.

$$(x-x_1)^2+(y-y_1)^2+(z-z_1)^2=(v_sT)^2 \tag{7.10}$$

$$(x-x_2)^2+(y-y_2)^2+(z-z_2)^2=(v_s(T+t_{12}))^2 \tag{7.11}$$

$$(x-x_3)^2+(y-y_3)^2+(z-z_3)^2=(v_s(T+t_{13}))^2 \tag{7.12}$$

$$(x-x_4)^2+(y-y_4)^2+(z-z_4)^2=(v_s(T+t_{14}))^2 \tag{7.13}$$

From a geometric perspective, the location of the PD source is described as the intersection of rotation symmetric hyperboloids [5].

In case of considering only the suspected area where the PD source is located, the number of the sensors can be reduced to two sensors. This technique is applied only for the transformers having two oil-fill valves by which the UHF sensors are mounted, as shown in Figure 7.11a.

The suspected area of the PD source can be determined by solving the following equations [6]:

$$D=t_a v_{\text{oil}} \tag{7.14}$$

$$t_a=\frac{t_{\text{diag}}-t_m}{2} \tag{7.15}$$

where D is the distance between the PD source to the respective UHF sensors, t_a is the propagation time from the PD source to the respective UHF sensors, t_{diag} is the propagation time from each sensor diagonally through the transformer, t_m is the propagation time difference between the signals as recorded by each of the two UHF sensors (Figure 7.11b). Considering different propagation times from the PD source to the respective UHF sensors denoted as t_b, t_1, and t_2, a line D will be constructed to represent the likely location of the PD sources. Figure 7.11 illustrates PD source location using two sensors at different positions and the time parameters used in the calculation.

7.4.3.2 Application of the Triangulation Method for PD Localization in a Transformer

To locate the PD source, the fixable or movable AE PD sensors can be used to detect PD signals. In the following example, four AE PD sensors are orthogonally placed at the suspected area on the transformer tank. A HFCT is also used to detect the electrical PD signal. The dimension of the investigated transformer, as shown in Figure 7.12, is 190×240×492 cm. The origin (0,0,0) in the Cartesian coordinate system is determined at the middle point of the transformer. The AE PD sensors are installed at various places on the exterior surface of the transformer, as shown in Figure 7.13a, b. The green AE PD sensor (AE1) is located at (−15, 50, 95).

The red AE PD sensor (AE2) is located at (45, 30, 95). The yellow AE PD sensor (AE3) and the blue AE PD sensor (AE4) are placed at (−35, 16, −95) and (−15, 15, 95), respectively. The HFCT was equipped around the ground wire of the transformer, as presented in

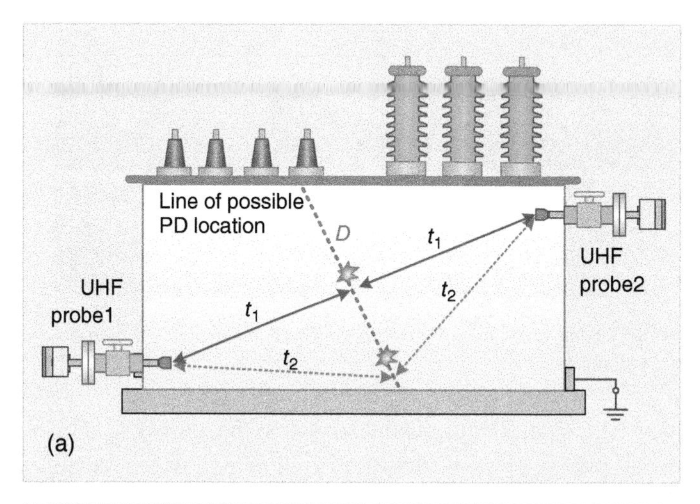

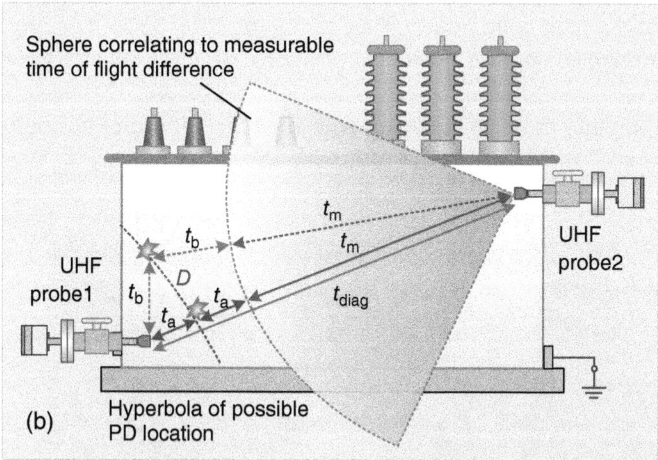

Figure 7.11 PD localization in a transformer with two UHF probe: (a) in case of the propagation times from the PD source to the respective UHF sensors are the same; (b) in case of the propagation times from the PD source to the respective UHF sensors are different [6].

Figure 7.12 The investigated transformer.

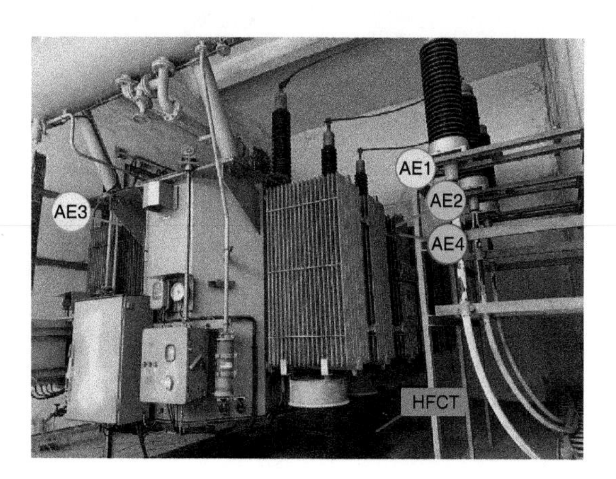

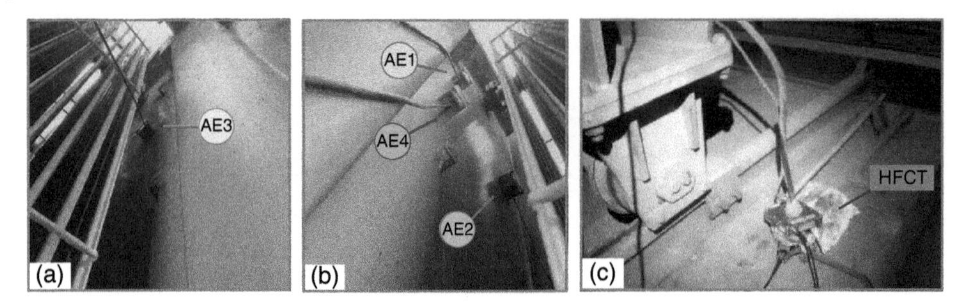

Figure 7.13 AE PD sensors and HFCT positions: (a) AE PD sensor 3; (b) AE PD sensor 1, 2, and 4; (c) HFCT clamping around the ground wire.

Figure 7.13c. When PD occurs, AE PD signals transmit and register at the sensors, as shown in Figure 7.14. Concurrently, the EE PD signal is captured by the HFCT, as exhibited in Figure 7.15. The triangulation technique is applied, and then the PD source location is computed and represented as the pink point, as displayed in Figures 7.16–7.17. Then, the investigated transformer is untanked and the PD source positions are revealed, as represented in Figure 7.18.

From this case study, the computer program did not provide the exact PD source position due to the complex structure of the transformers and complex characteristics of the

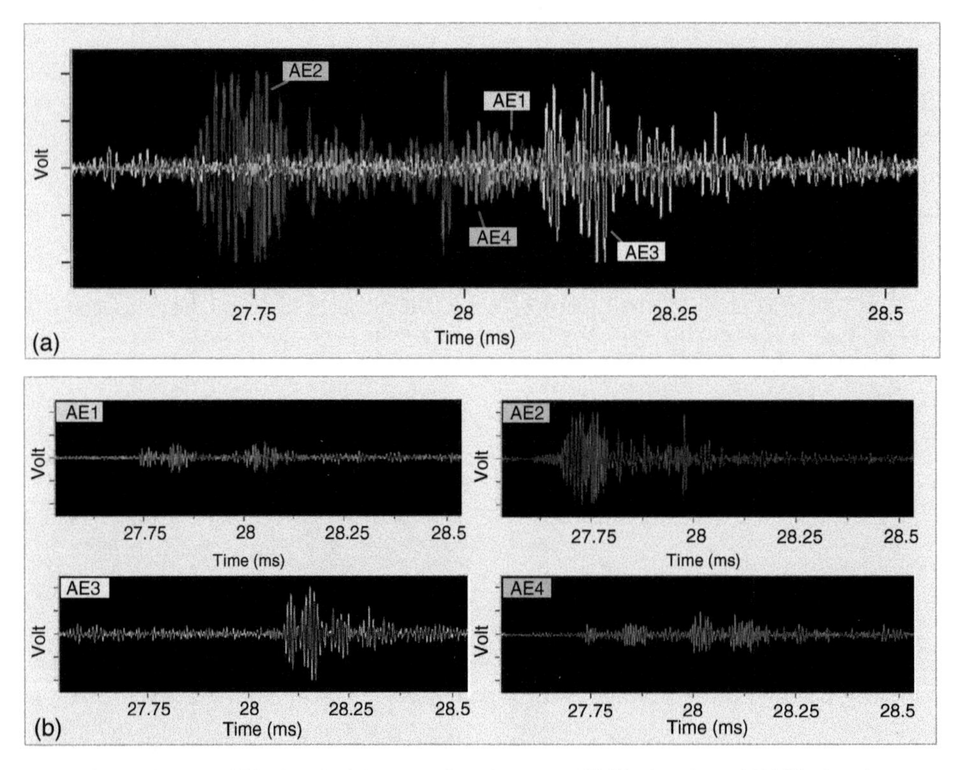

Figure 7.14 Detected PD signals: (a) comparison between AE PD signals and EE PD signal obtained from AE PD sensors and the HFCT; (b) AE PD signals from AE PD sensors 1–4.

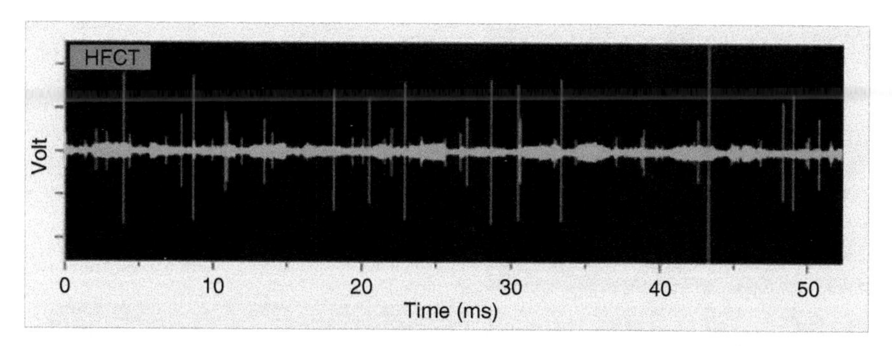

Figure 7.15 EE PD signal detected by the HFCT.

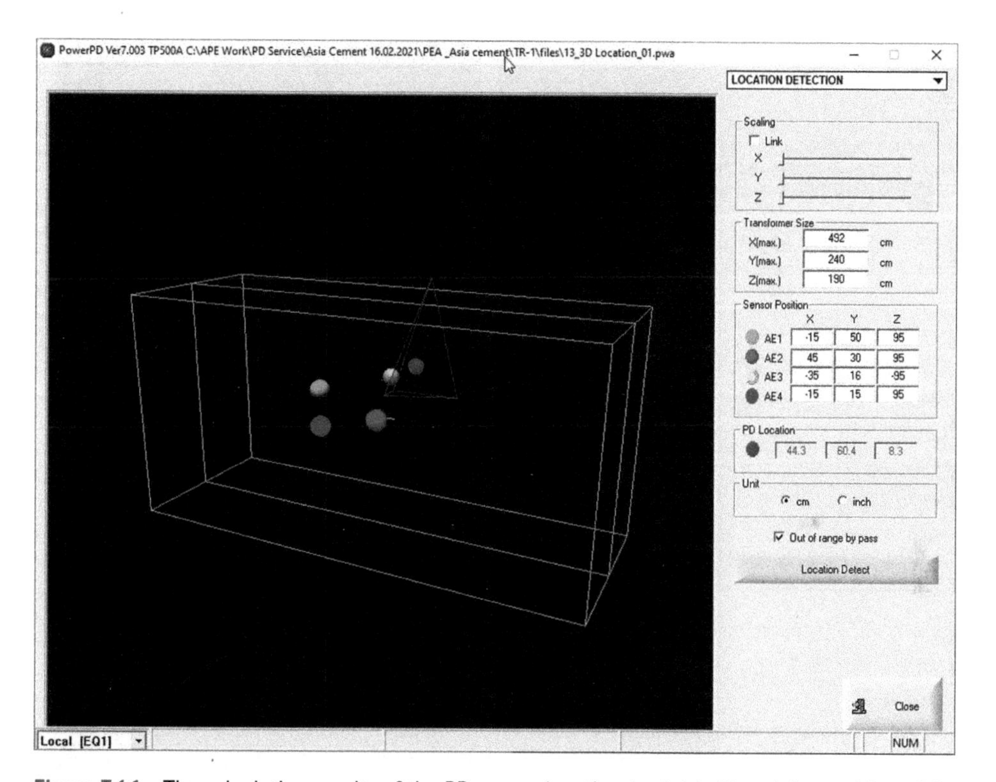

Figure 7.16 The calculation results of the PD source location (a pink ball) and the position of the sensors (a green ball, a red ball, a yellow ball, and a blue ball).

acoustic signals. Applying acoustic PD detection method based on the time arrival at three or more AE PD sensors orthogonally placed on the transformer tank with the assumption of a uniform propagation velocity usually suffers from strong uncertainty and partly misleading results for not considering the tank wall signals. Besides, it is very difficult to apply AE PD sensors to distinguish between mechanically induced clicking and ticking sounds and the sound originating from the PD sources within the transformer. Moreover, the AE signal processing requires industrious and elaborated work [7].

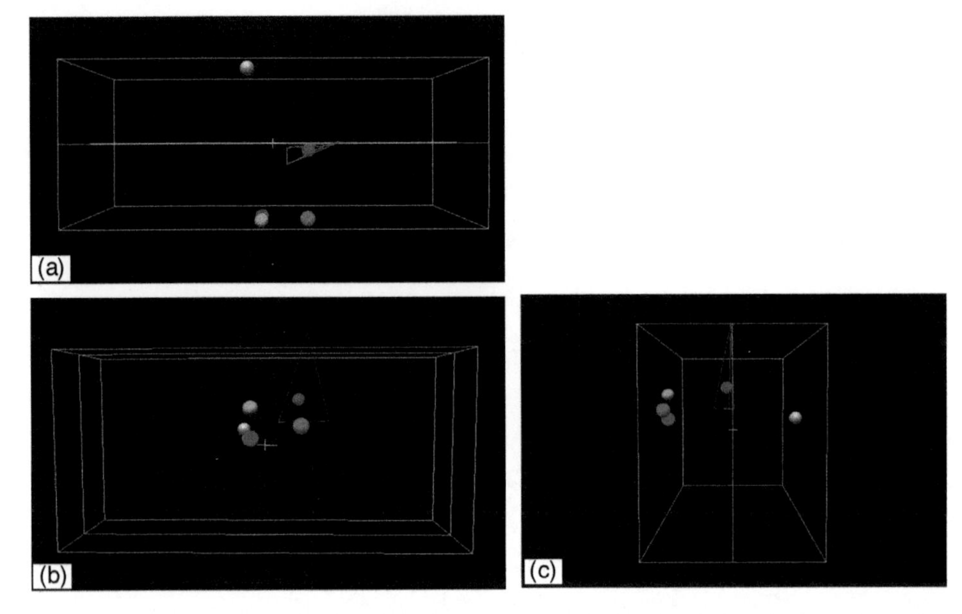

Figure 7.17 Third angle projection of the triangulation technique for PD localization: (a) top view; (b) front view; (c) side view.

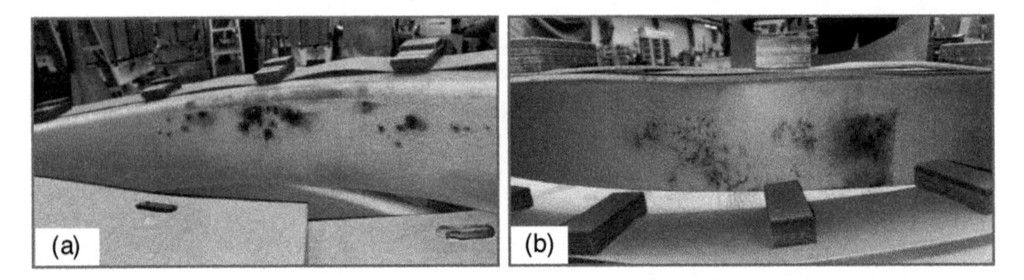

Figure 7.18 PD source locations in the investigated transformer (a) discharge area #1; (b) discharge area #2.

Based on the mentioned problem, another PD localization technique called the *flat problem* is proposed to reduce the large uncertainties of PD source location obtained from acoustic methods. First, to roughly locate the PD source, the suspected area of the transformer is scanned by replacing one or several AE PD sensors on the tank wall to find AE PD signal correlated to electrical PD signal detected by an electrical PD sensor such as an HFCT. The AE PD sensors have to move to find the strongest zone especially when only one AE PD sensor is used for the first step. The signals obtained from three AE PD sensors in this step are shown in Figure 7.19a. Second, to locate the PD source, three AE PD sensors are placed in a straight row on the transformer tank to reduce the triangular problem to a two-dimensional flat problem. If possible, the sensors are evenly spaced. The distance between the sensors corresponds to the initially found distance to the source (Figure 7.19b). After AE PD signals and electrical PD signals are captured, the sensors are placed vertically to capture the PD signals, as shown in Figure 7.20. By assisting with a developed software tool, a three-dimensional position result can be created to demonstrate the PD source location, as depicted in Figure 7.21 [7].

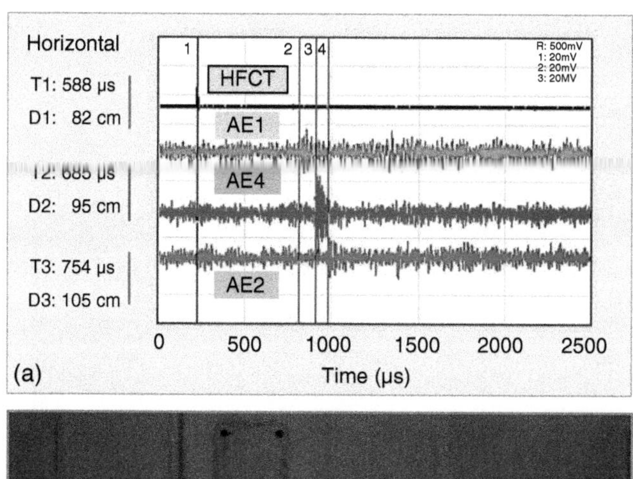

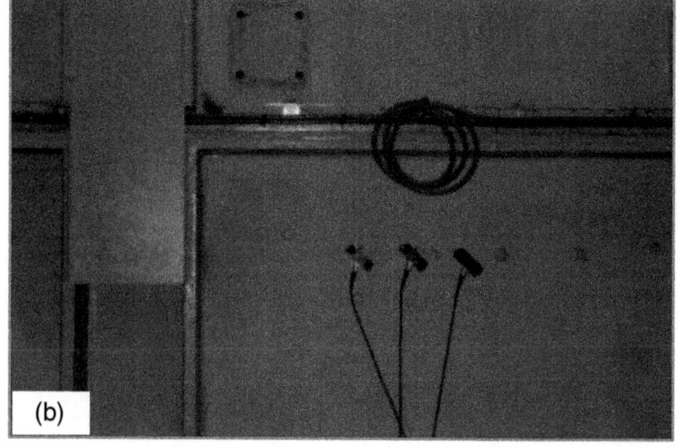

Figure 7.19 PD signals and the sensor installation (a) the AE PD signals (a green line, a blue line, and a red line) and EE PD signal (a black line) for roughly PD source location [7]; (b) a horizontal placing three AE PD sensors [8].

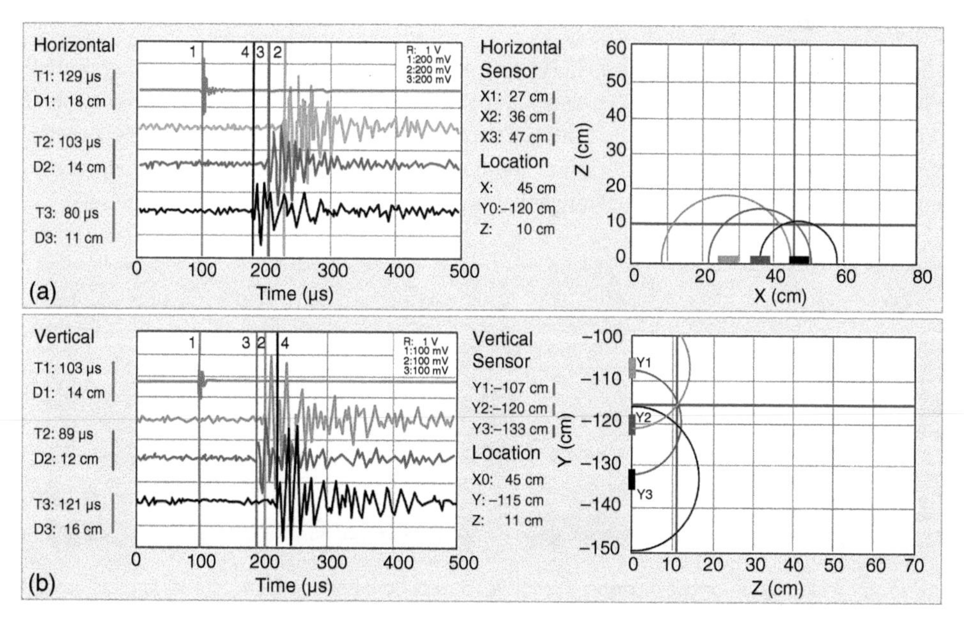

Figure 7.20 EE and AE PD signals obtained from (a) horizontal AE PD sensors; (b) vertical AE PD sensors including the results from the software program [8].

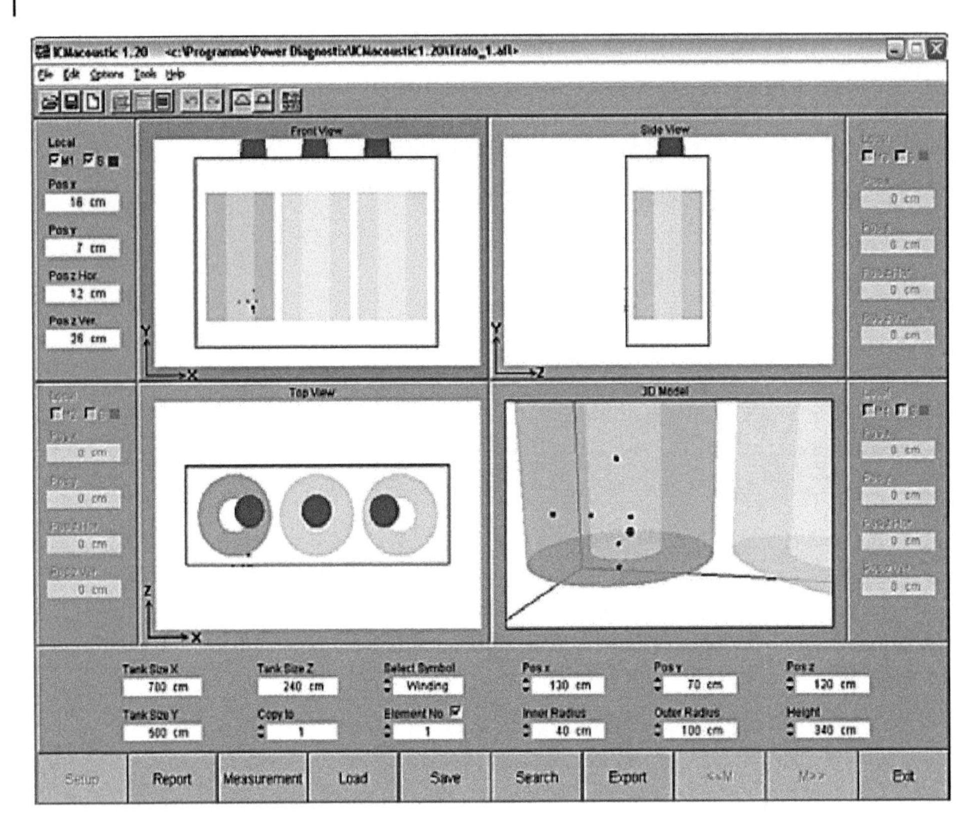

Figure 7.21 The examples of PD source location obtained from the flat problem technique [7].

7.4.4 Bouncing Particle Localization Method

To detect AE PD signal generated from a bouncing particle, an ultrasonic probe with a pointed tip is used to press on the GIS or GIL external surface to pick up AE PD signals, as illustrated in Figure 7.22. The position of the tip is moved to pick up the strongest AE PD signal where the suspected particle is located. The ultrasonic frequency range of the detected ultrasonic wave is then downed to the audio frequency range that is fed to the headphone for interpretation.

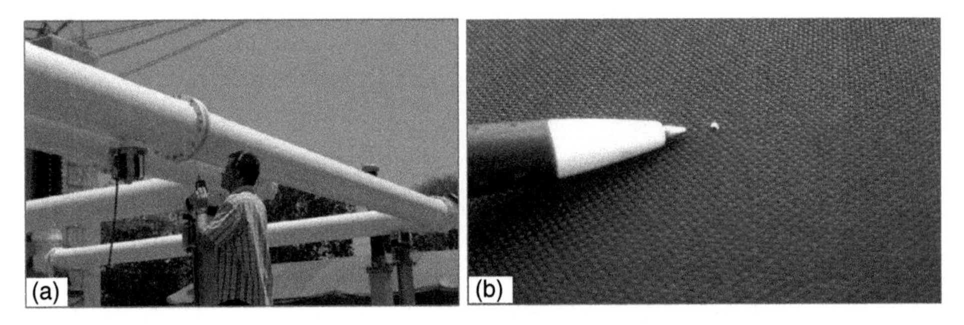

Figure 7.22 Application of the ultrasonic probe for detecting a bouncing particle (a) application of ultra-probe with a pointed tip used for field test; (b) zoom of particle found in GIL.

7.5 PD Localization Techniques for the External Insulation

For external insulation, the PD localization can be achieved by using a corona camera or an airborne acoustic probe. The optical PD measurement utilizing the corona camera or the acoustic PD measurement employing the airborne acoustic sensor is superior to the conventional electrical PD measurement. This is because applying the coronal camera or the airborne acoustic sensor for corona discharge and surface discharge detection needs simple procedures to locate the PD source. Besides, utilizing the corona camera is unaffected by (EM) interference, which is particularly advantageous for onsite and on-line testing.

7.5.1 Application of the Corona Camera

Corona inspection can be performed from the ground, from the insulated truck-mounted aerial platform, or from the air using a helicopter or unmanned aerial vehicle (UAV) for aerial inspections. The distance between an observer and the high-voltage equipment is determined by personnel safety considerations. Using a corona camera aimed at the desired object enables the existing corona discharge or the occurring surface discharge to be detected and recorded with UV emissions shown as an overlay on the visible image. With UV radiation shown in white, red, or blue, the corona camera provides an efficient method for locating the PD source location. Figures 7.23–7.24 show a field test using the corona camera to check for corona discharges and surface discharges.

7.5.2 Application of the Airborne Acoustic Probe

An airborne ultrasonic probe is also widely used to locate a PD source of the corona discharge or surface discharge of the external insulation, as represented in Figure 7.25. A user has to point the probe at the suspected PD source, providing the strongest AE PD signal. In the case of long distance between the ultrasonic probe and the target, the ultrasonic wave concentrator will be used with the ultrasonic probe to increase its sensitivity. Then the ultrasonic frequency range of the detected ultrasonic wave is downed to the audio frequency range for interpretation.

Figure 7.23 Corona discharge inspection in a field test performing from (a) the ground; (b) the insulated truck-mounted aerial platform; (c) a drone; (d) a helicopter [9, 10].

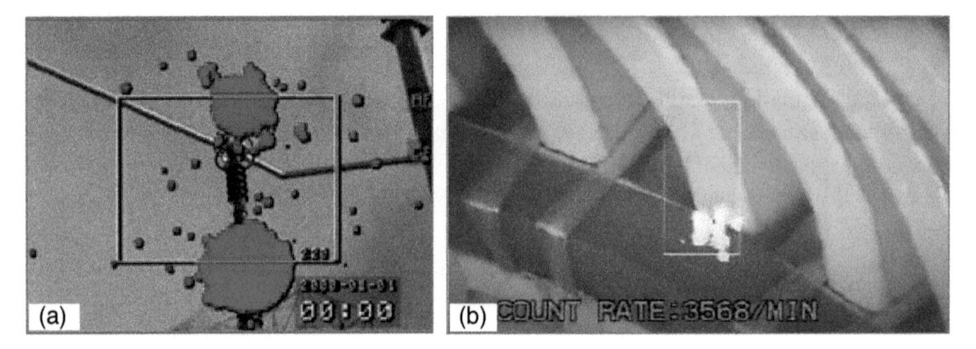

Figure 7.24 Corona activities detected by the corona camera (a) at an outdoor substation (in red); (b) inside a generator (in white) [11, 12].

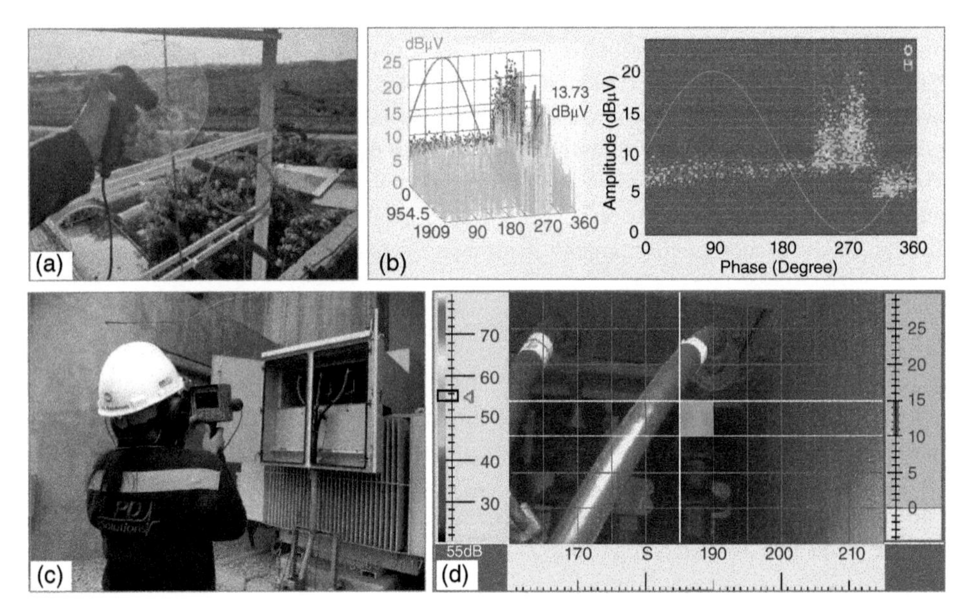

Figure 7.25 Application of the airborne acoustic probe for PD source localization: (a) corona discharge detection at the high-voltage part of the supporting insulator, (b) corona discharge pattern, (c) surface discharge detection at the cable terminations, (d) identifying the PD source.

References

1 International Electrotechnical Commission (2012). *IEC 60034: Rotating Electrical Machines – Part 27-2: On-Line Partial Discharge Measurements on the Stator Winding Insulation of Rotating Electrical Machines*. Geneva, Switzerland: IEC.

2 International Electrotechnical Commission. (2019). IEC 60071: insulation co-ordination – Part 1: definitions, principles and rules. International Electrotechnical Commission, Geneva, Switzerland.

3 Conseil International des Grands Réseaux Électriques. (2016). Guidelines for partial discharge detection using conventional (IEC 60270) and unconventional methods. *Technical Brochure. Ref. 662.*

4 Conseil International des Grands Réseaux Électriques (2012). High-voltage onsite testing with partial discharge measurement. *Technical Brochure. Ref. 502.*

5 Conseil International des Grands Réseaux Électriques (2017). Partial discharges in transformers. *Technical Brochure. Ref. 676.*

6 Coenen, S. (2012). Measurement of partial discharges in power transformer using electromagnetic signals. PhD dissertation. University Stuttgart.

7 Gross, D.W. and Soeller, M. (2012). On-site transformer partial discharge diagnosis. In: *Proceedings of the 101 IEEE International Symposium on Electrical Insulation, June 10–13.* San Juan, PR: IEEE.

8 Gross, D. (2016). Acquisition and location of partial discharge – especially in transformers. PhD thesis. Technischen Universität Graz.

9 CoprelecSpA High Techonology. (2018). OFIL unaempresa de altatecnología. https://www.coprelec.cl/ofil-ciq4 (accessed 16 December 2018).

10 HLD Highline Division. (2018). Corona, infrared & HD inspection. https://www.highhttp://linedivision.com/hld/php/services/servicesCorona.php (accessed 30 May 2021).

11 Genutis, D. A. (2017). Corona imaging, http://www.halcotestingservices.com/coronaimaging.html (accessed 30 May 2021).

12 Ofilsystems (2012). Corona on generator, https://www.youtube.com/watch?v=b1AFgNcrDAE (accessed 30 May 2021).

8

PD Measurement Under Direct and Impulse Voltage Stress Conditions

8.1 Introduction

The previous chapters and, therefore, the main content of this book, are related to PD measurement under AC power frequency conditions. This comes from the point of view that the physical PD behavior of electrical insulations and all related subjects should be discussed in terms of their practical applicability, which until recently has mainly been AC power frequency voltage stress. Related standards such as IEC 60270 have also focused on those conditions as well. However, in the last decades, applications such as DC and impulse voltage stress conditions have increased and require quality testing methods and procedures to provide reliable operation of applied components within electrical power systems. At impulse voltage stress, besides the introduced testing methods and procedures for switching (SI) or lightning (LI) impulse conditions, there is special interest in the quality assessment at repetitive voltage impulses with very short rise times and high repetition frequency generated by applied switching components like IGBT- or SiC-converters [1, 2]. That was also promoted by the increasing number of applied power electronic components in electrical systems.

To ensure the reliability and operation safety of electrical insulation under these electrical conditions, appropriate quality assessment tools and procedures are required. Thus, there is a worldwide endeavor to develop adequate requirements for PD measuring procedure and methods based on knowledge of PD behavior under such alternative stress conditions.

From a physical point of view, PD phenomena are dependent only on fulfillment of two conditions, as already mentioned (see Chapter 2):

- Local electrical field strength E_{loc} (given by voltage stress) in that part must exceed the (intrinsic) dielectric field strength E_d of the insulation material and, therefore, fulfilling the equation

$$E_{loc} \geq E_d \tag{8.1}$$

- Available free charge carrier(s), mainly electron(s), enable the initiation of any ionization process.

Partial Discharges (PD) - Detection, Identification, and Localization, First Edition. Norasage Pattanadech, Rainer Haller, Stefan Kornhuber, and Michael Muhr.

It is obvious that the "fulfilling" of those conditions should be influenced by the type of applied voltage, insulation material, and many other parameters such as temperature. This chapter discusses the options and issues for electrical PD measurement under DC and impulse stress conditions.

8.2 PD Measurement at Direct Voltage Conditions

Before discussing the PD behavior and derived requirements and conditions at assumed direct voltage (DC), it should be noticed that at insulation stress under DC not only the steady-state DC field conditions have to be considered. Moreover, the steady-state condition for electrical field within the insulation will be reached only after a certain so-called transition time after DC switched operations (on/off) or possible polarity reversals. The duration of that time might be in range of some seconds, minutes, or even more, mainly depending on insulation dielectric parameters such as polarization ability, conductivity, and temperature [3]. Additionally, in some practical cases the (pure) steady DC field might be superimposed with small alternating AC components (ripple). At all these cases the electrical field conditions within the insulation are not in a steady state and, therefore, possible PD activities during that transition time are related more to the already-described PD behavior under AC conditions. Hereafter the focus is on PD measurement issues at steady-state DC field conditions. Note, that there is a large variety in PD behavior for various insulation materials, which requires adequate PD measurement and quality assessment procedures.

Concerning the PD phenomena in gases, it can be stated that in ambient air the discharge activity (e.g. occurring in air gaps under DC at both polarities) are in principle comparable to those occurring under AC, which might be illustrated by discharges of Trichel type (Figure 8.1, see Chapter 2, Section 2.2) [4].

As one can see, the pulse magnitude is very similar in both cases, but the pulse number at DC is much higher than at AC. In principle, that confirmed that at AC, the electrical strength condition (Eq. 8.1) is reached only in a certain time slot of 50 Hz test voltage – here in approximately (4.8–5.2) ms – but at DC, no such limitation exists. However, that behavior could be changed if the gas is enclosed, as is the case for gas-insulated-systems (GIS).

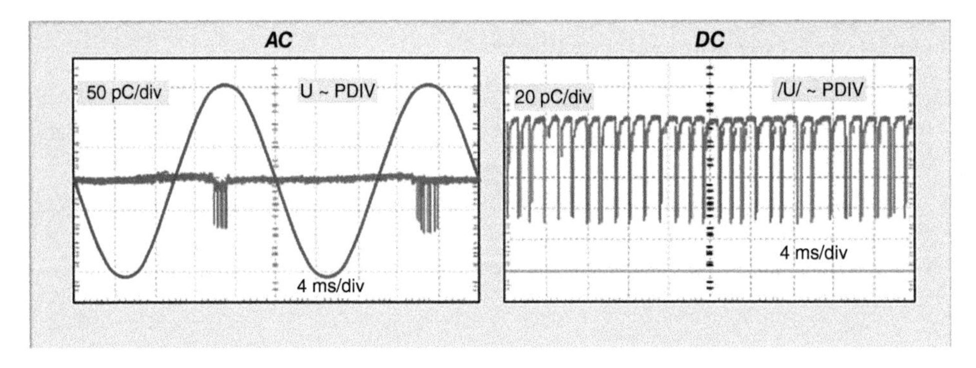

Figure 8.1 Trichel pulses at point-to-plane arrangement in air under AC- and DC-conditions.

Such insulation systems are mainly designed as hybrid-type, which means that besides the (pure) gas insulation, some interfaces mostly of polymeric materials are located between electrodes. For such hybrid insulation, those interfaces might be charged even below the measurable inception voltage of the (pure) gas insulation, which could influence the inception field strength of the whole system [5].

More difficult and until now not fully understood is the PD behavior at DC conditions for liquid and solid insulation, which according to the introduced classification were termed as *internal discharges*. To explain the PD occurrence at internal discharges, usually the modeling of cavity discharges is used.

Approaches for modeling such discharges was described in Chapter 3. Here we discuss the electrical field strength behavior in principle for a simplified cavity arrangement under DC conditions (Figure 8.2a). After switching on the DC and reaching a *steady-state condition,* the electrical field strength over the cavity will be increased to the value $E_{cav,max}$ if no discharges occur (Figure 8.2b). Now it should be assumed that for the cavity insulation, a certain electrical (intrinsic) field strength is valid at which the condition according to Eq. (8.1) might be fulfilled ($E_{inc,min}$). At the same time, it will be assumed that no free electron is available, with the consequence that the electrical field strength over the cavity E_{cav} will be further increased. The free electron should be available at first at E_{inc}, what means that a certain time delay in discharge inception must be considered, expressed by the value t_L and, at the same time, accompanied by an "overshoot" in electrical field strength expressed by ΔE. It should be noticed that the value t_L often termed as *statistical time lag* is mainly determined by the dielectric characteristic of the insulation material (e.g. conductivity, permittivity, polarization ability as well as temperature, etc.).

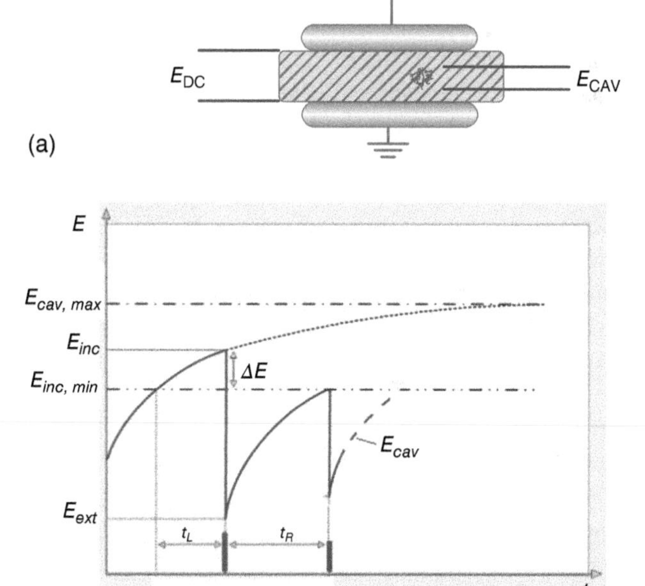

Figure 8.2 Electrical field strength over a cavity in solid insulation at DC (schematically): (a) basic arrangement for cavity discharge; (b) electrical field strength behavior over the cavity.

In case of practical insulation, the value of t_L lies in the range of seconds to minutes or even some hours [6]. At reaching the value of E_{inc} the electrical field over the cavity collapses to a residual value E_{ext} caused by an internal discharge and leading to a current pulse in a connected external circuit. After the extinction of the discharge, the same procedure might be repeated if, after an elapsing or recovery time t_R, the condition (8.1) is fulfilled again and free electrons are available. This leads to a consecutive discharge and to the next current pulse and so on (see Chapters 2 and 3 for more detailed description of discharge behavior in a cavity).

It should be noticed that the value of electrical field strength as well as the availability of free electrons are of stochastic character; therefore, in practical cases all termed values are scattering over a wide range so e.g. the recovery time t_R could be from minutes up to hours [7]. It is obvious that such a PD behavior has consequences for performance of practical PD measurement and relevant testing procedures.

Concerning the measurement requirements for PD detection under DC conditions, it can be stated there is no significant difference to the situation at AC power frequency. The measurable PD signals caused by the current pulses in an external circuit are of similar frequency characteristic as measured at AC conditions and, therefore, all requirements for basic measuring circuits – including coupling units and measuring systems as well as calibrating procedures as specified in related standards for PD testing under AC voltage – are applicable [8], as discussed in Chapter 4.[1] That means at a PD test circuit under DC conditions, the measuring PD system might be connected by coupling unit[2] via coupling capacitance C_K or directly with the test-sample C_a (Figure 8.3a,b). Especially in case of PD measurement under difficult noise conditions, often the *balance circuit arrangement* is used (Figure 8.3c).

As already mentioned, the PD behavior of internal discharges under DC conditions might be characterized by extremely large recovery time, which leads to a very low PD pulse repetition rate. That behavior has consequences for the measuring procedure, measurable parameter, and PD quantities. At first it should be noticed that the evaluation of the PD inception voltage (PDIV) is very problematically, because the rise time on the test voltage (ramp) after switching on (e.g. the beginning of electrical field stress within the insulation) has a significant influence on measurable PD activity. For that reason, at PD testing procedure of various HVDC-equipment, different test voltage ramps with respect to voltage steps are recommended [9–11]. For some practical applications (e.g. at PD testing of HVDC converter transformers), a minimal number of PD pulses over a certain threshold for apparent charge is required [12]. That is in contrast to the common understanding of PD inception, at which the first occurring PD pulses are counted. Additionally, if by some reasons the insulation material was already pre-stressed, any remaining space charges within the insulation material should be considered. In contrast to an alternating electrical field in which polarity changes in few milliseconds at power frequency these space charges are likely concentrated on imperfections in the insulation and could influence the PD activity on test results [13]. For example, it was investigated

1 This is also valid for *pulseless* discharges.
2 Sometimes also termed as *measuring impedance* Z_m.

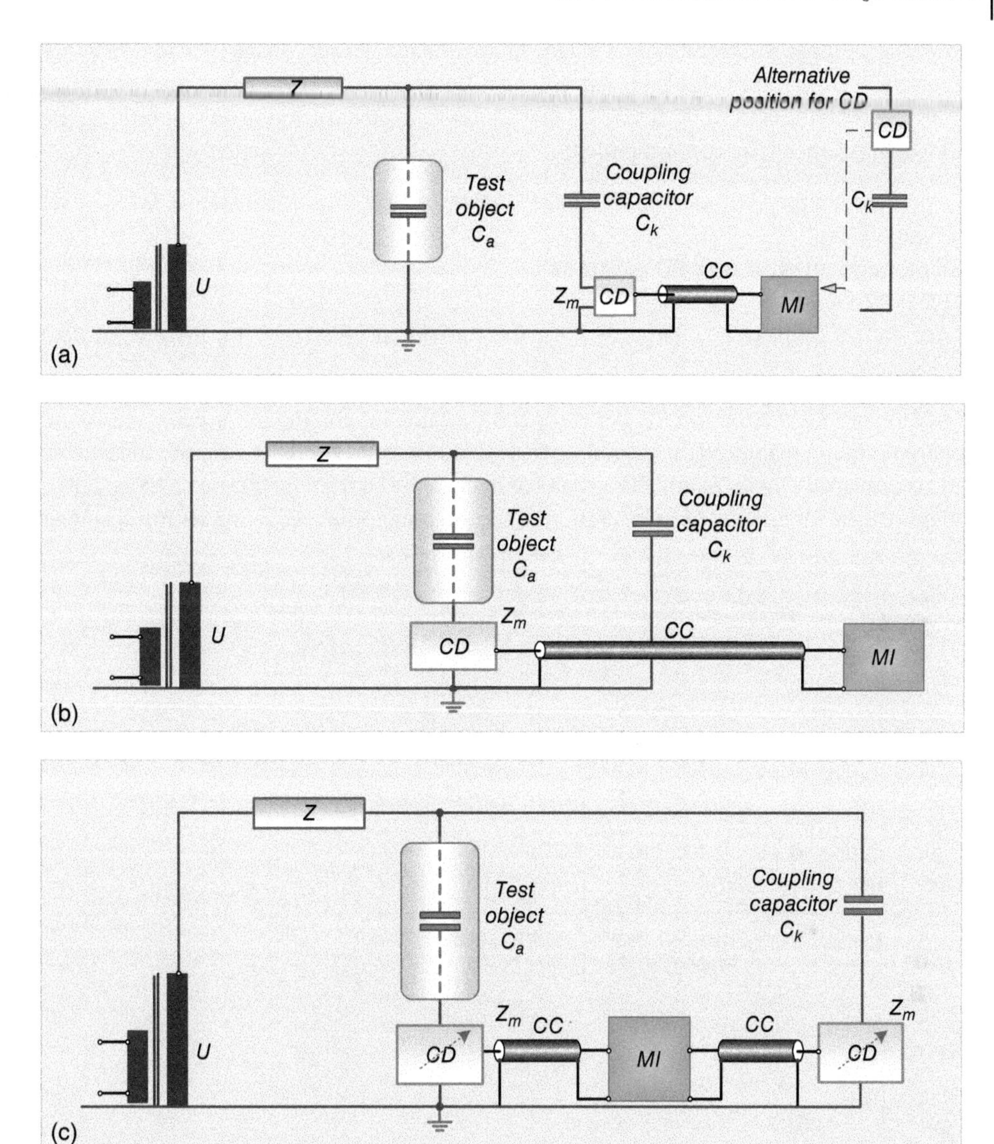

Figure 8.3 PD measurement circuit under DC conditions: (a) coupling device CD in series with the coupling capacitor C_k; (b) coupling device CD in series with the test object (C_a); (c) balanced circuit arrangement [8].

that the tree-growing behavior inside polymeric material was extremely influenced by such space charges. Finally, especially under on-site PD testing conditions, one cannot be sure that the measured signal is really caused by the expected PD pulse and not by any other external noise activity.

It might be concluded that in case of DC testing, the electric field stress during test voltage rise and decrease (switch on/off) differs from that during the period when the voltage is constant and in steady-state conditions. Therefore, PD testing procedures described for alternating voltage to determine the PD inception and extinction voltage (PDIV, PDEV) are generally not applicable for PD measurement at DC conditions [14].

Despite that and according to the related standard for measuring of PD quantities, the following parameter are recommended:

- Apparent charge Q of each individual PD pulse vs. the *time* at constant DC test level (Figure 8.4a)
- Time between successive PD pulses Δt
- PD pulse count m
- Accumulated apparent charge Q_{cum} of the individual pulses vs. the measuring *time* (Figure 8.4b)

Additional information on PD behavior can be gained if the PD pulse count m vs. the apparent charge magnitude q exceeding specified threshold levels during the measuring time is displayed (Figure 8.5a). This graph has been deduced from the PD pulse train shown in Figure 8.4a. Moreover, the presentation of the pulse counts m occurring within specified

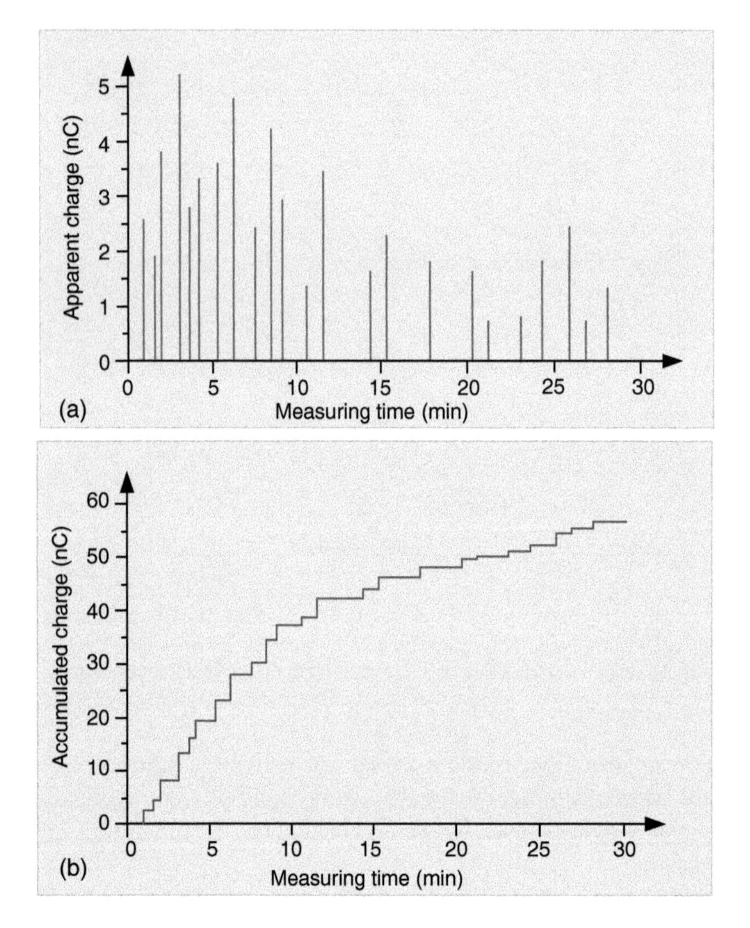

Figure 8.4 Measurable PD quantities recommended by IEC 60270:2000 [15]: (a) apparent charge of individual PD pulses; (b) accumulated apparent charge.

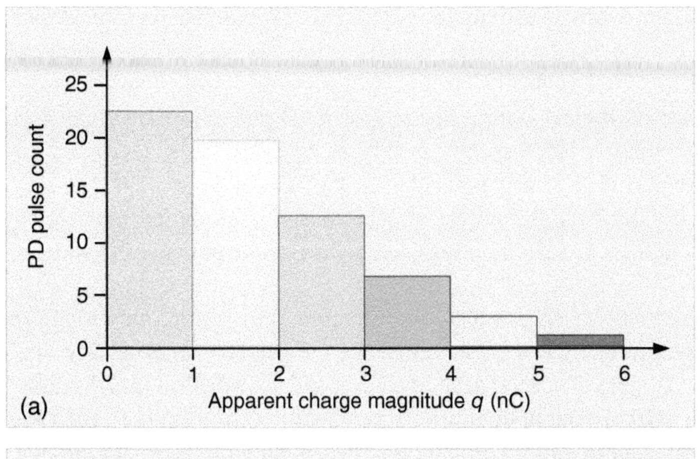

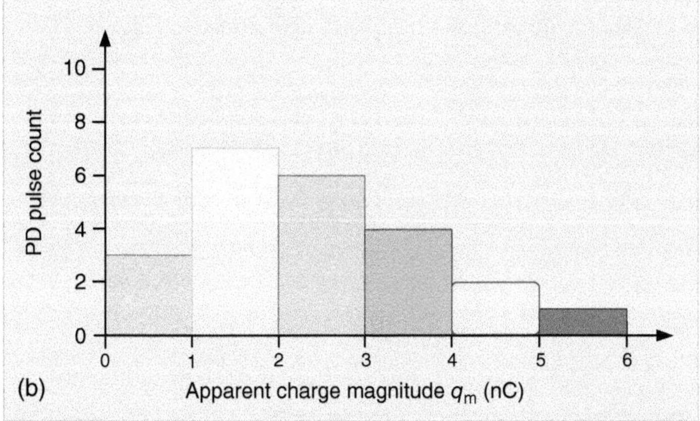

Figure 8.5 Derived PD quantities recommended by IEC 60270:2000 [15]: (a) PD pulse count m exceeding the following thresholds for the apparent charge magnitude q: {0, 1, 2, 3, 4, 5} nC; (b) PD pulse count m occurring within the apparent charge intervals (classification) q_m: {(0−1)/(1−2)/(2−3)/(3−4)/(4−5)} nC.

limits of the apparent charge magnitude seems useful for assessing the PD activity during direct voltage measurement.

As the PD pulses are often scattering over an extremely wide range, which refers to both the magnitude and the repetition rate, a statistical analysis of the significant PD quantities is highly recommended [16]. For that purpose, the PD measuring system should be equipped with an appropriate postprocessing unit, which enables statistical procedures – number of PD pulses m exceeding certain threshold levels of apparent charge magnitude q_m and depicted as $m = f(q_m)$ dependency (like an empirical *density function*) (Figure 8.5b).

Due to the lack of knowledge for PD behavior of internal discharges at DC stress conditions and their harmfulness, recent investigations are mainly focused on detection and recognition of PD sources as well as their location within the insulation, respectively. Besides the above-mentioned statistical analysis, there are some PD evaluation procedures with future perspectives for applying even at DC conditions – pulse sequence analyzing

(PSA) and advanced signal processing (ASP) methods [17, 18]. It should be mentioned that digital measuring PD devices equipped with appropriate postprocessing units are needed to apply of all these methods.

The PSA method is based on the evaluation of successive PD pulses in *magnitude* and *occurrence time* and related differences, respectively (Figure 8.6). Originally developed for PD measurement at AC, this method is also applicable under DC conditions [19]. The estimated differences in ΔQ and Δt allow different graphical interpretation in form of related charts and, in the same manner, in certain cases conclusions about possible PD sources (Figure 8.7) [20].

However, also at PSA method two issues must be considered: the extraordinary long PD measuring time (due to the extremely low pulse rate) at DC conditions and necessary suppression of external noise. Despite of special developed processing algorithm [21] or modified graphical description (3D) of related quantities e.g. $\Delta Q/\Delta t = f(\Delta Q, \Delta t)$ [22], the PSA approach needs further application and investigation to become a more accepted method for the evaluation of partial discharge quantities during tests with direct voltage.

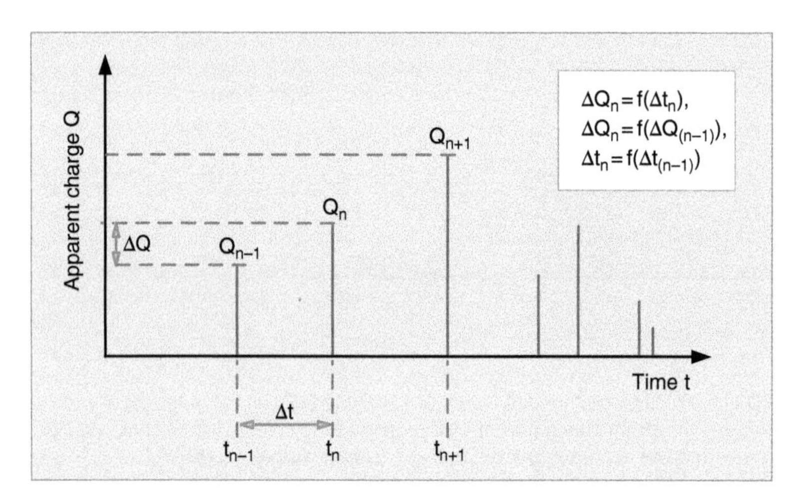

Figure 8.6 Principle of pulse sequence analyzing.

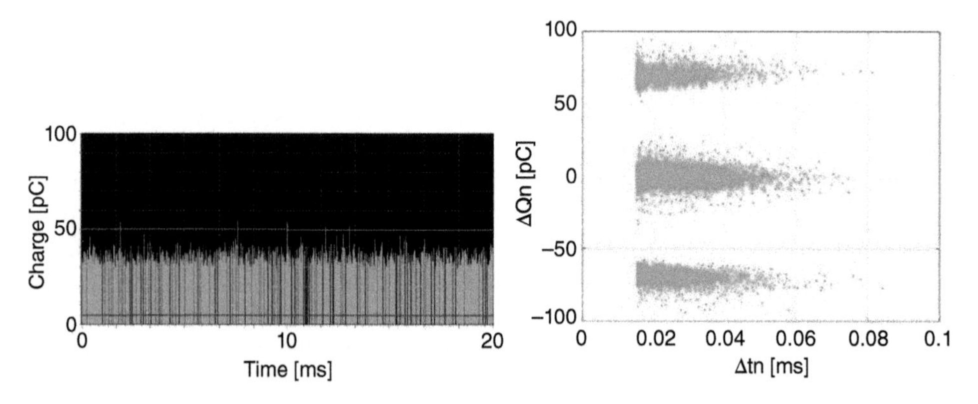

Figure 8.7 Pulse sequence analyzing of Trichel pulses.

Recently, ASP methods have been adapted for DC conditions. This technique was mainly accompanied by the development of fast electronic circuits as FPGA as well as advanced computer technology and appropriate PD measurement systems. At the same time, special procedures or algorithms, respectively, were developed to recognize the noise environment, suppress the interference signals, and/or to separate it from measurable (PD) pulses [23, 24]. In some cases, also the application of neural networks, together with special processing algorithms like wavelet methods, was investigated as an option for recognizing the noise environment and to separate it from measurable (PD) signals [25, 26]. Without going into detail (see also Chapter 4), it should be mentioned that the main obstacle for application of that method is the time for adaption on the noise environment – *learning time* – which must be as short as possible, preferably in the range of few μs. Another approach seems to be the application of spectral analyzing methods to assign the detected signals to the origin PD sources and/or disturbances by clustering algorithms [27, 28].

Nevertheless, it could be stated that the major obstacles for PD measurement under DC conditions are the possible evidence of very large recovery time of PD activity and, therefore, increased efforts for separation of noisy and desired PD signals. The long recovery time leads to low pulse rate and is strongly dependent on the dielectric material characteristic, possible space charge impact, especially during the electrical stressing or prestressing time as well as ambient parameter like temperature and so on. The suppression of electromagnetic (EM) disturbances or its elimination for the measuring processing seems to be possible by applied ASP methods, but that is not generally solved for practical applications.

We can conclude that the broader application of PD measurement methods and procedures under DC stress conditions should be different for various type of insulation with different dielectric material respectively and thus need further investigation.

8.3 PD Measurement at Impulse Voltage Conditions

Previous chapters focused on PD measurement for high-voltage insulation, which is stressed by continuously applied test voltage as AC and DC. But, as already mentioned in the introduction, many new electrical components operate under voltage stress conditions that differ from those introduced at AC power frequency as well as under DC conditions. Besides the "classical" impulse voltage stress conditions, as for example switching (SI) – or lightning (LI) impulse voltage as well as test voltages with very fast front time (VFF), a new class of voltage stress conditions is of growing interest: Repetitive voltage impulses with very short rise times and high repetition rate are generated by fast switching components like IGBT- or SiC-converters [1, 2]. The latter one was also promoted by the increasing number of applied power electronic components in electrical power systems, especially by pulse width modulation (PWM) motor-driven parts. To ensure the reliability and operation safety of electrical insulation even under such different electrical conditions, appropriate PD measuring and quality assessment tools must be developed.

8.3.1 PD Measurement at Classical Impulse Voltage Conditions

The standardized classical impulse test voltages – lightning-, switching-, or oscillating impulse – are introduced in quality-testing procedures for high-voltage insulation equipment [14]. It is obvious that PD should be measured even under such conditions as a well-proven quality assessment tool. However, conventional PD measurement systems and methods are not commonly suitable for that kind of measurement. In principle, two main issues must be solved when PD measurement under impulse conditions should be performed:

1) The EM disturbance caused by the impulse generator must be suppressed. The working principle of impulse generators is commonly based on fast switching discharge of capacitances, which leads to a residual (displacement) current in the whole test circuit dependent on the "slew rate" of test voltage (dV/dt). The impact of such currents on the measuring signals might be different, dependent on the decoupling method for the PD signals. According to Chapter 4 there are different options for that coupling: direct, capacitive, inductive, and EM (antenna) methods. For each of them there are different ways for suppression of disturbance signals caused by the impulse generator and mainly transferred as displacement current pulses. A typical example for such disturbance signal is depicted in Figure 8.8, which was measured in the connection lead from decoupling unit to the measuring device. It can be seen that the noise signal occurs during the front time of the voltage test impulse, which is characterized by maximal gradient of test impulse voltage $(dV/dt)_{max}$. For PD detection, it is necessary to suppress or at least minimize such noise signals.

The suppression of noise signals at classical impulse voltage seems to be more difficult because each impulse is a single event and not part of a pulse train, as is the case at repetition pulses e.g. generated by power electronic components. In the latter case, the noise signals could be suppressed by special filter techniques and/or ASP procedures (e.g. correlation methods). At classical impulse voltage, one option for noise suppression might be realized by special electrode design to avoid the displacement (noise) current on the measurement path. As an example, for measurement of PD current signal at a basic arrangement (tip-plane) for an oil-barrier insulation stressed by lightning impulse voltage the tip

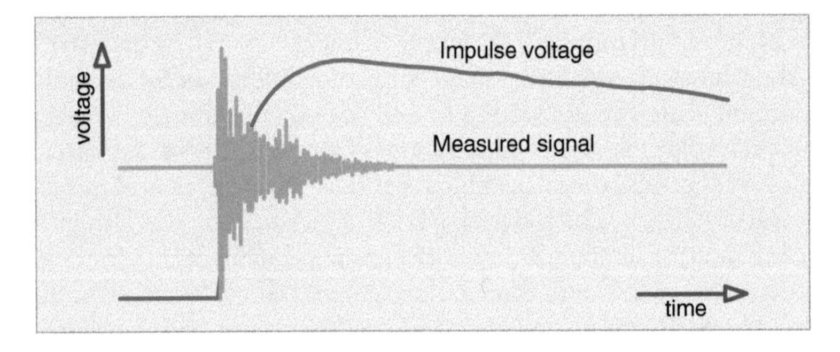

Figure 8.8 Measured signal via coupling unit superimposed with disturbances caused by the impulse voltage generator.

electrode is designed with two parts – guard and sensing part (Figure 8.9). With such a design, it is possible under the given conditions to minimize the displacement (noise) current and significantly increase the signal-to-noise (S/N) ratio. Note: This technique was applied in the past at basic investigations of dielectric behavior of gaseous, liquid, and solid insulation [29–31].

Due the described requirements (e.g. special electrode design), that method is mainly preferred for basic investigations performed in research institutions. Obtaining the PD current signal, other PD quantities like apparent pulse or accumulated charge are also available if appropriate post-processing procedures such as time integration or summarizing are provided.

PD activity starts after the maximum of test voltage – i.e. after the "disturbance" period (Figure 8.9). This might also be used for suppression procedure, for example, if the measuring path (coupling unit + measuring device) will be "opened" only after the voltage maximum e.g. by an electronic switch synchronized with the test impulse voltage signal (see also Chapter 4 on gating). That PD measuring technique is applied at oscillating impulse testing, especially for onsite PD testing of medium voltage power cable [32]. However, a

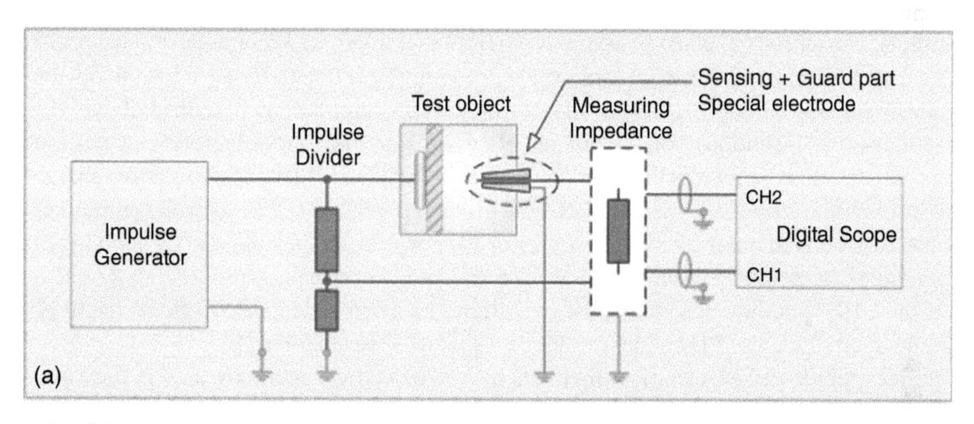

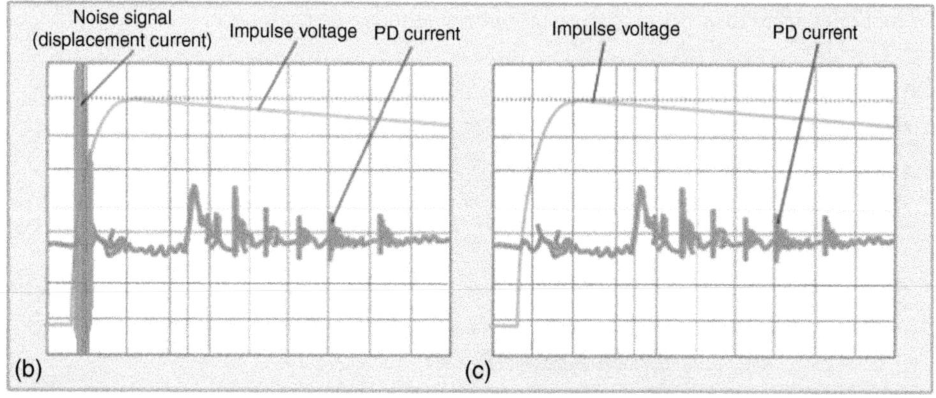

Figure 8.9 Measured PD signal at lightning impulse voltage without and with special electrode configuration: (a) PD test circuit; (b) PD current without guard part; (c) PD current with guard part electrode.

possible lack of information of the whole PD activity should be consider, if during the "disturbance" time slot some PD signals occur.

Commonly introduced for PD measurement of encapsulated electrical equipment like GIS or power transformers under AC testing procedure, also the UHF measurement technique could be applied even under the given impulse voltage conditions. That is possible if the frequency spectrum of test impulse voltage inclusive disturbances (displacement currents, electromagnetic interferences) differs compared with the spectrum of PD signals as it is schematically shown in Figure 8.10, where the PD spectrum exceeds the noise spectrum at a critical frequency f_{crit}. The typical value of f_{crit} is in the range of a few hundred MHz, whereas the PD spectrum extends up to at least 1 GHz. The PD activity might be detected if the PD detector is equipped with a bandpass filter having a lower cutoff frequency above f_{crit}.

Without going in detail (see Chapter 5), at such kind of UHF measurement and dependent on the applied filter characteristic, the full information about the PD pulse might not be obtained. Therefore, those methods allow qualitative PD measurement like PD detection only, but for quantitative measurement like evaluation of charge values, etc. they are not appropriate; nevertheless, it is very useful to measure PD under such conditions. The required equipment for that measurement could be realized by different options – for example, capacitive (coaxial) or inductive (HFCT) couplers with connected appropriate filters, EM couplers, or UHF antennas. The measurable PD parameter depends on the applied coupling option. In case of measuring current impulses as described above, the instantaneous (inception) voltage at first PD event and via time integration of current impulse the pulse (apparent) charge could also be obtained. Likewise, the pulse charge might be summarized to a "accumulated pulse charge" related to the test voltage impulse event. Also derived quantities as PD energy or PD power could be obtained by applying of adequate postprocessing procedure.

If the UHF technique is applied, the repetition rate of the detected PD signals might be obtained, but the pulse PD charge cannot be evaluated (see Chapter 5).

Besides the described electrical methods also nonelectrical methods for PD measurement under impulse voltage conditions like optical or ultrasonic measurement are applied. For more details to those kind of measurement methods, see Chapter 6.

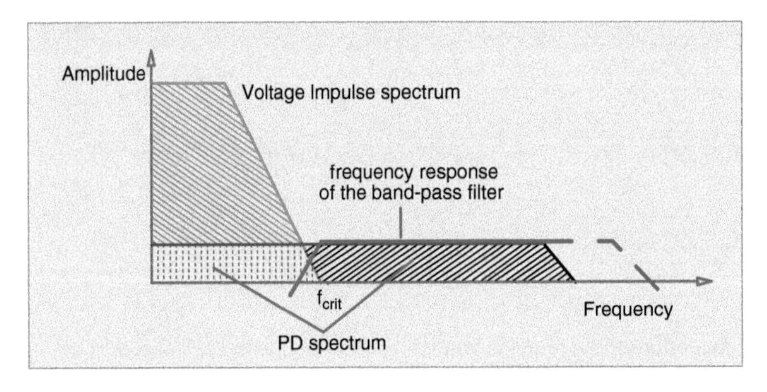

Figure 8.10 Frequency spectra of impulse voltage and PD pulse (schematically).

8.3.2 PD Measurement at Repetitive Pulse Voltage Conditions

As already mentioned, a novel kind of electrical stress on insulation is generated by applied power electronic devices especially in connection with rotating machines or similar applications fed by pulse width modulation (PWM) inverters. That technique is based on fast solid-state switching operations and might be characterized by a repetitive voltage pulse train with very short rise times (switching edges) and/or large slew-rates at pulse frequencies up to some tens of kHz (Figure 8.11).

Those voltage pulses occur at the terminals of fed equipment and might lead especially at winding insulation to overvoltage and oscillations distributed over the winding turns. In the case of excited PDs by such phenomena, the exposed insulation will be accelerated aged and an unexpected outage must be considered.

Therefore, the recognizing and assessment of PDs under such conditions is more and more required. The difficulties and issues at such PD measurement are similar, as already described for classical impulse conditions (see Section 8.3.1), namely the effective suppression or, at least, minimizing of EM disturbances generated by the impulse voltage source gathering a sufficiently high signal-to-noise (S/N) ratio. The new challenge compared with Section 8.3.1 is the repetitious character of the stress voltage (pulse train), including disturbances combined with extremely fast rise and decay times of pulse signals, respectively. The repetitive noise signals allow the application of adaptive filter technique as well as ASP procedures – for example, neural network methods often combined with wavelet algorithms [33, 34].

All these procedures need a certain time, often termed as "training" or "learning" time, to create or adapt effective algorithms for eliminating or, at least, minimizing the noise in relation to the measured signals. It should be noticed that for such purposes, fully digitized measuring systems with adequate postprocessing units and algorithms respectively are required.

For decoupling of PD signals from the measuring circuit, the already discussed options like direct, capacitive, inductive, and EM coupling methods are usable. Especially for the first three methods, the decoupled signal must be filtered to recognize the desired PDs and suppress the noise signals. As an example, a typical PD measurement on a PWM-fed

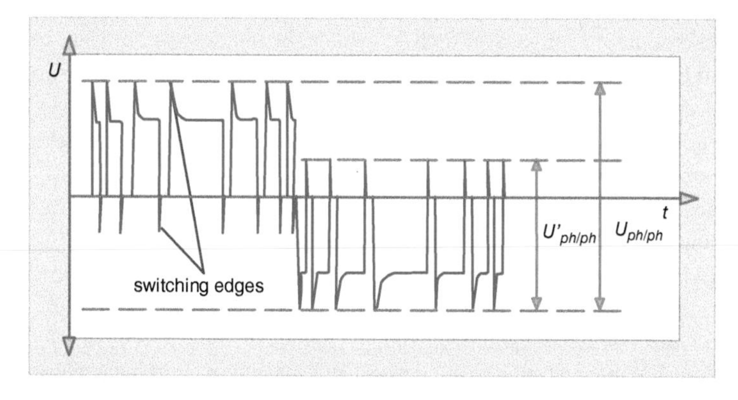

Figure 8.11 Phase/phase voltages ($U_{ph/ph}$) at the terminals of a machine fed by a PWM converter (schematically).

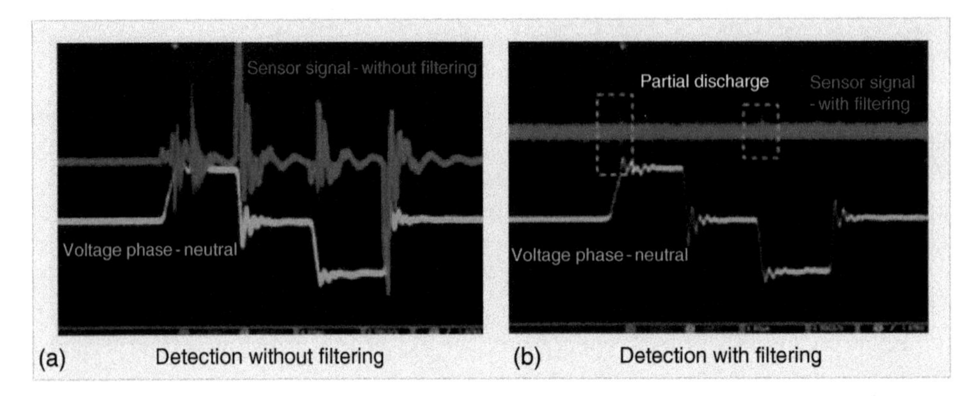

Figure 8.12 PD detection at a PWM fed motor measured via a capacitive sensor and connected filter: (a) detection without filtering; and (b) detection with filtering (pink: decoupled signal, green: terminal voltage) [35].

motor [35] underlines the necessity of filtering. This is also demonstrated in Figure 8.12 at the detected signal decoupled by a capacitive (coaxial) sensor and with and without appropriate filter [35].

In case of applied UHF technique and by PD detection with EM coupling the filter is already provided by the limited bandwidth of the applied antenna, which means that no signals out of the measurable frequency range of the antenna will be detected. According to design and performance of the applied antenna, the measurable frequency and, therefore, the detectable part of PD spectrum is in the range of about 0.5–3 GHz. A typical arrangement for such a measurement is depicted in Figure 8.13a, at which the PD emission of a sample is detected if the sample (twisted pair) is stressed by an electronic converter generating PDs [36].

Usually, the received signal must be elaborated by some postprocessing procedures to separate the noise and the PD signal. In the described example here, the noise was subtracted from the received signal by appropriate postprocessing [36]. The results depicted in Figure 8.13b show clear separated signals (noise, PD) after that procedure. But some issues at this type of measurement should be noticed. At first it can be seen that the noise spectrum has certain parts in frequency range over 1 GHz, which means that in contrast to Figure 8.10, the noise and the PD spectra are not significantly separable. That could be an obstacle for the choice of appropriate filters. Second, the magnitude of detectable PD signals is influenced by geometric parameters as type of applied antenna including distance to the sample as well as its size with respect to volume. This fact leads in many PD measurements to additional application of nonelectrical methods like optical or acoustic measurement to confirm the evidence of PDs in the investigated test object.

In this context it should be noticed that the evaluation of PD inception and in the same manner the estimation of the partial discharge inception voltage (PDIV) is one of most difficult issues, especially in case of repetitive impulse voltage conditions. Besides the impact of statistical time lag as was discussed in previous chapters, this chapter provides evidence of strong noise environment, which might lead to an ambiguous situation at detected signals – noise or PD. For this reason, some publications recommend a special defined

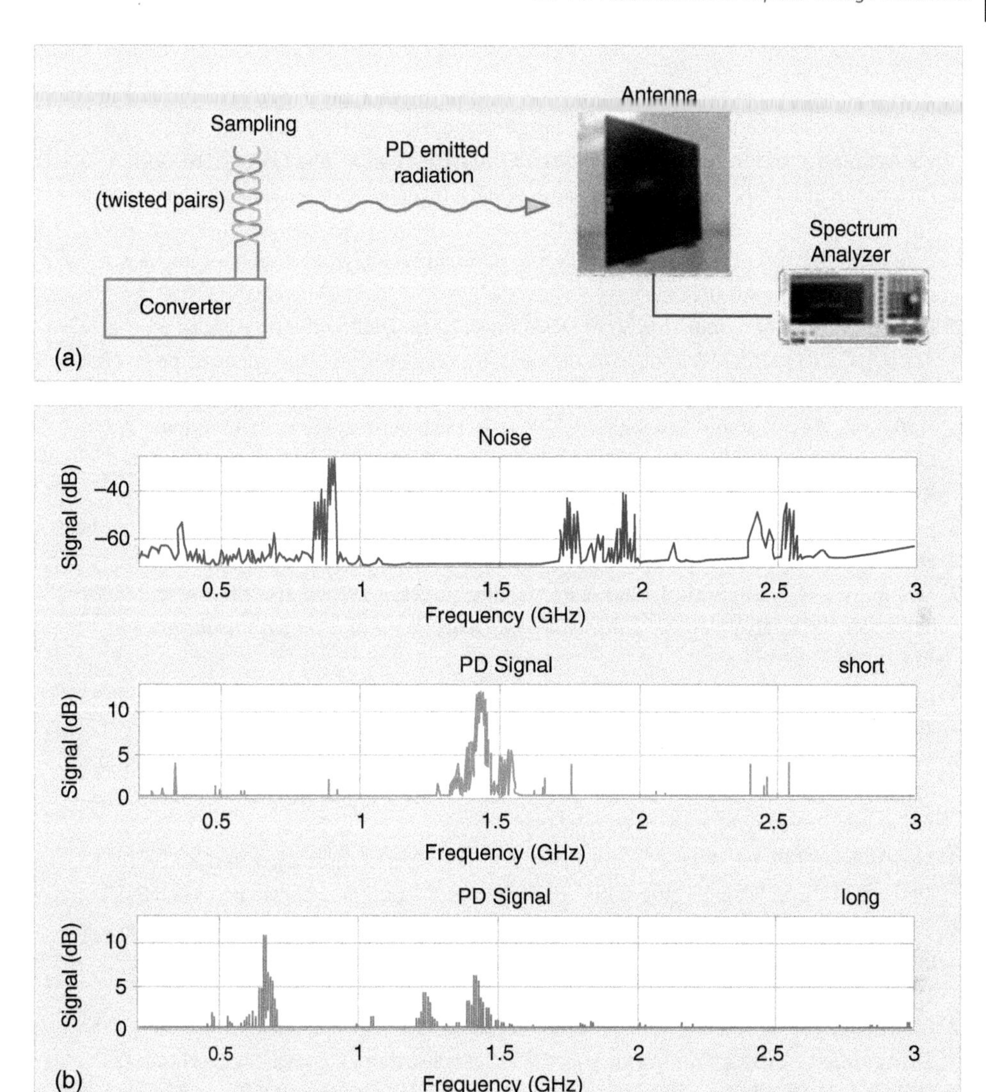

Figure 8.13 PD detection at a winding sample with different length (twisted pairs – long, short) fed by SiC-inverter and measured via a log-periodic antenna and a spectrum analyzer: (a) Measuring arrangement; (b) detected signals [36].

inception voltage – the so-called repetitive partial discharge inception voltage (RPDIV) [37, 38]. The procedure for evaluation of RPDIV is not unified – some authors prefer a certain and various number of detected PD pulses in relation to the applied pulses, so this topic is not further examined in this book.

In conclusion, measurable parameters are mainly the magnitude and the repetition rate of the detected PD signals, including the difficulties at evaluation of PD inception. Further PD quantities such as pulse charge or another derived parameter cannot be evaluated (see Chapter 5).

References

1. Moriyasu, S. and Okuyama, Y. (1999). Surge propagation of PWM-inverter and surge voltage on the motor. *IEEE Transactions on Industry Applications* 119 (4): 508–514.
2. International Electrotechnical Comission (2008). *IEC 60034-18-42: Rotating Electrical Machines – Qualification and Acceptance Tests for Partial Discharge Resistant Electrical Insulation Systems (Type II) Used in Electrical Machines Fed from Voltage Converters.* Geneva, Switzerland: IEC.
3. Kreuger, F.H. (1995). *Industrial Hgh DC Voltage.* Delft: Delft University Press.
4. Trichel, G.W. (1938). Mechanism of the negative point-to-plane corona near onset. *Physical Review* 54: 1078.
5. Haller, R., Pihera, J., and Svoboda, M. (2014). PD behavior at AC and DC voltage. In: *Conference on Condition Monitoring and Diagnosis (CMD)*, OC 3-01. Jeju, South Korea: IEEE.
6. Fromm, U. (1995). Partial discharge and breakdown testing at high DC voltage. PhD thesis. Delft University of Technology.
7. Morshuis, P.H.F. and Smit, J. (2005). Partial discharges at DC voltage: their mechanism, detection and analysis. *IEEE Transactions on Dielectrics and Electrical Insulation* 12 (2): 328–340.
8. International Electrotechnical Commission (2000). *IEC 60270: High-Voltage Test Techniques – Partial Discharge Measurements.* Geneva, Switzerland: IEC.
9. Van Brunt, R.J. and Misakian, M. (1980). Comparison of DC and 60 Hz AC positive and negative partial discharge inception in SF_6. *Conference on Electrical Insulation & Dielectric Phenomena - Annual Report 1980* 461–469.
10. Morshuis, P.H.F., Jeroense, M., and Beyer, J. (1997). Partial discharges part XXIV: the analysis of PD in HVDC equipment. *IEEE Electrical Insulation Magazine* 13 (2): 6–16.
11. Cavallini, A., Montanari, G., Tozzi, M., and Chen, X. (2011). Diagnostic of HVDC systems using partial discharges. *IEEE Transactions on Dielectrics and Electrical Insulation* 18 (1): 275–284.
12. International Electrotechnical Commission (2001). *IEC 61378-2, 2001: Converter Transformers – Part 2: Transformers for HVDC Applications.* Geneva, Switzerland: IEC.
13. Shihab, S. (1972). Teilentladungen in Hohlräumen von Polymeren Isolierstoffen bei hoher Gleichspannung (in German). PhD thesis. Technical University Braunschweig.
14. Hauschild, W. and Lemke, E. (2014). *High-Voltage Test and Measuring Techniques.* Berlin Heidelberg: Springer-Verlag.
15. International Electrotechnical Commission (2016). IEC 60270: 2000 + AMD1: High-voltage test techniques – Partial discharge measurements.
16. Fromm, U. and Gulski, E. (1994). Statistical behavior of internal partial discharges at DC voltage. *Proceedings of 1994 4th International Conference on Properties and Applications of Dielectric Materials (ICPADM)*, Brisbane (AU), 670–673. IEEE.
17. Hoof, M. and Patsch, R. (1994). Analyzing partial discharge pulse sequences: A new approach to investigate degradation phenomena. In: *IEEE Symposium on Electrical Insulation*, 327–331. Pittsburgh: IEEE.
18. Roy, N.B. and Bhattacharya, K. (2020). *Application of Signal Processing Tools and Artificial Neural Network in Diagnosis of Power System Faults*, 1e. Boca Raton, FL: CRC Press.

19 Pihera, J., Hornak, J., Kupka, L. et al. (2018). Modified pulse sequence analysis for PD measurement at DC. In: *IEEE 2nd Int. Conference on Dielectrics*, 101–104. Budapest: IEEE.

20 Pirker, A. (2020). Messung und Darstellung von Teilentladungen bei Gleichspannung zur Identifikation von Defekten gasisolierter Systeme (in German). PhD thesis. Graz University of Technology.

21 Hoof, M., Freisleben, B., and Patsch, R. (1997). PD source identification with novel discharge parameters using counterpropagation neural network. *IEEE Transactions on Dielectrics and Electrial Insulation* 4 (1): 17–32.

22 Pirker, A. and Schichler, U. (2019). Application of NoDi* pattern for UHF PD measurement on HVDC GIS/GIL. In: *5th Int. Conf. on Condition Monitoring, Diagnosis and Maintenance.* Bucharest, Romania: Graz University of Technology.

23 Ardila-Rey, J.A., Cerda-Luna, M.P., Rozas-Valderrama, R.A. et al. (2020). Separation techniques of partial discharges and electrical noise sources: a review of recent progress. *IEEE Access* 8: 199449–199461.

24 Hochbrückner, B. Spiertz, M., Zink, M. et al. (2019). Digital filtering methods for interferences on partial discharges under DC voltage. *XXIst International Symposium on High Voltage Engineering (ISH)*, Budapest. Springer. Cham.

25 Pattanadech, N., Nimsanong, S., Potivejkul, S., et al.2015. Partial discharge classification using probabilistic neural network model. *18th International Conference on Electrical Machines and Systems (ICEMS)* Pattaya City, Thailand. IEEE.

26 Ganguly, B., Chaudhuri, S., Biswas, S. et al. (2021). Wavelet kernel-based convolutional neural network for localization of partial discharge sources within a power apparatus. *IEEE Transactions on Industrial Informatics* 17 (3): 1831–1841.

27 Koltunowicz, W. and Plath, R. (2008). Synchronous multi-channel PD measurements. *IEEE Transactions on Dielectrics and Electrical Insulation* 15 (6).

28 Hochbrückner, B. Spiertz, M., Zink, M., et al. (2019). Comparison of algorithms for clustering of partial discharge signals under DC voltage. *2nd International Conference on High Voltage Engineering and Power Systems (ICHVEPS)*, Bali. IEEE.

29 Lemke, E. (1967). Breakdown mechanism and characteristics of non-uniform field electrode configuration in air for switching surge (in German). PhD thesis. Dresden, Technical University.

30 Hauschild, W. (1970). About the breakdown of insulating oil in a non-uniform field at SI voltages (in German). PhD thesis. Dresden, Technical University.

31 Densley, R.J. and Salvage, B. (1971). Partial discharges in gaseous cavities in solid dielectrics under impulse voltage conditions. *IEEE Transactions on Electrical Insulation* EI-6 (2): 54–62.

32 Omicron (n.d.) MPD 800: Universal partial discharge measurement and analysis system, www.omicronenergy.com/MPD800 (technical data).

33 Satish, L. and Nazneen, B. (2003). Wavelet-based denoising of partial discharge signals buried in excessive noise and interference. *IEEE Transactions on Dielectrics and Electrical Insulation* 10 (2): 354–367.

34 Cusidó, J., Rosero, A., Romeral, L. et al. (2006). Fault detection in induction machines by using power spectral density on wavelet decompositions. In: *IEEE 37th Power Electronics Specialists Conf.* Jeju, South Korea: IEEE.

35 Billard, T., Abadie, C., and Lebey, T. (2017). Recent advances in on-line PDs detection in power conversion chains used in aeronautics. In: *IEEE Proc. Workshop Elect. Mach. Des., Control Diagnosis*, 281–289. Nottingham: IEEE.

36 Rumi, A., Righetti, L., Cavallini, A., and Seri, P. (2022). Electromagnetic UHF Emission of Motor Turn-Turn Insulation Samples fed by SiC-Converter. In: *IEEE Conf. on Electr. Insul. and Dielectric Phenomena (CEIDP)*, 1–4. Denver, CO: IEEE https://doi.org/10.1109/CEIDP55452.2022.9985312.

37 Cavallini, A., Fabiani, D. Hiroshi, H. et al. (2017). Insulation degradation under fast, repetitive voltage pulses. *CIGRE TB 703, WG D1.43*

38 International Electrotechnical Commission (2011). *IEC 61934 ED3/TS: Electrical Insulating Materials and Systems – Electrical Measurement of Partial Discharges (PD) under Short Rise Time and Repetitive Voltage Impulses*. Geneva, Switzerland: IEC.

9

Monitoring of PD Behavior

9.1 Introduction

Partial discharge (PD) measurement is accepted worldwide as a technique to evaluate the insulation integrity of high-voltage equipment. In a high-voltage equipment manufacture, PD measurement is used for assessing the quality of the new insulation system by which the localized defects can be detected. Then such equipment has to be transported and assembled on-site. The insulation defect may occur during the transportation and installation process. PD detection can also be applied. Once the high-voltage equipment is operated, the aging process of the insulation system begins. For reliable operation, the condition of aged insulation and the aging process of the insulation of the service high-voltage equipment have to be observed. The presence of PD in the insulation system is one of the significant problems that can lead to the failure of high-voltage equipment. Therefore, the early detection of insulation problems due to PD is essential. Inarguably, PD monitoring is a crucial task that shows an effective way to increase the operational reliability of high-voltage equipment, as reported in [1–4].

9.2 PD Monitoring

PD monitoring is the process of observing a PD behavior occurring in the insulation system. The aim of PD monitoring is to detect any changes in insulation conditions during operation at the very early stage of the change. Generally, PD monitoring can provide localized defect information in the insulation system of most high-voltage equipment types. For rotating machines, PD will provide information on the presence of the aging process and the aging effect as well. To obtain PD signal, the off-line PD measurement and/or the on-line PD measurement need to be performed. Both methods have advantages and disadvantages with respect to one another. In general, the off-line PD measurement is periodically performed. For on-line PD measurement, periodically or continuously PD measurement is possible. Generally, the insulation deterioration needs time to process, depending on the types of insulation material, PD types, the service stresses, and the operating conditions. After detecting the harmful PD signals, maintenance should allow sufficient time to take action to avoid failure.

Partial Discharges (PD) - Detection, Identification, and Localization, First Edition. Norasage Pattanadech, Rainer Haller, Stefan Kornhuber, and Michael Muhr.

9.2.1 Off-Line and On-Line PD Measurement

PD measurement can be divided into off-line and on-line PD measurement. Off-line PD measurement is performed with the high-voltage equipment isolated from the power system. In contrast, on-line PD measurement is conducted when the high-voltage equipment is normally operating. Some important information, especially partial discharge inception voltage (PDIV) and partial discharge extinction voltage (PDEV), is mainly obtained from the off-line PD measurement, whereas other information exhibiting the PD behavior during the operation of high-voltage equipment under the operating stresses is only acquired from on-line PD measurement.

9.2.1.1 Off-Line PD Monitoring

Since high-voltage equipment is isolated from the power system with off-line PD monitoring, a separate power supply is needed. PD measurement is usually performed periodically according to a time-based maintenance strategy. However, if the measurement shows significant insulation deterioration, the off-line PD measurement must be carried out more frequently. Off-line PD monitoring is also conducted to investigate unusual results acquired from any on-line PD monitoring. The advantages and disadvantages of off-line PD monitoring are summarized here.

Advantages:

- It is possible to measure the PD activity at different voltage levels and voltage waveforms.
- Information such as PDIV, PDEV can be obtained, and they are very useful in the assessment.
- Off-line monitoring provides a fingerprint of PD activity, which is a basis for further trend analysis.
- A higher signal-to-noise ratio results in increased sensitivity and selectivity.
- Reliable criteria are provided to support the interpretation of the PD test result and diagnosis.

Disadvantages:

- An external voltage supply is needed.
- Testing is time consuming.
- It is expensive.
- The obtained PD signals are not influenced by the operating condition of high-voltage equipment.
- An outage is required.

Off-line PD testing and monitoring can be performed with various types of voltage waveforms available for various types of high-voltage equipment. Table 9.1 summarizes the off-line – on-site PD test techniques with different voltage waveforms for high-voltage apparatus. Figure 9.1 presents the influence of the test voltage waveforms on PD patterns. The PD measurements were implemented on the solid insulation with an air cavity. Tables 9.2–9.3 illustrate PD activities and the observed insulation deterioration obtained from off-line PD measurement of a rotating machine during 2014–2017.

Table 9.1 Test voltage used on off-line – on-site PD test for various HV equipment.

Equipment to be tested PD on-site		Cables: Oil-paper cables	Cables: Extruded cables MV	Cables: Extruded cables HV	GIS	Instrument transformers	Power transformers	Rotating machines
Very low frequency voltage	VLF	–	PD	–	–	–	–	–
Alternating Voltage (AC)	ACTC	–	–	–	PD	PD	PD	PD
	ACRL	PD	PD	PD	PD	PD	PD	PD
	ACRF	PD	PD	PD	PD	PD	PD	PD
Damped alternating voltage	DAC	PD	[a]PD	[a]PD	–	–	–	PD

Abbreviations: ACTC, transformer circuit for AC voltage generation, ACRL, inductance-tuned resonant circuit for AC voltage generation, ACRF, frequency-tuned resonant circuit for AC voltage generation, MV, medium voltage, HV, high voltage

[a] PD, mainly for PD diagnostics.

Source: modified according to [5].

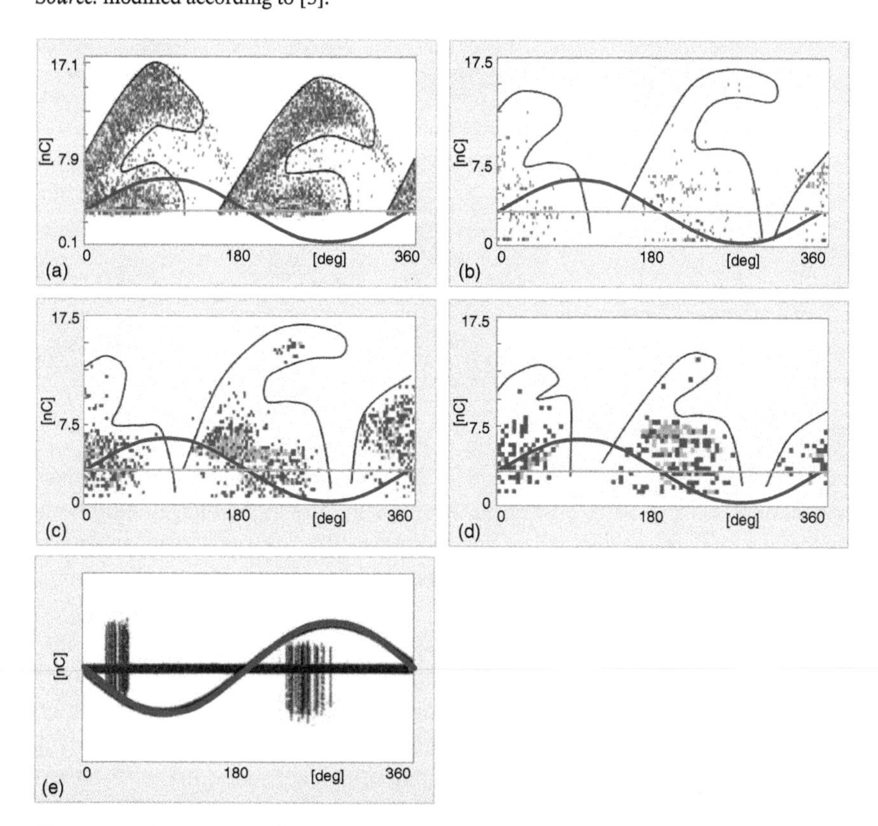

Figure 9.1 Influence of the test voltage waveforms on PD patterns experimented on the solid insulation with an air cavity [6]. Where: (a) two minutes 50 Hz AC voltage testing, (b) 20 shorts 265 Hz DAC voltage testing, (c) 20 shorts 520 Hz DAC voltage testing, (d) 20 shorts 930 Hz DAC voltage testing, (e) three minutes 0.1 Hz VLF voltage testing.

Table 9.2 PD activities obtained from off-line PD measurement [7].

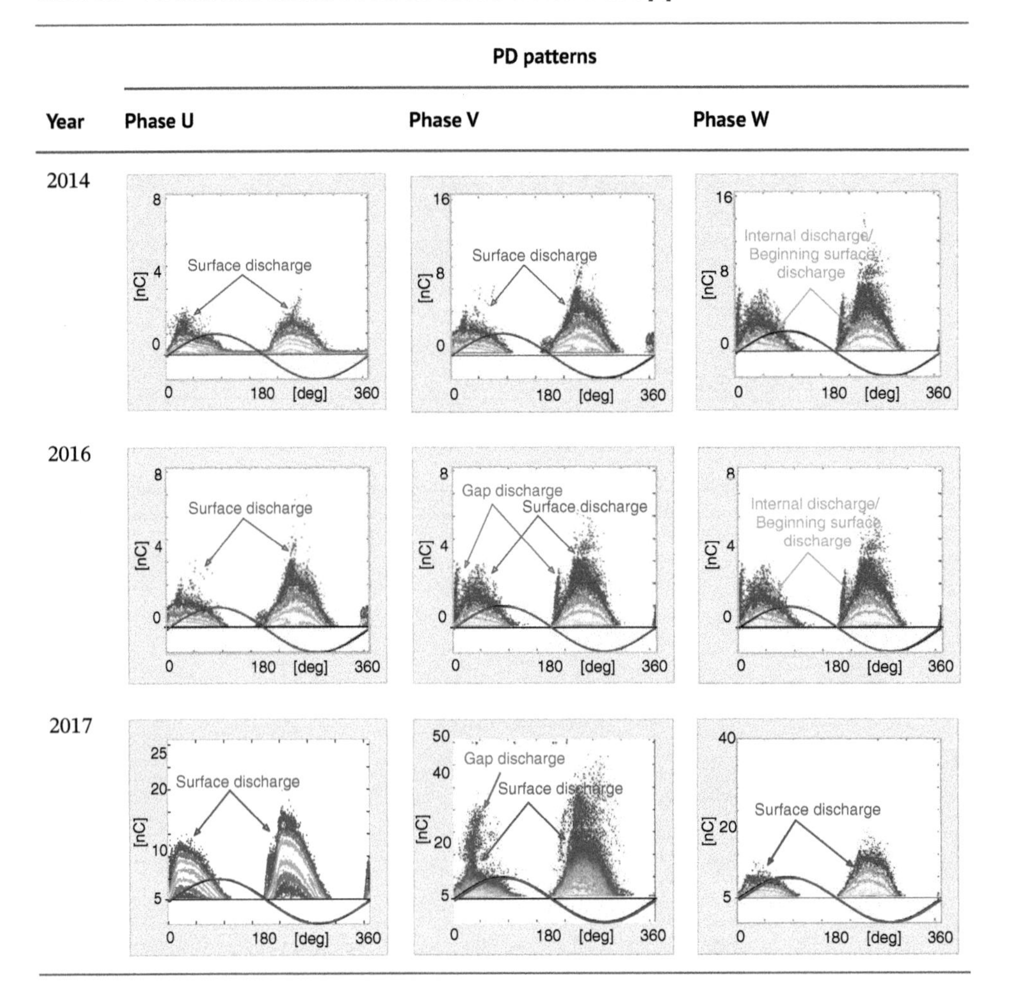

9.2.1.2 On-Line PD Monitoring

On-line PD measurement has gained widespread acceptance. A PD measuring system may be installed permanently. To perform on-line PD monitoring, the high-voltage equipment is operating normally and connected to the power system. The detected PD signals are influenced by the operating stresses – thermal stress, electrical stress, mechanical stress, and atmospheric condition.

On-line PD monitoring can be conducted in two ways, i.e. periodic and continuous monitoring. The periodic on-line PD measurement will be performed on the high-voltage equipment at regular intervals depending on either the time base planning or the condition of the high-voltage equipment. In comparison, the continuous on-line PD measurement will continuously be performed on the high-voltage equipment to acquire the stream PD data. Periodical PD measurement is quite suitable for the case of the slow process of insulation deterioration or PD being the symptom of the failure of the insulation system, such as the PD activity in the rotating machines. Figure 9.2 demonstrates the impact of the thermal stress and the load conditions on the PD behavior of the rotating machines. It is clear that

Table 9.3 Contaminated stator bar insulation and the growing deterioration of the rotating machine associated with the PD test results in Table 9.2 [7].

Year	Physical appearance under inspection
2014	 
2016	 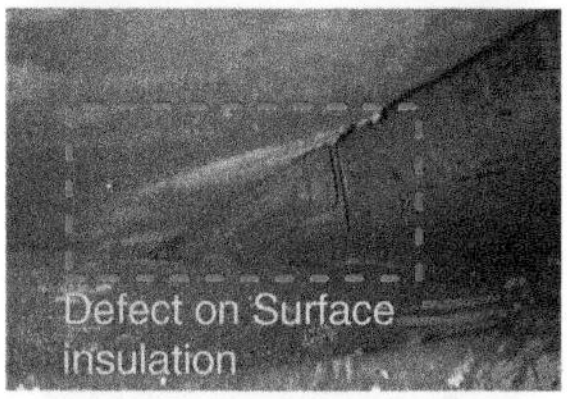
2017	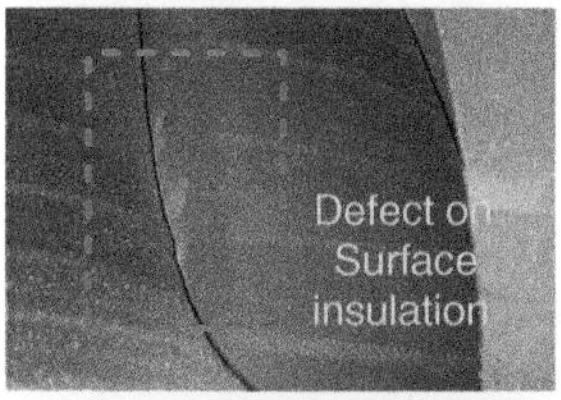 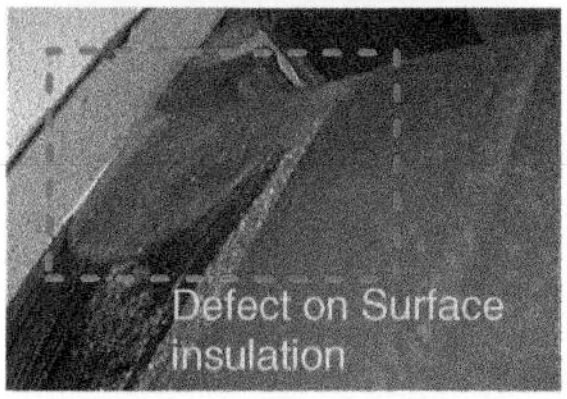

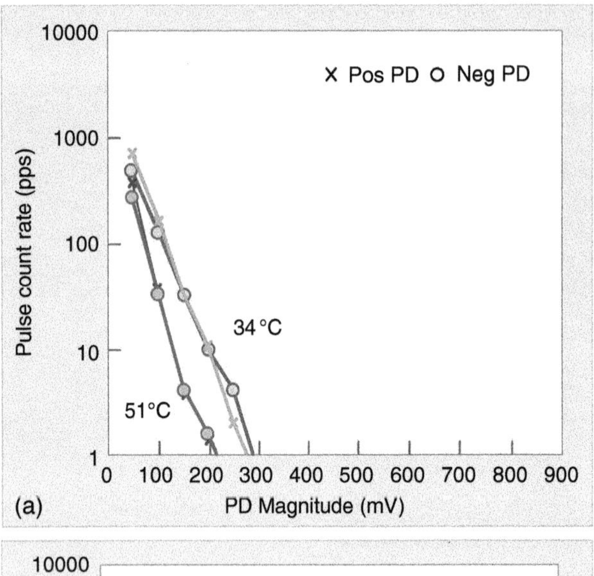

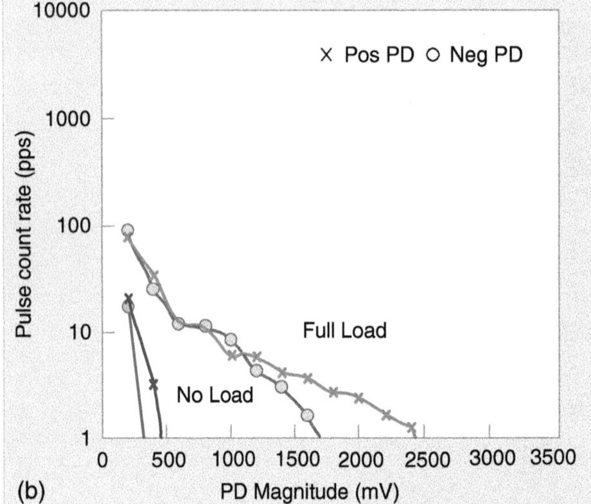

Figure 9.2 The impact of (a) thermal stress and (b) load condition on PD behavior of the rotating machines [8].

PD magnitude depends on the stator temperature and the load of the rotating machine. Figure 9.3 reveals that some PD sources were active with the higher temperature Figure 9.4 presents the on-line PD trend to failure on a 10 kV motor. The PD peak increased from 6.5 to 13.5 nC in one day. Then the PD amplitude was quite steady with a consistent increase for a further nine weeks before the failure of the motor occurred.

Generally, an on-line PD monitoring device comprises a PD sensor and data acquisition. Within the data acquisition unit, a signal transducer converts the PD detected signal to an information stream, and software characterizes the PD signal parameters. Furthermore, special software is employed in the data acquisition unit to monitor and trend the derived PD parameters presented in a display. The monitoring device will generate an automatic warning if preset limits of the derived PD parameters or trends exceed the threshold.

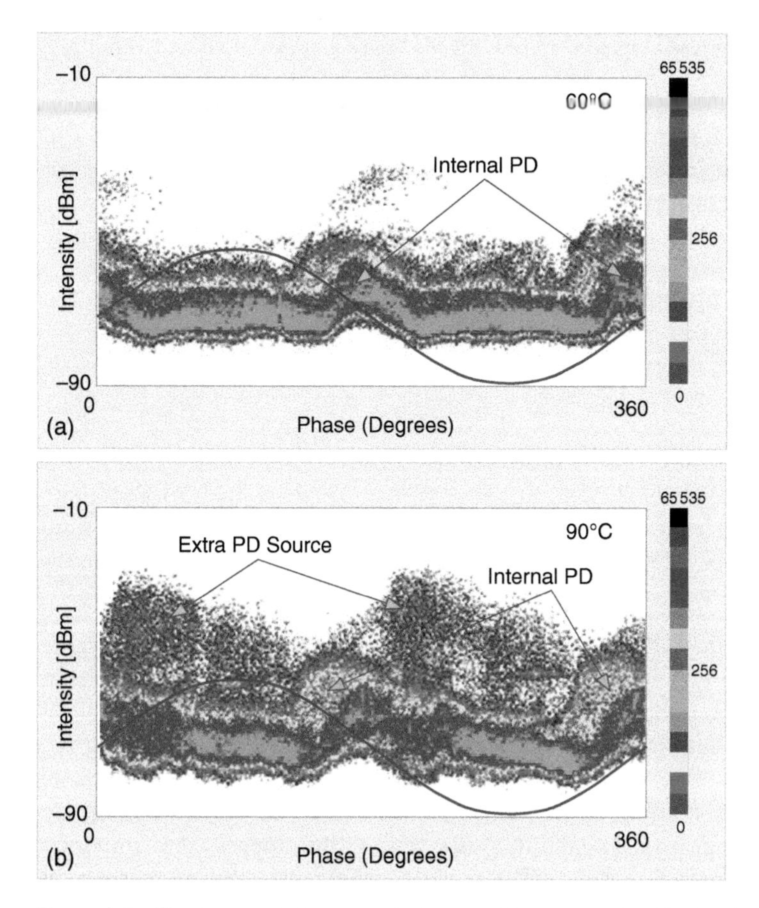

Figure 9.3 The impact of the thermal stress on PD source activation in the insulation system of the rotating machine (a) 60 °C (b) 90 °C [9].

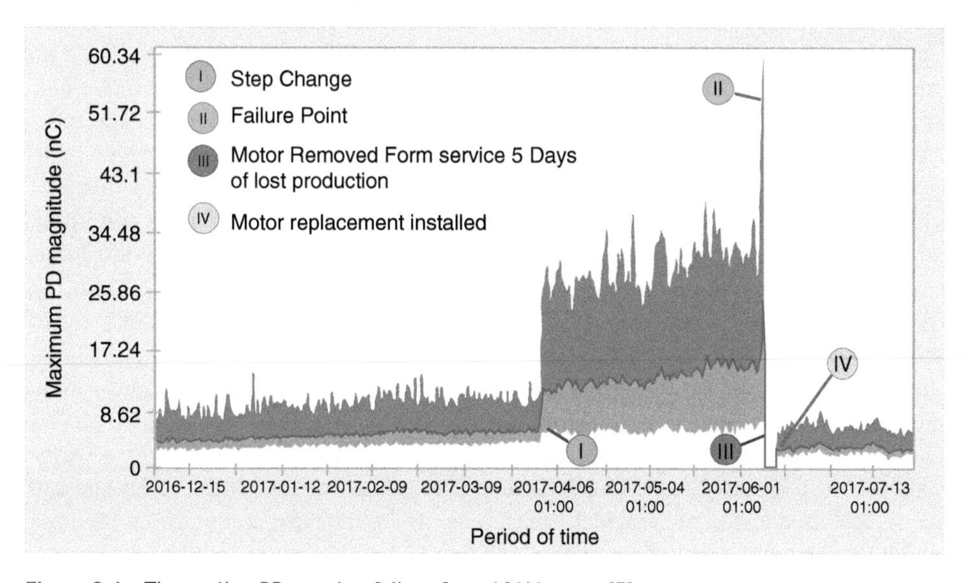

Figure 9.4 The on-line PD trend to failure for a 10 kV motor [3].

The advantages and disadvantages of on-line PD monitoring are summarized as follows:

Advantages:

- It is suitable for a power plant located remote site difficult to access.
- The risk is reduced for in-service high-voltage equipment operated in a hazardous gas (Ex/ATmosphères EXplosible, or ATEX) zone or other hazard areas.
- It is easier to trend PD data over time to achieve real-time condition assessment, which allows detecting any changes in PD activity.
- The initial fingerprint PD of the machine in operating condition is provided.
- Because there are no external power sources, there is no service interruption.
- It is cost-effective.

Disadvantages:

- Noise and disturbances may impact the detected PD signal.
- It may not acquire PDIV and PDEV.
- Different PD test result analyses may be obtained from the same PD source when the on-line PD monitoring uses different PD sensor types or different PD analysis and diagnosis techniques.

Obviously, noise and disturbance occurring in the on-line PD monitoring have more impact on the PD test result analysis; however, with the advanced analog circuit and software algorithms, the noise and disturbance do not seem like an insurmountable problem. Figure 9.5 represents on-site PD measurement with phase-resolved PD (PRPD) pattern and noise separation.

Not only periodic and continuous on-line PD measurement, CIGRE working group for gas-insulated switchgear (GIS) condition assessment also proposes the temporary continuous PD measurement, which offers some advantages as the comparison in Table 9.4.

Currently, on-line PD measurement is by far developed. The experience gained from PD monitoring is now increasing steadily. However, the PD diagnosis obtained from only on-line PD measurement is not enough to assess the insulation condition. The data acquired from the off-line PD measurement, in particular PDIV and PDEV, are also essential to combine with stream on-line PD data measurement to yield the high reliability of the PD diagnosis. Moreover, other information such as the insulation type relating to the PD mechanism, the load profile, the dielectric test results from different techniques, and so on are needed to elevate the accuracy of the insulation condition. The data from the on-line PD monitoring is regularly updated based on the latest information from the on-line monitoring, while the PD data from the off-line PD monitoring will be updated only when new data is available, e.g. following the next scheduled testing.

To analyze and diagnose the same PD problem, various types of on-line PD sensors and technologies utilized for PD acquisition and analysis may not produce comparable PD diagnostic results, which have different interpretations of the defect type or severity. Therefore, the users necessitate understanding the accuracy of the PD test technique and analysis tool used to determine which measurements are the most accurate; and then use the more accurate test results for a critical decision.

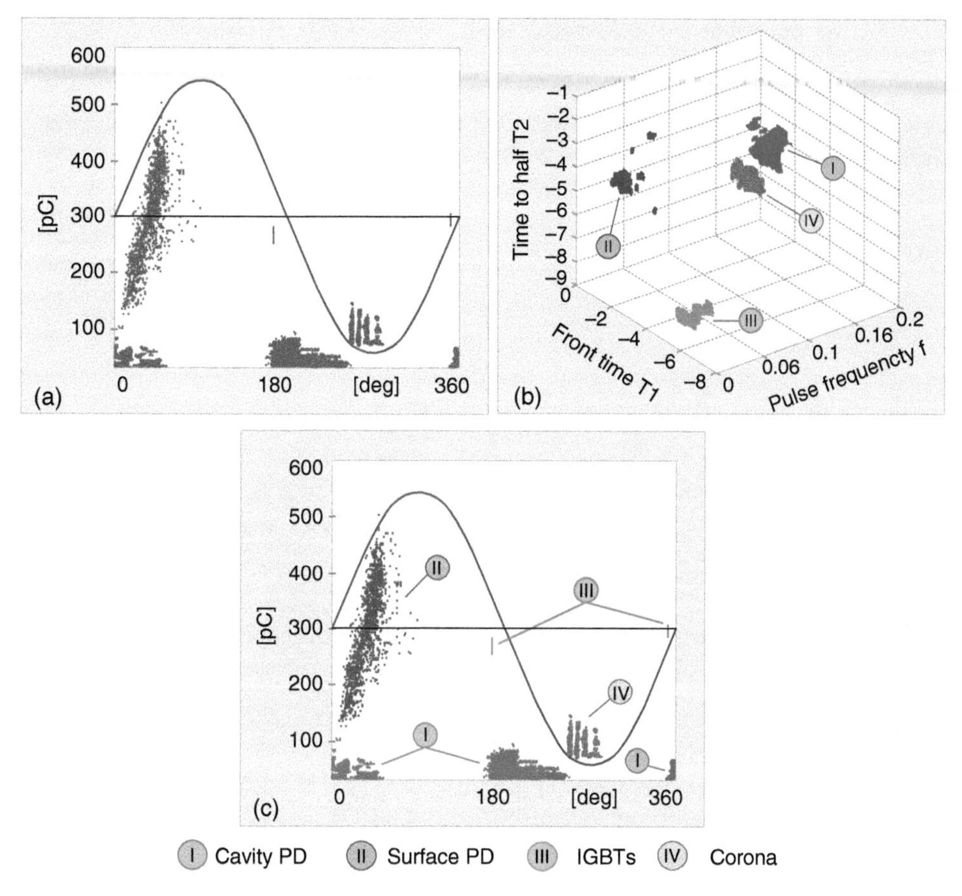

Figure 9.5 PRPD patterns: (a) PRPD pattern without noise separation, (b) identification of PD cluster, (c) PRPD pattern after applying noise separation algorithm [10].

9.2.2 PD Monitoring System

A PD monitoring system is needed to fulfill the estimation of the insulation condition. Generally, PD monitoring is composed of at least a PD sensor or multi-PD sensors and a data acquisition unit. A PD monitoring device may be installed permanently on the high-voltage equipment allowing the user to access the periodic PD data. Besides, the stream PD data may be sent to the PD analysis center for diagnosis and prognosis with the expert system. This kind of monitoring generally notifies users if a threshold value is reached. Details of the PD monitoring components are described in this section.

9.2.2.1 PD Sensor

There are many types of PD sensors used for the PD monitoring device. These sensors are usually permanently installed, either on or in high-voltage equipment. The PD sensors should be suitable and easily installed with the high-voltage equipment target. These sensors must not be the cause of the failure of high-voltage equipment. In principle, PD signals can be detected from conducted or electromagnetically radiated pulses generated by a PD source. Different PD

Table 9.4 Overview of different time-based strategies for on-line PD measurement.

Time strategy	Description	Advantages	Drawback	Time scale	Relative cost
Periodic	Representing a snapshot in time on a limited number of simultaneously connected location	• Flexible use	• No PD diagnosis between two campaigns • Risk of nondetection of intermittent PD • A high level of human expertise required	Some hours – some days	Lowest initial costs (depends on the strategic role of the substation)
Temporary continuous	Representing the development on a few connected locations for a limited time	• Flexible use • Tracking of every discharge occurring in a time window • Time history • Trend analysis • Alarm generation If occurring in a time window • Remote control	• Specific algorithms are required to eliminate noise interference • Expert system software is required	Some days – 1 yr	In the mid-level
Continuous	Representing the development for all locations without limited time	• Tracking of every discharge • Time history • Trend analysis • Alarm generation If occurring in a time window • Remote control	• Specific algorithms are required to eliminate noise interference. • Expert system software is required for diagnosis.	Lifetime	Highest initial cost

Source: Modified according to [11].

sensors or PD detection methods such as capacitive couplers (CCs), high-frequency current transformer (HFCT)s, Rogowski coils (RCs), directional coupler sensor (DCS), and antennas can be used, depending on the expected frequency range of the PD signal to be measured. High sensitivity, modern communication, and cost-effectiveness of PD sensors are also considered. A multimodal PD sensor system may be used to enhance the efficiency of PD monitoring. Generally, on-line PD measurement is performed based on nonconventional PD measurement. Tables 9.5–9.8 summarize the advantages and disadvantages of conventional and unconventional PD measurement methods for high-voltage equipment.

Table 9.5 Comparison of conventional and unconventional PD measurement applied to GIS.

Conventional (IEC 60270)	Unconventional	
	UHF method (antenna)	Acoustic method (AE sensor)
Calibrated in pC	Charge calibration not possible: look CIGRE sensitivity check	Charge calibration not possible
High sensitivity (< 1 pC)	High sensitivity (< 5 pC)	High sensitivity (< 5 pC)
Coupling capacitor ($C_C \ll C_{EUT}$)	Built-in and external radio frequency (RF) sensor	Typically, piezoelectric acoustic energy
Electromagnetic interference (EMI) difficult to suppress	EMI relative unproblematic	No EMI but background vibration
Not possible to locate PD	Allows implementing time of pulse time arrival method for locating PD	Excellent for locating many types of PD (manually of pulse time arrival, auscultatory, and triangulation method)
Type testing, final factory test	On-site commissioning, assessment, monitoring	Useful for both factory and on-site

Source: Modified according to [12].

Table 9.6 Comparison of conventional and unconventional PD measurement applied to transformer.

Conventional (IEC 60270)	Unconventional	
	UHF method (antenna)	Acoustic method (AE sensor)
Calibrated in pC	Charge calibration not possible	Charge calibration not possible
Coupling capacitor ($C_C \ll C_{EUT}$)	Built-in and external RF sensor	Typically piezoelectric acoustic energy
EMI difficult to suppress	EMI relative unproblematic	No EMI but background vibration
Not possible to locate PD	Allows implementing time of pulse time arrival method for locating PD	Allow for locating of PD (manually of pulse time arrival, auscultatory, and triangulation method)
Type testing, final factory test	Useful for on-site monitoring	Useful for on-site monitoring

Source: Modified according to [12].

9.2.2.2 Data Acquisition

PD signals and noise obtained from PD sensors must be separated before the PD signals are amplified and characterized. Then, PD analysis, diagnosis, and even prognosis are performed and communicated to the user as a condition of the insulation. Details of each part of data acquisition are discussed in this section.

Table 9.7 Comparison of conventional and unconventional PD measurement applied to cables.

	Unconventional	
Conventional (IEC 60270)	**HF measurement (use of coupling capacitor and HF coupling devices)**	**HF/VHF measurement (use of HFCT)**
Calibrated in pC	Calibrated readings in pC not possible	Calibrated readings in pC not possible
Difficult to suppress EM interference on-site	Better applicability under on-site conditions	Applicable under on-site condition but strongly influenced by defect position, especially at VHF frequency
High sensitivity in low noise environment	Distributed measurements possible at both sides of the cable	Sensitivity influenced by the measurement frequency and the type and the position of sensors
PRPD reference patterns are available for different PD defects	PRPD patterns similar to conventional measurements	PRPD patterns similar to conventional measurements
Applicable off-line, not on-line	Applicable off-line and on-line	Applicable off-line and on-line
Not possible to locate PD	With time-domain reflectometry, PD localization possible, for long cable lengths, measurements on both ends of the cable recommended	PD localization possible on the cable accessories, multiple sensors on various locations being needed
Standardized procedures for type testing, factory testing, and on-site testing	No standardized procedure for on-site commissioning, assessment, or monitoring	No standardized procedure for on-site commissioning, assessment, or monitoring. Comparison of different detection systems not possible

Source: Modified according to [12].

Input Amplifier and Noise and Interference Reduction

Under on-site PD measurement, noise and interference are unavoidable. PD sensors are always used to detect the weak PD signal and noise. Therefore, noise cancelation algorithms are a necessity to separate the PD signals from noise. Various types of noise filtering methods such as adaptive digital filtering, program-controlled bandpass filtering, or wavelet denoising detailed in [13] can be employed for this task. The PD signals need then to be amplified before being sent to the signal-processing unit.

Signal-Processing Technique

In this process, the amplified PD signal will be characterized. The PD amplitude, PD polarity, and the number of PD events, including AC phase position, are analyzed. Because the detected PD is not the real PD occurring in the insulation system, the attenuation, and dispersion of the PD pulse during propagation to the PD sensor, including the sensitivity of the PD sensor, should be considered to achieve an actual PD value.

Table 9.8 Comparison of conventional and unconventional PD measurement applied to rotating machine.

	Unconventional	
Conventional (IEC 60270)	**HF measurement (use of capacitive coupler and HFCT)**	**UHF measurement (use of strip antenna)**
Calibrated in pC	Calibrated readings in pC not possible	Calibrated readings in pC not possible
Difficult to suppress EM interference on-site	Applicable under on-site condition but influenced by defect position	Applicable under on-site condition but strongly influenced by defect position
High sensitivity in low noise environment	Sensitivity influenced by the measurement frequency and the type and the position of sensors	Sensitivity influenced by the measurement frequency and the type and the position of sensors
PRPD reference patterns are available for different PD defects	PRPD patterns similar to conventional measurements	PRPD patterns similar to conventional measurements
Applicable off-line, not on-line	Applicable off-line and on-line	Applicable off-line and on-line
Not possible to locate PD	With time-domain, PD localization possible, two sensors on given locations are needed.	PD localization currently not possible
Standardized procedures for type testing, factory testing, and on-site testing	Standardize procedure for on-site, both off-line and on-line, PD measurement using CC	No standardized procedure for on-site, on-line PD measurement. Comparison of different detection systems not possible

Evaluation and Diagnosis

The PD information is generated via the evaluation and diagnostic process. The PD amplitude and intensity, PD trend, the type of PD source, and the PD source location are evaluated to get a degree of insulation failure risk due to PD. A continuous PD monitoring system can be integrated into a supervisory diagnostic system at a remote diagnostic center for the insulation condition assessment. The appropriate communication link is needed and plays an important role. So far, specialized personnel like experts are very important for diagnosing PD data. Additionally, due to the rapid development of high-technology and computer technology, dedicated diagnostic software based on artificial intelligence and knowledge-based expert systems is now available. However, due to various types of PD sensors and diagnosis systems that may apply for PD signal detection and different criteria for PD test result analysis may be performed, the assessment of the PD data should not be comparable. The correct diagnosis for the PD test data obtained from various PD measuring systems requires enormous effort and time. Basically, the doubling rule used for evaluating the risk of PD event is that when the PD quantity doubled within six months (or one year), it shows that the insulation is at great risk of failure. Unfortunately, this rule is not effective with a low-amplitude PD value. Besides, the doubling rule is sometimes also not valid for the high amplitude but stable PD value by which the insulation may suddenly fail without a doubled increase of PD. The principle of PD measurement with direct A/D conversion of the input PD pulses and with integration at a bandpass filter and subsequent A/D conversion is presented in Figures 9.6 and 9.7.

Visualization

PD parameters, i.e. PD magnitude, phase-resolved partial discharge (PRPD) pattern, and other derived PD parameters will be displayed. PD magnitude can be represented in term of apparent charge [pC], voltage [mV], and intensity [dB]. PRPD pattern (or Ø, q, n pattern) is widely used for PD visualization. PRPD is composed of the PD magnitude (q_i) on the ordinate, and the phase angle ($Ø_i$) on the abscissa for each individual of PD pulse, whereas the intensity of PD occurrence (n) within each phase/magnitude window is visualized by a suitable color code whose scale may be visualized by the side of the plot [15]. Figure 9.8 shows the principle of the PRPD analysis (PRPDA) system. Figure 9.9 depicts various types of PRPD patterns. Besides, additional PD parameters, such as the derived PD parameters (the statistical PD quantities) and the trending PD parameter over a given period are manifested.

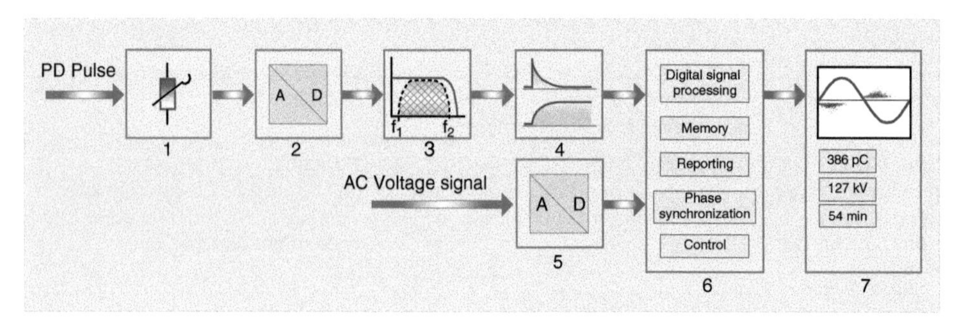

Figure 9.6 PD instrument with direct A/D conversion of the input PD pulses. (1) input PD pulses with attenuator, (2) A/D converter for PD pulses, (3) digital bandpass filter, (4) numerical integrator, (5) A/D converter for AC, (6) an acquisition unit, and (7) and evaluation and visualization unit [14].

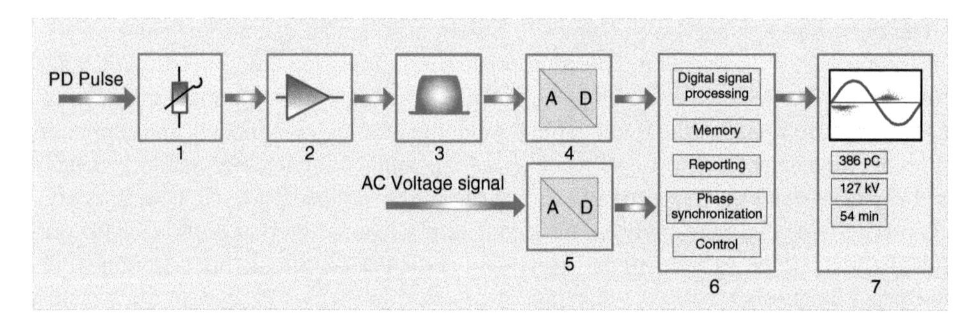

Figure 9.7 PD instrument with integration at a bandpass filter and subsequent A/D conversion. (1) attenuator, (2) amplifier, (3) bandpass filter, (4) A/D conversion PD, (5) A/D conversion AC, (6) acquisition unit, (7) evaluation and visualization unit [14].

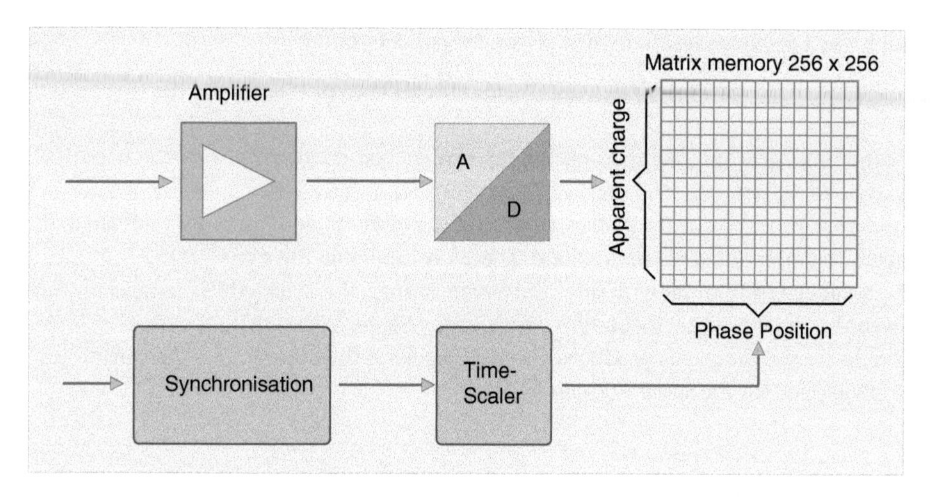

Figure 9.8 Principle of the PRPDA system [16].

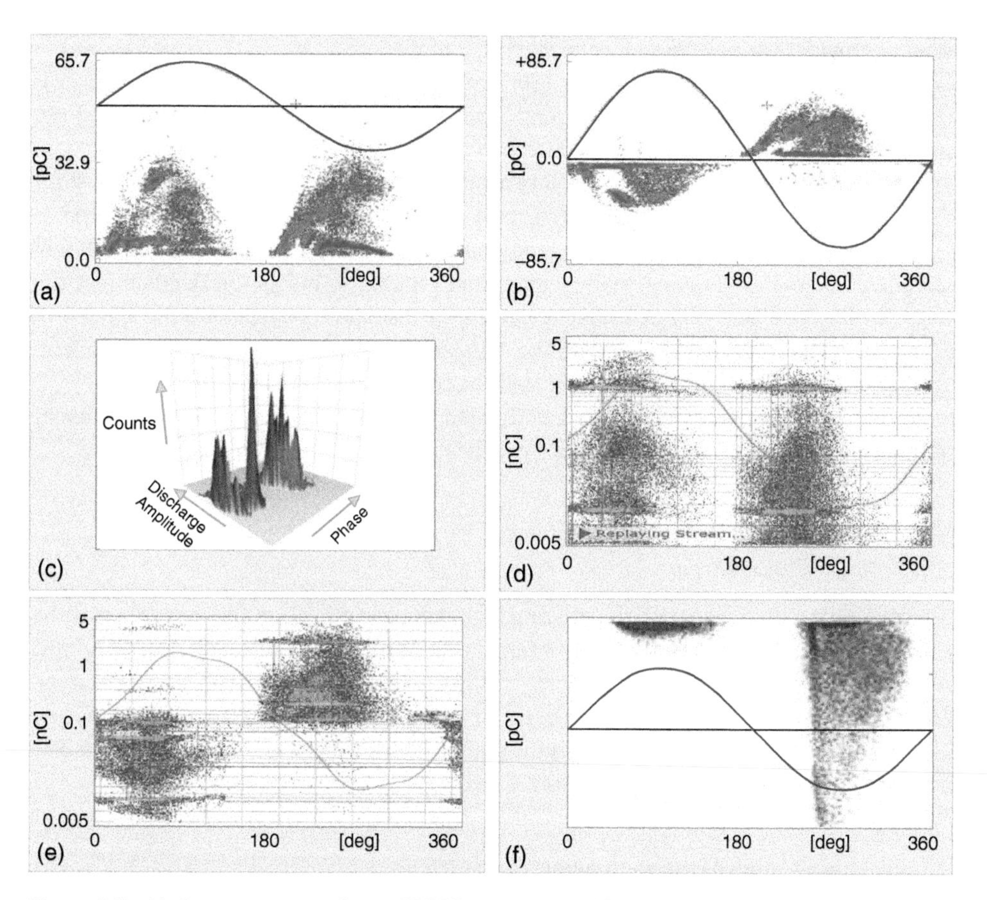

Figure 9.9 Various representations of PRPD patterns: (a) linear unipolar, (b) liner bipolar, (c) three-dimensional, (d) logarithmic unipolar (increasing amplitude from bottom to top), (e) logarithmic bipolar, (f) linear unipolar (increasing amplitude from top-down) [16].

To apply the PD diagnosis efficiently, please keep this in mind:

- PD measurement cannot detect all insulation-related problems in the insulation system [17].
- PD test results are only comparable under similar test conditions with the same PD measurement device.
- The detected PD signal varies to the structure of the high-voltage equipment related to its capacitance and inductance parameters. Therefore, applying the same PD measurement for high-voltage equipment with different power ratings, the detected PD values may differ, even though the actual insulation damage may be the same [18].
- Generally, the PD diagnosis produces different results if the different PD measuring systems are applied for the same PD activity [18].

9.3 Application of PD Monitoring

PD monitoring is now widely accepted to be used with high-voltage equipment. PD measurement provides various types of information, mainly depending on the material used in high-voltage equipment. For purely organic material mainly used in transformers, instrument transformers, underground cable, and so on, PD test result analysis provides the local condition of the insulation system used in such equipment.

The presence of PD can lead to the failure of such high-voltage equipment within a short time. However, insulation deterioration caused by PD is usually a relatively slow process for inorganic material or composite organic material, which is typically used in rotating machines. Therefore, PD is normally a symptom relating to insulation degradation, the presence of aging effects and potential defects in the insulation system, and the aging process of the insulation [17, 18]. Based on the material type used in high-voltage equipment, the interpretation of PD levels and trends may be completely different.

The application of PD monitoring can be applied for the existing high-voltage equipment and the new equipment supporting smart grid technology.

9.3.1 Application of PD Monitoring for the Existing High-Voltage Equipment

Performing PD monitoring with the existing high-voltage equipment, the first choice is the periodic off-line PD measurement that is familiar to most high-voltage engineers. Usually, existing high-voltage equipment has been in service for a period of time in the electrical system, and therefore, the aging process of the insulation is unavoidable. Performing off-line PD measurement, the test voltage level is the point that should give particular attention. For application on-line PD measurement, the following issues need to be considered [19]:

- Importance of the equipment
- Maintenance history of the equipment
- Potential repair time
- Potential cost of the outage
- Potential damage from the failure
- Potential cost of the damage

- Effect on reputation
- Effectiveness of the PD monitoring system adopted
- Availability of monitoring tools and trained personnel
- Cost of monitoring

9.3.2 Application of PD Monitoring for the New Equipment Supporting Smart Grid

Smart grid is aimed to increase the levels of system reliability, efficiency, security, and quality of electrical energy service. A smart grid must have intelligent monitoring and control of the power equipment from distant locations. Accordingly, smart on-line PD monitoring will play an essential role in a smart grid [20–22]. In a smart grid network, equipment must be closely monitored to avoid failure.

A smart PD monitoring supporting the smart grid is rapidly developed due to the evolving and modern technologies in the three main areas:

1) *Evolution of PD detection and communication technology:* New PD sensors with smarter, smaller, and more features are designed to interoperate with smart ubiquitous IoT technology and mobile devices for the advanced function supporting. Smart PD sensors can communicate on-line to a remote location with high efficiency and high reliability.
2) *Evaluation of data analytics and processing technology:* Artificial intelligence (AI) deals with a huge stream of data and data mining, expert knowledge, and back-end information technology systems that can support PD analysis, diagnosis, and prognosis of the condition of the insulation system.
3) *Evolution of high-voltage equipment:* New designs can incorporate built-in sensors and smart devices with remote control.

A smart PD monitoring supporting the smart grid should fulfill the following concepts:

- Supporting large data size obtained from high sampling rate PD detection and multi-PD sensors.
- High sensitivity to detect weak PD, which has a low magnitude and stochastic nature
- High denoising capability, especially the discrimination of PD and noise
- High ability and variety to operate with modern communication devices and systems, especially mobiles and ubiquitous IoT equipment
- Low power consumption

A smart PD monitoring device is toward higher reliability, higher intelligence, higher performance, increasing maintenance intervals, or decreasing maintenance costs and more cost-effectiveness.

Internet of Things (IoT) plays a crucial role in the dramatic development of the smart grid system. IoT also affects the enhancement of a smart on-line PD monitoring device. Principally, the IoT system is composed of IoT devices, IoT gateway, and IoT platform. IoT device is cooperated with a PD sensor to transmit PD signals such as PRPD in real-time. Typically, the small or weak PD needed to be detected at an early stage. Small PD may be rapidly developed to be harmful, requiring a short time for pure organic insulation. To improve the PD measurement, various types of PD sensors and/or multi-PD sensors may be applied. All detected PD signals from each PD sensor can be multiplexed before feeding it to the communication network. The integrated and processed signals can be monitored using the communication links (IoT gateway), which is fast enough to provide information about the dynamic state of the insulation. The data scanning rate and latency for PD on-line and remote condition monitoring of a smart grid should be less than 20 ms. Considering this point, the data monitoring and refreshment rate through the remote terminal unit (RTU) and power line carrier (PLC) system may not be suitable. The smart grid may fail to monitor some dynamic rate of the power device [23]. Then, the PD data from the IoT gateway will be saved and managed in the IoT platform. The PD diagnosis and prognosis will be performed and then alarm the user if the diagnosis result falls within a warning range. IoT system requirements for a smart on-line PD measuring system are as follows: high reliability, high interface utilization, high immunity that can withstand a severely noisy environment, non-intrusive measuring instrument, low power consumption, and cost-effectiveness.

Wireless sensor network (WSN) has become an emerging development in technological advancement within remote monitoring. WSNs technology applications include infrared, Wi-Fi, Bluetooth, IEEE 802.15.4 or ZigBee, and etc. detailed in Table 9.9. Application wireless technology for PD measurement is highly potential. Furthermore, digital signal processing techniques could be utilized and integrated with a WSN to enhance a smart on-line PD monitoring device. A wireless network may be deployed close to high-voltage equipment and automatically organize itself to form an ad hoc multi-hop network for communicating with one another and allowing the dynamic flow of monitor data to support the operation of a smart on-line PD monitoring [22, 24, 25].

Application wireless network for PD monitoring, two issues need to be considered:

1) PD signals may interfere with the WSN communication signals causing poor communication in the WSN network.
2) WSN signal might increase the measured radiometric PD power.

9.4 Challenges for PD Monitoring in Future

There are some challenging issues for the application of PD monitoring:

- Extracting the information from the detected PD signal is needed much more correctly for evaluation of the insulation condition and the remaining lifetime of high-voltage equipment.
- PD measuring and interpreting the PD test result for high-voltage motors fed by a variable-speed pulse width modulation drive are not clearly understood [26–29].
- For most unconventional PD measurements, the application and interpretation of the PD test results are not yet standardized, and the related recommendations or criteria to manage the test results may vary. Accordingly, the benefit of PD diagnosis and prognosis sometimes seems not tangible to users [17, 30, 31].
- Suppose the different PD test methods and analysis produce comparable results, the user's confidence in the data increases. However, various PD measurements and diagnosis techniques are employed in practice and provide different interpretations of the defect type or severity. Users would then need to clearly understand the PD measurement techniques to determine whether measurements and results are the most accurate [17].

Table 9.9 Wireless sensor network application in the industry [25].

Wireless sensor network	Technical summary	Typical radio band	Transmission range	Data rate	Applications
Wi-Fi	It is probably the most widely used wireless local area network (WLAN) technology based on the IEEE 802.11 series of standards.	2.4, 5 GHz	150 m	54 Mbps	Video and monitoring based applications, smart home
Bluetooth	Bluetooth low energy technology is a global standard, which enables devices with coin cell batteries to be wirelessly connected to standard Bluetooth enabled devices and services.	2.4 (v1.x, v4), 5 GHz (v3.0)	10–150 m	1 , 24 Mbps	Remote access, sports and fitness, indoor positioning (HAIP), smart phone based applications
ZigBee (IEEE 802.15.4)	A well-defined protocol stack for WSN with features of self-deployment, low complexity, low data rate and low cost, etc. based on IEEE 802.15.4 standards.	780 (China), 868, 915 MHz, 2.4 GHz	100–300 m	20, 40, 250 Kbps	Smart energy, home automation, building automation, health care, remote control, retail services, etc.
RFID	A fast-developing radio technology used to transfer data from an electronic tag, which includes identification, information collection, etc.	125 kHz (LF), 13.56 (HF), 433 MHz (UHF), 2.4 GHz (MW)	<10 cm, < 1 m, 4–20 m, 60–100 m	1–5 Kbps, 6.62–26.48 Kbps, 40–640 Kbps, 200–400 Kbps	Logistic, E-car license, one pass card
433 MHz enabled proprietary solutions	Proprietary solutions by using one of the most commonly used ISM (industrial, scientific, and medical) radio band in China	433 MHz	300–1500 m	<10 Kbps	Home security, environment monitoring, etc.

References

1 Koltunowicz, W., Badicu, L., Broniecki, U., and Belkov, A. (2016). Increased operation reliability of HV apparatus through PD monitoring. *IEEE Transactions on Dielectrics and Electrical Insulation* 23 (3): 1347–1354.

2 Leibfried, T. (1998). Online monitors keep transformers in service. *IEEE Computer Applications in Power* 11 (3): 36–42.

3 Renforth, L.A., Giussani, R., Mendiola, M.T., and Dodd, L. (2019). Online partial discharge insulation condition monitoring of complete high-voltage networks. *IEEE Transactions on Industry Applications* 55 (1): 1021–1029.

4 Dodd, L., Giussani, R., McPhee, A., and Burgess, A. (2019). Online partial discharge condition monitoring of complete networks: challenges and solutions explained through case studies for the pulp and paper industry. *IEEE Industry Applications Magazine* 25 (5): 18–26.

5 Conseil International des Grands Réseaux Électriques. (2012) High-voltage on-site testing with partial discharge measurement. *Technical Brochure. Ref. 502.*

6 Wester, F.J. (2004). Condition assessment of power cables using partial discharge diagnosis at damped AC voltages. PhD thesis. Delft University of Technology.

7 Phloymuk, N., Phumipunepon, N., and Pattanadech, N. (2018). Partial discharge behaviors of a surface discharge problem of the stator insulation for a synchronous machine. In: *12th International Conference on the Properties and Applications of Dielectric Materials (ICPADM)*, 385–388. IEEE.

8 Lloyd, B.A., Campbell, S.R., and Stone, G.C. (1999). Continuous on-line partial discharge monitoring of generator stator windings. *IEEE Transactions on Energy Conversion* 14 (4): 1131–1138.

9 Conseil International des Grands Réseaux Électriques. (2003) Knowledge rules for partial discharge diagnosis in service. *Technical Brochure. Ref. 226.*

10 Garnacho, F. Sánchez-uran, M. Ortego, J. et al. (2010). Partial discharge monitoring system for high voltage cables. CIGRE Paper B1–306. Paris.

11 Conseil International des Grands Réseaux Électriques. (2017). Benefits of PD diagnosis on GIS condition assessment,. *Technical Brochure. Ref. 674.*

12 Conseil International des Grands Réseaux Électriques. (2016) Guidelines for partial discharge detection using conventional (IEC 60270) and unconventional methods. *Technical Brochure. Ref. 662.*

13 Sriram, S., Nitin, S., Prabhu, K.M.M., and Bastiaans, M.J. (2005). Signal denoising techniques for partial discharge measurements. *IEEE Transactions on Dielectrics and Electrical Insulation* 12 (6): 1182–1191.

14 International Electrotechnical Commission (2015). *IEC 60270:2000+AMDI: High-Voltage Test Techniques – Partial Discharge Measurements*. Geneva, Switzerland: IEC.

15 International Electrotechnical Commission (2017). *IEC 60034: Rotating Electrical Machines – Part 27-1: Off-Line Partial Discharge Measurements on the Winding Insulation*. Geneva, Switzerland: IEC.

16 Conseil International des Grands Réseaux Électriques. (2017) Partial discharges in transformers. *Technical Brochure. Ref. 676.*

17 Stone, G.C. (2012). A perspective on online partial discharge monitoring for assessment of the condition of rotating machine stator winding insulation. *IEEE Electrical Insulation Magazine* 28 (5): 8–13.

18 IEC 60034-27-2 (2012). *Rotating Electrical Machines – Part 27–2: On-Line Partial Discharge Measurements on the Stator Winding Insulation of Rotating Electrical Machines*. Geneva, Switzerland: IEC.

19 Conseil International des Grands Réseaux Électriques. (2013). Guideline to maintaining the integrity of XLPE cable accessories. *Technical Brochure. Ref. 560*.

20 Montanari, G.C. and Cavallini, A. (2013). Partial discharge diagnostics: from apparatus monitoring to smart grid assessment. *IEEE Electrical Insulation Magazine* 29 (3): 8–17.

21 Montanari, G.C., Hebner, R., Seri, P., and Ghosh, R. (2021). Self-assessment of health conditions of electrical assets and grid components: a contribution to smart grids. *IEEE Transactions on Smart Grid* 12 (2): 1206–1214.

22 Gaouda, A.M. (2013). Adaptive Partial Discharge monitoring system for future smart grids. In: *IECON 2013 – 39th Annual Conference of the IEEE Industrial Electronics Society*, 4982–4987. IEEE.

23 Baki, A.K.M. (2014). Continuous monitoring of smart grid devices through multi protocol label switching. *IEEE Transactions on Smart Grid* 5 (3): 1210–1215.

24 Sheng, Z., Mahapatra, C., Zhu, C., and Leung, V.C.M. (2015). Recent advances in industrial wireless sensor networks toward efficient management in IoT. *IEEE Access* 3: 622–637.

25 Mohamed, H. (2018). Partial discharge detection and localization using software defined radio in the future smart grid. PhD thesis. University of Huddersfield.

26 Stone, G.C., Sedding, H.G., and Chan, C. (2016). Experience with on-line partial discharge measurement in high voltage inverter fed motors. In: *Petroleum and Chemical Industry Technical Conference (PCIC)*, 1–7. IEEE.

27 Tozzi, M., Cavallini, A., and Montanari, G.C. (2010). Monitoring off-line and on-line PD under impulsive voltage on induction motors – part 1: standard procedure. *IEEE Electrical Insulation Magazine* 26 (4): 16–26.

28 Tozzi, M., Cavallini, A., and Montanari, G.C. (2011). Monitoring off-line and on-line PD under impulsive voltage on induction motors – part 2: testing*. *IEEE Electrical Insulation Magazine* 27 (1): 14–21.

29 Tozzi, M., Cavallini, A., and Montanari, G.C. (2011). Monitoring off-line and on-line PD under impulsive voltage on induction motors – part 3: criticality. *IEEE Electrical Insulation Magazine* 27 (4): 26–33.

30 Gross, D. (2016). Acquisition and location of partial discharge in transformers. PhD thesis. TU Graz.

31 Hauschild, W. and Lemke, E. (2019). *High-Voltage Test and Measuring Techniques*, 2e. Springer Nature Switzerland AG.

10

Evaluation of PDs

10.1 Introduction

Failures in technical insulations cannot be totally eliminated in spite of quality control, tests, or improvement of production. Such inhomogeneities result in local breakdowns (partial discharges, PD) and start the destruction of insulation materials. These destructions lead to a change of material properties and especially to an aging process. Therefore, it is very important to know not only the dielectric, thermal, and mechanical behaviors of the insulation material but also the reaction of this material to the stress created by PDs. These PDs have a dangerous effect on the insulation properties.

In gas insulation, the stresses by field changes are high, which can lead to PDs. In outdoor systems, PDs are caused by local field increase at metallic sharps or edges. These PDs have more influence to noise, interference, and chemical reactions. But there is a free exchange of the insulating medium (air). In gas-insulating systems (GIS) a regeneration of gas is not possible and reaction products are expected. In GIS, particles are also a great problem because they produce PDs, which reduce the dielectric strength of the gas insulation. Charges can also accumulate at solid insulations.

Solid insulating materials regarding PD are divided in two groups: (i) PD-resistant materials like mica, porcelain, and even glass; and (ii) polymer plastic likes PE, XPLE, and others that are PD sensitive. Stresses by PDs inside of these materials lead – in a shorter or longer time – to a complete breakdown. Also, surface discharges lead to chemical reactions of the materials. Besides these chemical reactions, there is also mechanical damage on the surface of the materials, which can also result in a complete breakdown. Therefore, in solid insulating materials, electrical discharges like PDs damage the insulation and finally cause a breakdown of the insulation.

In liquid-insulating materials, metallic particles, cellulose fibers, and dissolved gases as well as water content have a great influence to the dielectric behavior. These imperfections lead to field distortions and create PDs. These PDs in liquid-insulating materials start to degrade the material by aging and contamination and also thermal stress.

In mixed dielectrics like oil-paper insulations, epoxy resins with inorganic filler materials, mica with bonding materials, and so on are in use for power equipment.

Partial Discharges (PD) - Detection, Identification, and Localization, First Edition. Norasage Pattanadech, Rainer Haller, Stefan Kornhuber, and Michael Muhr.

The combination of different dielectrics restricts mostly the effect of PDs to only one part and avoids the disadvantages of a single dielectric.

Due to all the measurements of PDs in insulating materials, it is very difficult if not impossible to make a general statement about the size and riskiness of PDs within the insulation materials, since the special stresses on volumes and surfaces as well as the coupling path of the PD signal cannot be defined. In single and simple dielectrics, it may be possible to come to a conclusion to a certain extent, but that is limited without knowing the location of the PD source. Therefore, assessing the lifetime of equipment based only on PD measurement is not possible.

In order to assess the riskiness of insulation failure, PD diagnostics are very important. The intensity, level, phase relationship, and pattern of PDs are of great significance. This applies to the stress with alternating voltage, but it is more complex for stress with direct voltage since the relationship to the phase is missing. In addition, the cause of the PD behavior under direct voltage has not yet been sufficiently researched.

10.2 In-House and On-Site PD Testing

PD testing of high-voltage equipment is essential in order to ensure that the equipment will operate with high reliably and at high efficiency. PD testing can be divided into in-house PD testing and on-site PD testing. The overview of PD testing for high-voltage equipment from developing a prototype until the end of life of equipment is shown in Figure 10.1.

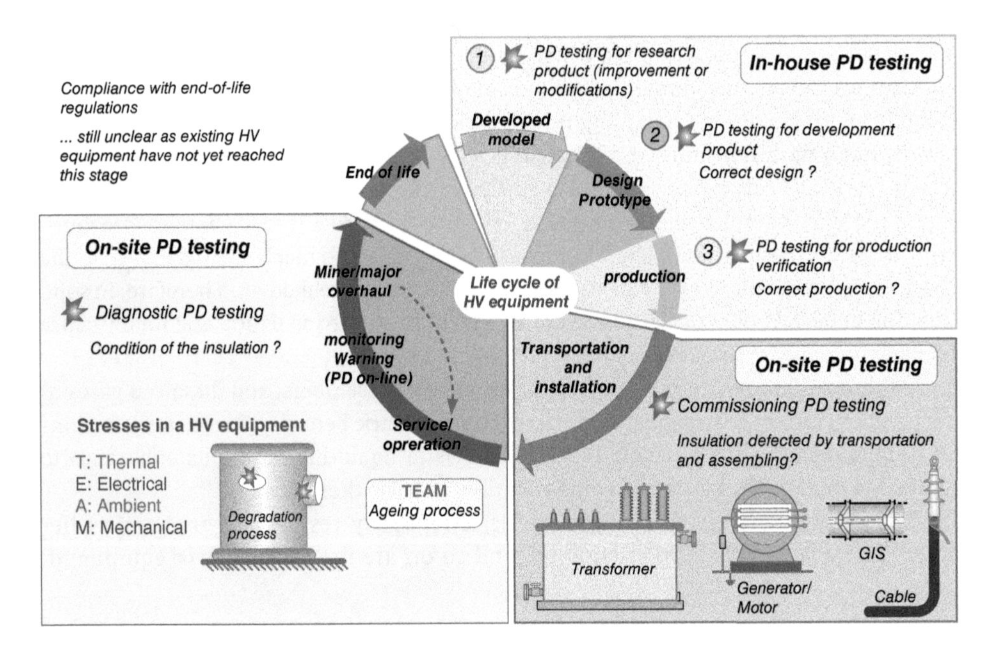

Figure 10.1 Overview of PD testing for a HV equipment.

10.2.1 In-House PD Testing

There are two reasons for performing in-house PD testing: product research and development, and for product quality verification:

1) *PD testing for product research and development.* Product research and development (R&D) is an integral part of the manufacturing process. During R&D, prototype high-voltage equipment is checked and adjusted as necessary in order to comply with operating parameters, and also to discover whether any improvements or modifications are required before commencing to manufacture the final product. For example, PD testing is used to determine the optimal impregnation time for the transformer manufacturing process, as shown in Figure 10.2.
2) *PD testing for product verification.* Product verification is an integral part of the in-house quality-control process for manufacturing the final product. PD testing is performed during the product verification phase in order to ensure that the final product will operate reliability and efficiency throughout its operating life.

10.2.2 On-Site PD Testing

On-site PD testing is generally classified as either commissioning tests or diagnostic tests:

1) PD commissioning tests are performed after on-site installation for the purpose of determining whether any significant damage has occurred during transportation and installation.

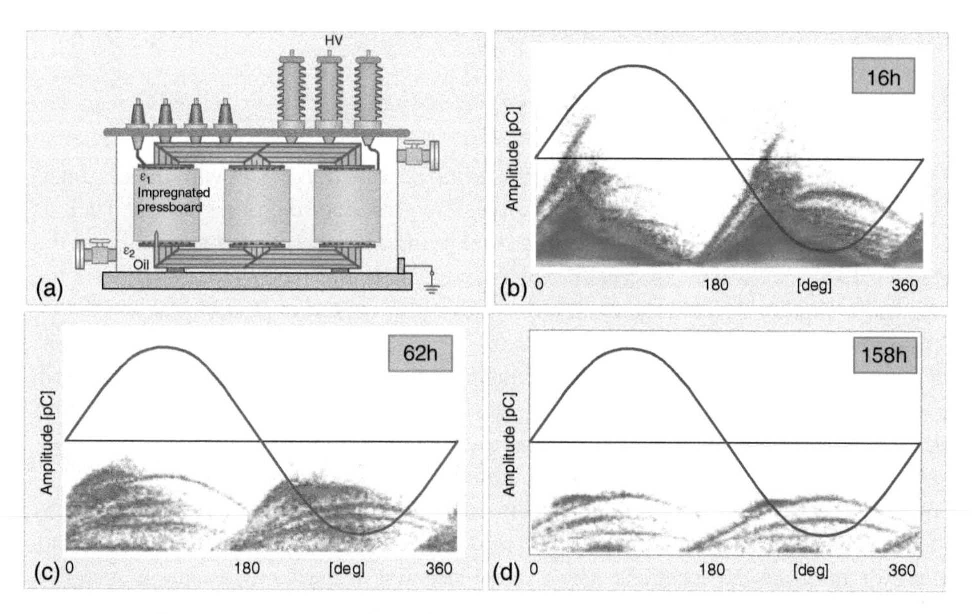

Figure 10.2 PD test results used to establish optimal impregnation time during transformer manufacture: (a) oil filled power transformer during manufacturing process; (b) PD patterns after drying and impregnation for 16 hours; (c) PD patterns after drying and impregnation for 62 hours; (d) PD patterns after drying and impregnation for 158 hours. [1].

2) PD diagnostic tests are performed during maintenance schedule and after any major repair work has been done. During maintenance program, PD testing is used to check the condition of the insulation. Normally, data from the first PD diagnostic test of newly installed equipment is used to determine a set of reference values (aka diagnostic fingerprint). After any major equipment repairs, diagnostic PD testing is required in order to confirm that the equipment is ready to go back in service.

10.3 How to Evaluate PD Test Results

Evaluation of in-house PD test results can be categorized as either PD testing for product R&D or PD testing for product verification. For in-house product R&D, the PD test result is valid, providing it at least equals design specification. Similarly, in-house production verification requires that the PD test results are within the limits of the relevant equipment standards. By contrast, evaluation of on-site PD test results can be categorized as either commissioning tests or insulation tests. For commissioning tests, the PD test results are required to be within limits set by the relevant operating standards and/or agreed to by the project manager, owner, and the end-user. However, because the nature of on-site PD testing to evaluate insulation condition is more problematical than performing the equivalent tests in-house, a full evaluation on-site is rarely possible. For example, if only PD test data is gathered, this gives only a limited understanding of the insulation condition, because a thorough explanation of the PD mechanism requires additional information, such as PD test result history (or fingerprints), PD location, trend analysis, and insulation properties. Moreover, a comprehensive understanding of insulation condition requires additional diagnostic test techniques, such as dissipation factor, polarization/depolarization current, and visual inspection.

10.4 Effect of PD on Insulation Degradation

In high-voltage equipment, the process of insulation degradation begins with the first operation of the equipment. The degradation process occurs due to various factors, such as defects in the insulation, the quality of the insulation used, operating conditions, and the aging mechanism. The aging process produces irreversible changes in the properties of the insulating structure due to four main stress factors: temperature stress, electrical stress, mechanical stress, and ambient or TEAM stress [1], [2]. Longer exposure to such stresses produces higher degradation of the motor insulation, as shown in Figures 10.3 and 10.4. Moreover, while the insulation system is in-service, it is subjected to the combined influence of such multiple aging factors as the above, which degrade insulation integrity, leading to insulation failure and ultimately equipment failure.

When PD occurs in the insulation system, although the destructive force released by individual PD events tends to be very small, the cumulative effect of serial discharge events in one or multiple PD locations has a direct negative impact on insulation integrity. Moreover, PD has an indirect effect on both the mechanical strength and thermal characteristics of the insulation material, and over time it is capable of causing significant degradation of the insulation system. The full impact of PD of insulation condition is a complex phenomenon:

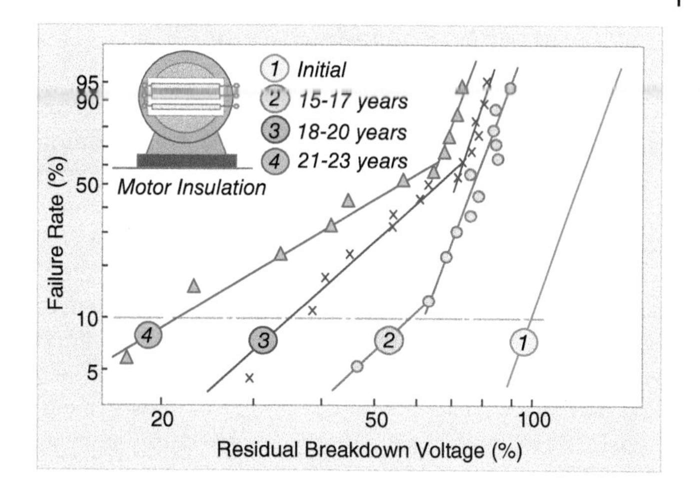

Figure 10.3 Insulation degradation as a function of operation time [3].

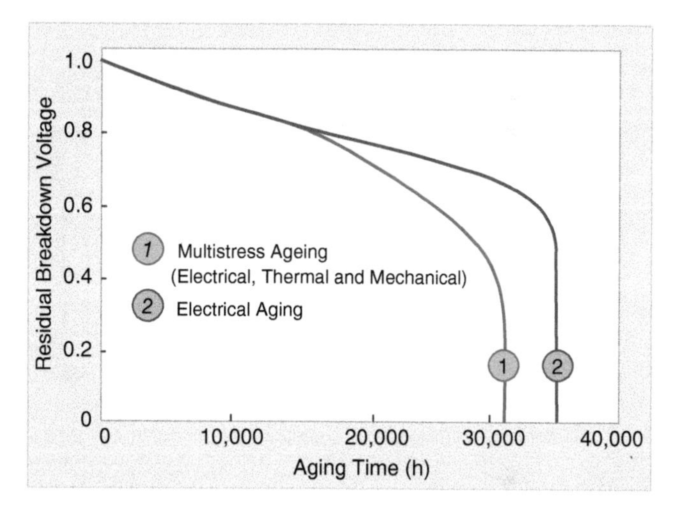

Figure 10.4 Insulation degradation as a function of experienced stresses [3].

1) *Impact of PD on organic and inorganic material.* High-voltage transformer insulation consisting of organic material (such as cellulose or mineral oil) is highly susceptible to degradation by PD activity, and a high level of degradation may occur due to even short exposure to PD activity, leading to rapid equipment failure. However, inorganic materials (such as mica) used in rotating machines have PD resistant characteristics, and thus PD is only symptomatic of insulation deficiency rather than leading directly to insulation failure. Furthermore, while PD may directly attack the insulation material and accelerate the influencing the aging process, failure probability (aka time-to-failure) tends to vary significantly due to other factors that may have no strong correlation to the level of PD activity per sec [4]. For example, the location of the PD source *vis a vis* the insulation material is an important factor in considering the degree and impact of aging.
2) *Impact of PD on insulation medium types.* PD activity in gas insulation (e.g. corona discharge in an air-insulated substation) is relatively harmless compared to PD activity in solid or liquid insulation (e.g. internal or surface discharge) [1].

10.5 Integrity of PD Measurement

PD amplitude and frequency contents as measured is almost always less than the original PD at the source due to attenuation of the PD signal during propagation from the PD source to the PD sensor. The degree of this attenuation is determined by the complexity of the internal structure of high-voltage equipment. Furthermore, the PD signal is also compromised by the sensitivity of the PD sensor, which varies according to the type of sensor used.

Nonconventional PD measurement techniques (as described in Chapters 5 and 6) prescribe various PD sensor types for taking PD measurements. PD measurement in accordant with nonconventional PD techniques tends to detect different PD phenomena than conventional PD measurement techniques, according to IEC 60270. This is implied by the fact that the two techniques use different units of measurement. That is, detection by conventional PD measurement is based on charge detection, whereas detection by nonconventional PD measurement is based on other PD phenomena detection, which provides more sensitivity to some kinds of PD defects, as illustrated in Figure 10.5 – meaning that such

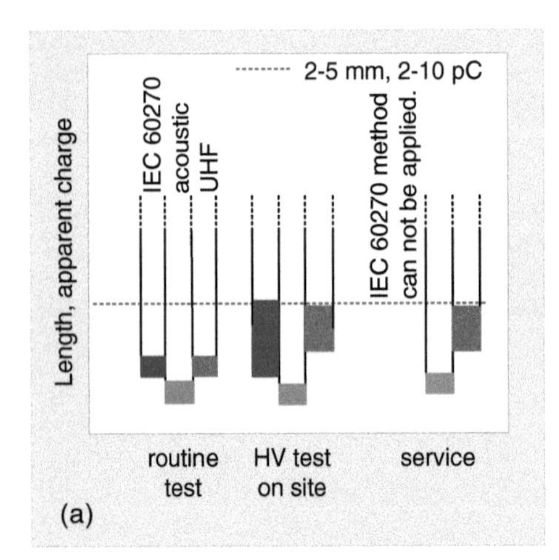

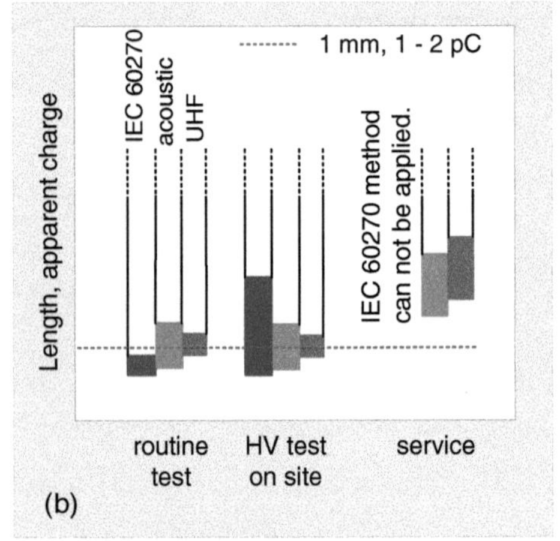

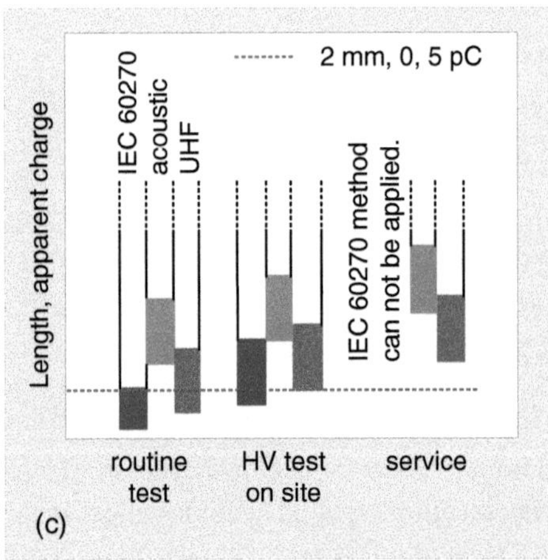

Figure 10.5 Sensitivity of PD diagnostic methods during routine test, on-site test and in service: (a) free metallic moving particle; (b) protrusion on HV conductor; (c) particle on insulation [5].

measurements are not directly comparable. Furthermore, charge calibration is only valid for conventional PD measurement.

Aside from the relevant experience of the person conducting PD test analysis, PD quantity and pattern are generally the two most useful to consider.

10.6 PD Quantity

There are two main types of PD quantity used in the analysis of PD test results – *basic PD quantity* and *derived PD quantity*. Moreover, there are three basic *PD quantities* (aka dimensional PD quantities) to be considered during analysis of insulation condition – *discharge magnitude, PD inception voltage* (PDIV), and *PD extinction voltage* (PDEV).

10.6.1 Discharge Magnitude

In conventional PD measurement, the *apparent charge* is measured and expressed in picocoulombs (pC). The *peak value* of the apparent charge (Q_p) and the largest repeatedly occurring PD magnitude recorded by a measuring system which has the pulse train response (Q_{IEC}) are expressed as the *PD magnitude* [6]. In nonconventional PD measurement, various types of PD *pulse magnitude* are designated, according to the sensor types used and the relevant industry standards. For example, the recommended standard for PD measurement in CC tests of rotating machines is the highest magnitude associated, with a PD pulse repetition rate of 10 pulses per second (Q_m) expressed in millivolts (mV) [4]. Similarly, for high-frequency current transformer (HFCT), Rogowski coil (RC), and UHF measurement of PD, the PD pulse amplitude is usually expressed in mV.

However, the amplitude of the PD spectra measured by the UHF technique is also expressed in decibel (dB). An increase in PD magnitude means higher PD intensity, which suggests a significant increase and/or acceleration of the insulation aging process. Increase in PD magnitude over time indicates that significant deterioration has occurred in the condition of the insulation. For example, a doubling of PD magnitude over a 6- or 12-month period provides a very strong indication of significant insulation deterioration during the same period. Moreover, increase in PD magnitude with the higher test voltage also indicates the degree of severity of the insulation problem, as demonstrated in Figure 10.6. However, increase in PD magnitude alone may not provide enough information to accurately evaluate the insulation condition, as shown in Figure 10.7. With the longer service time or the higher test voltage, after a steady increase of PD magnitude by a period, the PD magnitude may decrease or even being constant before failure of the insulation occurs.

10.6.2 PDIV, PDEV, and Other PD Quantities

PD inception voltage and *PD extinction voltage* are used as indicators of insulation condition, as higher PDIV and PDEV indicate the better insulation condition. They are classified as follows:

1) *PDIV:* The lowest applied voltage at which the magnitude of a PD pulse quantity becomes *equal to or higher than* a specified low value is designated the PDIV [6].
2) *PDEV*: The lowest applied voltage at which the magnitude of a PD pulse quantity becomes *equal to or lower than* a specified low value is designated the PDEV [6].

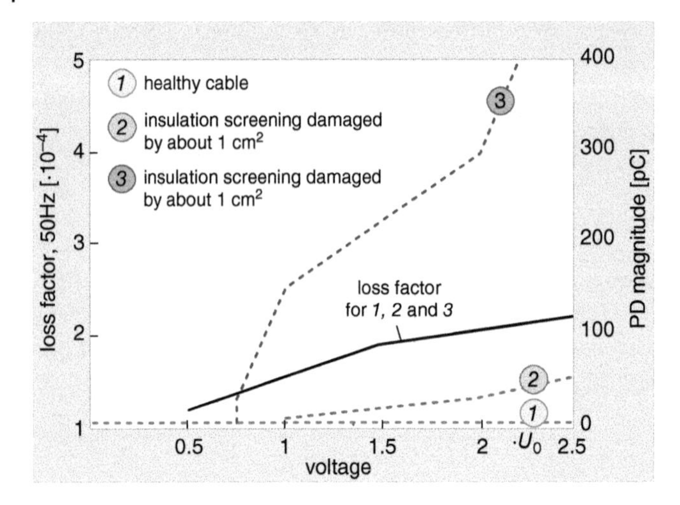

Figure 10.6 Application of PD values and loss factor for demonstrating insulation problem of XLPE cable [7].

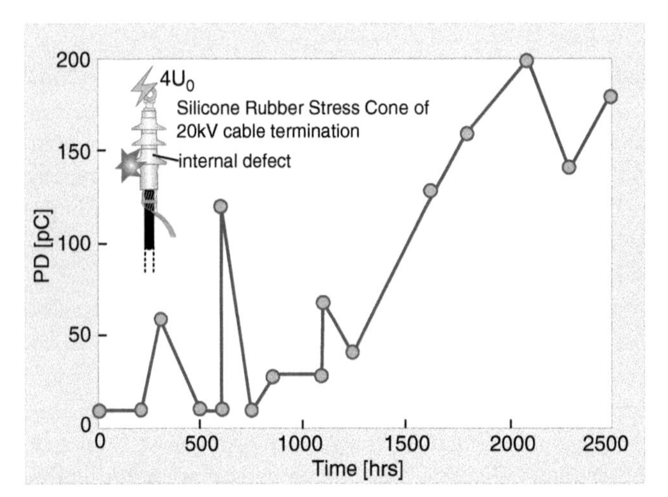

Figure 10.7 Example of PD development at $4U_0$ until breakdown caused by internal defect "grooves on the insulation core" of silicone rubber stress cone of 20 kV cable termination [8].

Other basic PD quantities, such as PD *discharge-phase*, PD intensity, and *time difference* between successive PD pulses, may apply for PD analysis as well. For derived PD quantities, which may display as the presence of time-dependent changes in PD behavior, the occurrence of the consecutive PD pulses in the function of power frequency cycle (at AC stress) and in time (at DC stress), as well as the intensity spectra of PD quantities, provide more information on discharge activity [8].

10.6.3 PD Quantity as Criteria for Evaluation of Insulation Condition

PD magnitude, PDIV, and PDEV may be used as the recommended values for interpretation and/or evaluation the insulation condition for some types of high-voltage equipment, both in the laboratory and on-site tests, as the following examples describe (Tables 10.1–10.3).

Table 10.1 Recommended PD quantities for interpretation of induced voltage PD test results obtained from a power transformer [8].

PD quantities	Condition	
PD magnitude	<1 nC	>1 nC
PD intensity (number of impulses)	low	high
PD pattern database recognition	regular	irregular
PD pattern change during and after enhanced voltage level	no significant change	significant change
PD magnitude during entire test	low	high
PD intensity during entire test	low	high
Number of the PD processes	one	several

Table 10.2 Recommended PDEV values for interpretation of commissioning test data obtained from a newly installed cable system [9].

	Withstand/conditioning (Monitored)			PD test
Voltage class (kV)	Test level (U_0)	Frequency range (Hz)	Duration (min)	PD pass/fail criterion
66–72	2.0	10–300	60	PDEV > 1.5 (No detectable PD at 1.5 U_0)
110/115				
132/138	1.7			
150/160				
275/285				
345/400				
500	1.5			

Table 10.3 Recommended PD criteria for interpretation of commissioning test data obtained from an in-service cable system [9].

			5–15 years		>15 years	
Voltage class (kV)	Frequency range (Hz)	Duration (min)	Test level (U0)	PD pass/fail criterion	Test level (U0)	PD pass/fail criterion
66–72	10–300	60	1.5	No detectable PD	1.1	No detectable PD
110/115						
132/138			1.4			
150/160						
220/230						
275/285						
345/400						
500						

10.7 PD Patterns

Two types of PD patterns – *phase-resolved PD* (PRPD) and *PD pattern with polarity count for problem analysis* – are generally used in PD problem analysis. The PRPD technique is generally the preferred analytical method, as it can be applied to basically all types of high-voltage equipment. By contrast, the PD pattern with polarity count for problem analysis is generally the preferred method for analyzing PD problems in rotating machines, whereby the number of PD pulses (both negative and positive) can provide an indication of the PD location in the insulation system.

10.7.1 Analysis of PD Patterns

The PRPD pattern consists of the apparent charge amplitude (q) and the phase angle (Ø) of the PD pulse at a specific test voltage [6]. The pattern is represented as a two-dimensional pattern described by the amplitude of the apparent charge in relation to the phase position. In recent times, a three-dimensional PRPD pattern has been proposed, consisting of the apparent charge amplitude (q), the phase position, and the number of PD pulses registered during the test, which is demonstrated by color. The three-dimensional PRPD pattern is shown in Figure 10.8, and the PD pattern with polarity count for problem analysis is illustrated in Figure 10.9.

Typical PRPD patterns (statistical behavior) represent the underlying physical phenomena of specific PD sources [1]. Analysis of PD patterns is an internationally accepted technique for assessing the insulation condition and is based on the assumptions that the shape of the PD pattern should depend only on each type of PD source (rather than the structure of the insulation system) and that such PD patterns measured by different PD detection techniques – both conventional and nonconventional – should be similar or comparable. Unfortunately, applying PD patterns for insulation problem analysis tends

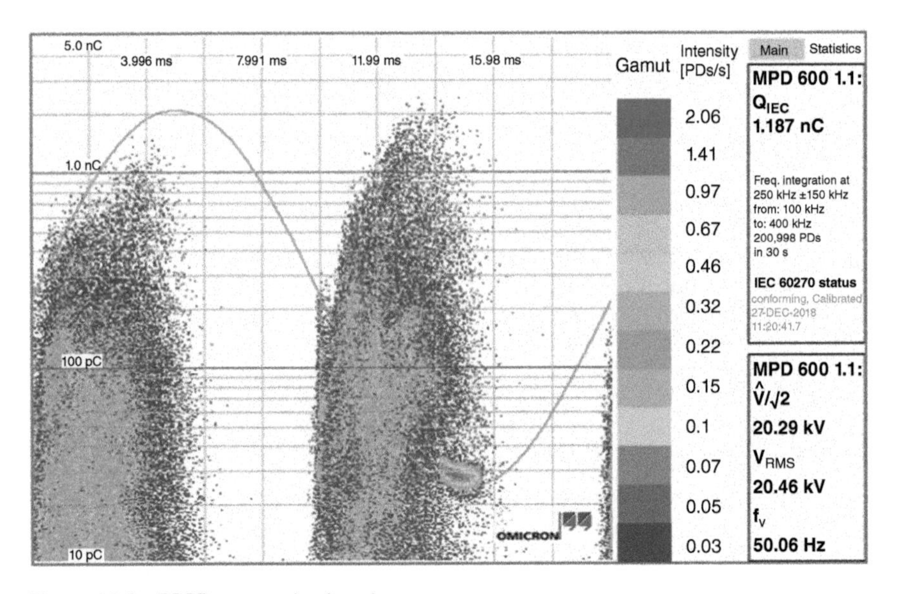

Figure 10.8 PRPD pattern (omicron).

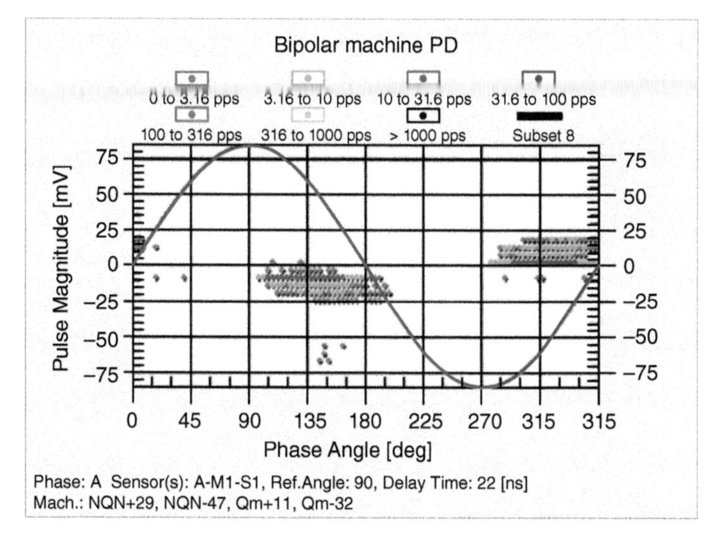

Figure 10.9 PD pattern with polarity count for problem analysis (IRIS).

to be rather difficult if there is more than one PD source activated at the time of PD pattern recording, or if there is high-frequency background noise present, because in these cases the recorded PD patterns will be a mixture of noise signal and all the PD patterns generated by each type of PD at the time. However, modern PD analysis software featuring PD recognition or PD-noise separation may be of assistance in PD pattern analysis.

10.7.2 PD Patterns as Criteria for Evaluation of Insulation Condition

Typical PD patterns occurring in transformers, rotating machines, and GIS are summarized in the Figures 10.10–10.15, respectively. Figure 10.10 illustrates the PRPD patterns which may happen in the insulation system of the transformers. The PRPD patterns obtained from the stator bar insulation testing in the laboratory and from onsite PD measurement are demonstrated in Figures 10.11 and 10.12 respectively. The typical PRPD patterns obtained from the offline PD GIS testing are shown in Figure 10.13.The typical PD spectra in GIS measured by UHF PD measurement technique are illustrated in Figure 10.14. The analysis of the PD problems taking place in the stator bars by comparing the numbers of positive and negative PD pulses detected is presented in Figure 10.15.

Nonetheless, PD patterns can be a useful indicator of insulation system deterioration, as shown by the example in Figure 10.16, which clearly indicates a steady increase in PD activity until week 9, when it dipped before increasing again. In week 12 of the aging process, the main insulation of one of the stator bar broke down. From this picture, it is confirmed again that just the PD level is not sufficient to track stator insulation degradation [8].

Figure 10.17 shows the PD development in a defect cable prior to breakdown. In Tables 10.4–10.5 PRPD patterns and PD spectra including their characteristics of the PD problems due to foreign particle in GIS are presented. PD patterns demonstrating PD problem development in the GIS insulation system caused by the protrusion form the inception stage to close to breakdown stage are shown in Figure 10.18.

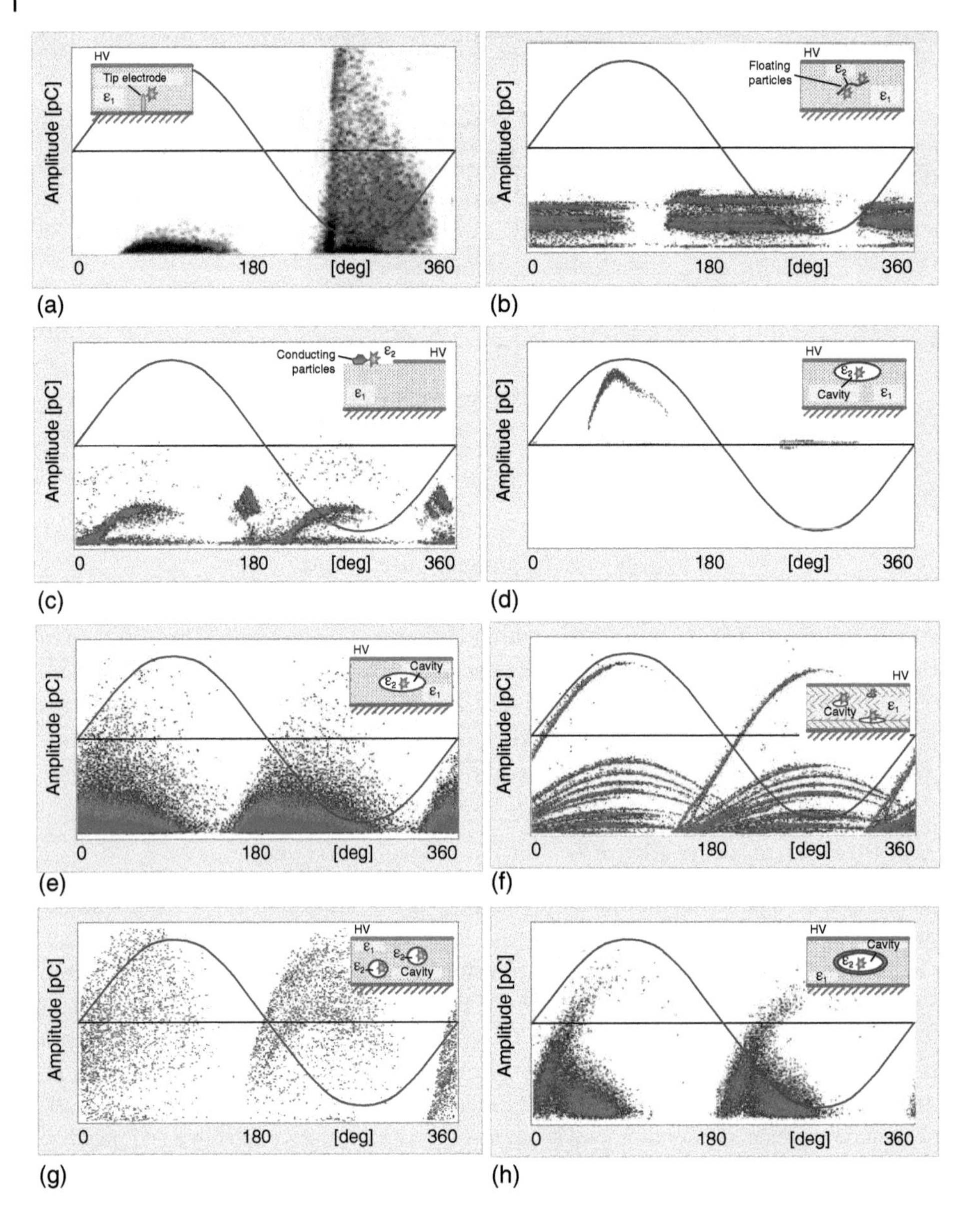

Figure 10.10 Typical PRPD patterns in transformers (conventional PD measurement) [1]: (a) Conducting material directly connected to the metallic electrode (tip electrode); (b) Conducting material without any contact to the metallic electrode (floating particle); (c) Conducting particles laying on the surface of the insulating material surface (surface discharge, creepage discharge); (d) Nonconducting material (cavity) with direct contact to the electrode; (e) Nonconducting material (cavity) without direct contact to the electrode #1; (f) Nonconducting material (cavity) without direct contact to the electrode #2; (g) Nonconducting material (cavity) without direct contact to the electrode #3; and (h) Nonconducting material (cavity) without direct contact to the electrode with interaction at the surface.

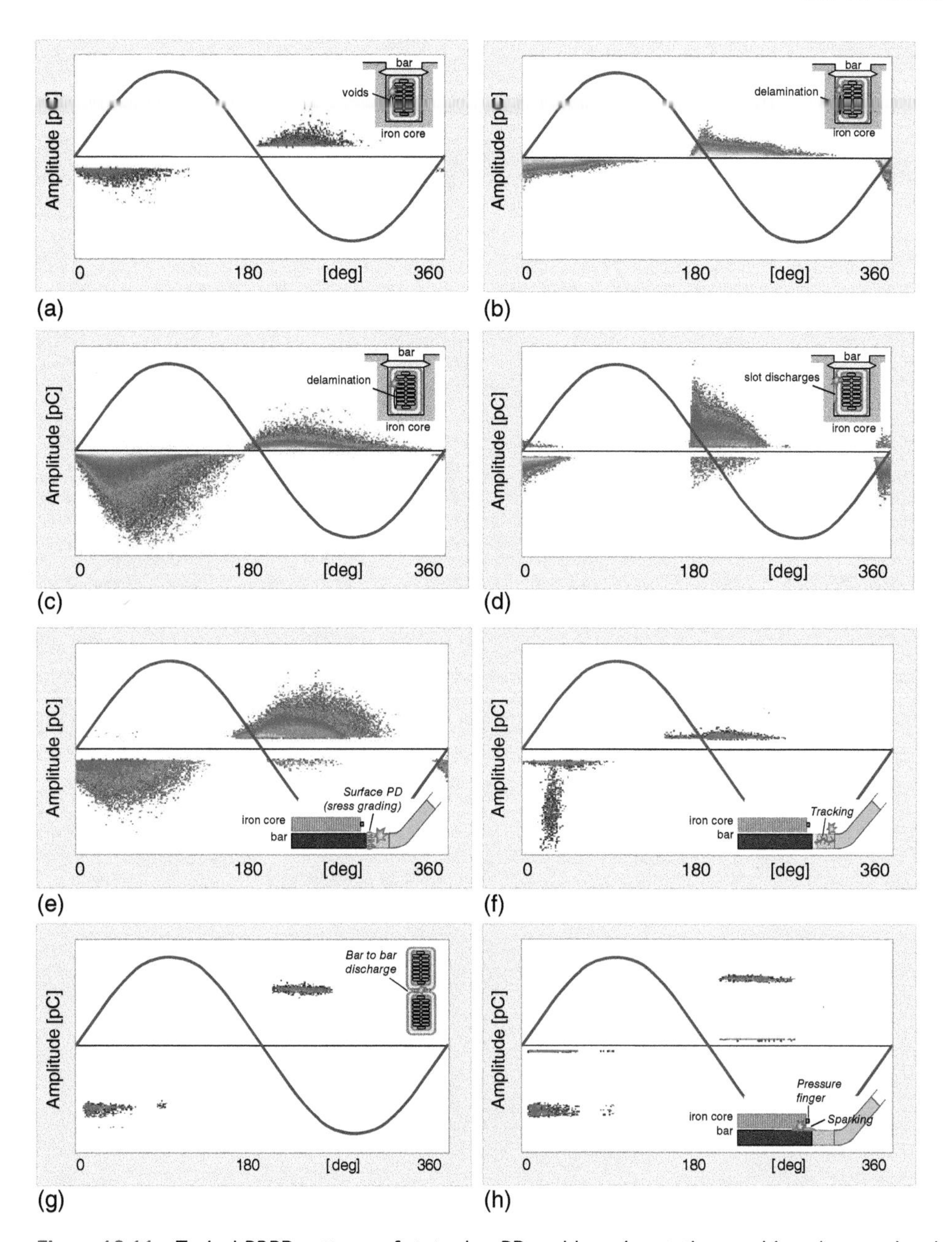

Figure 10.11 Typical PRPD patterns of stator bar PD problems in rotating machines (conventional PD measurement) [4]: (a) Internal void discharge PRPD pattern; (b) Internal delamination PRPD pattern; (c) Delamination between conductor and insulation PRPD pattern; (d) Slot partial discharge activity; (e) Corona activity at the S/C and stress grading coating; (f) Surface tracking activity along the end arm; (g) PD activity between two bars; and (h) PD activity between a bar and the press finger of the stator core.

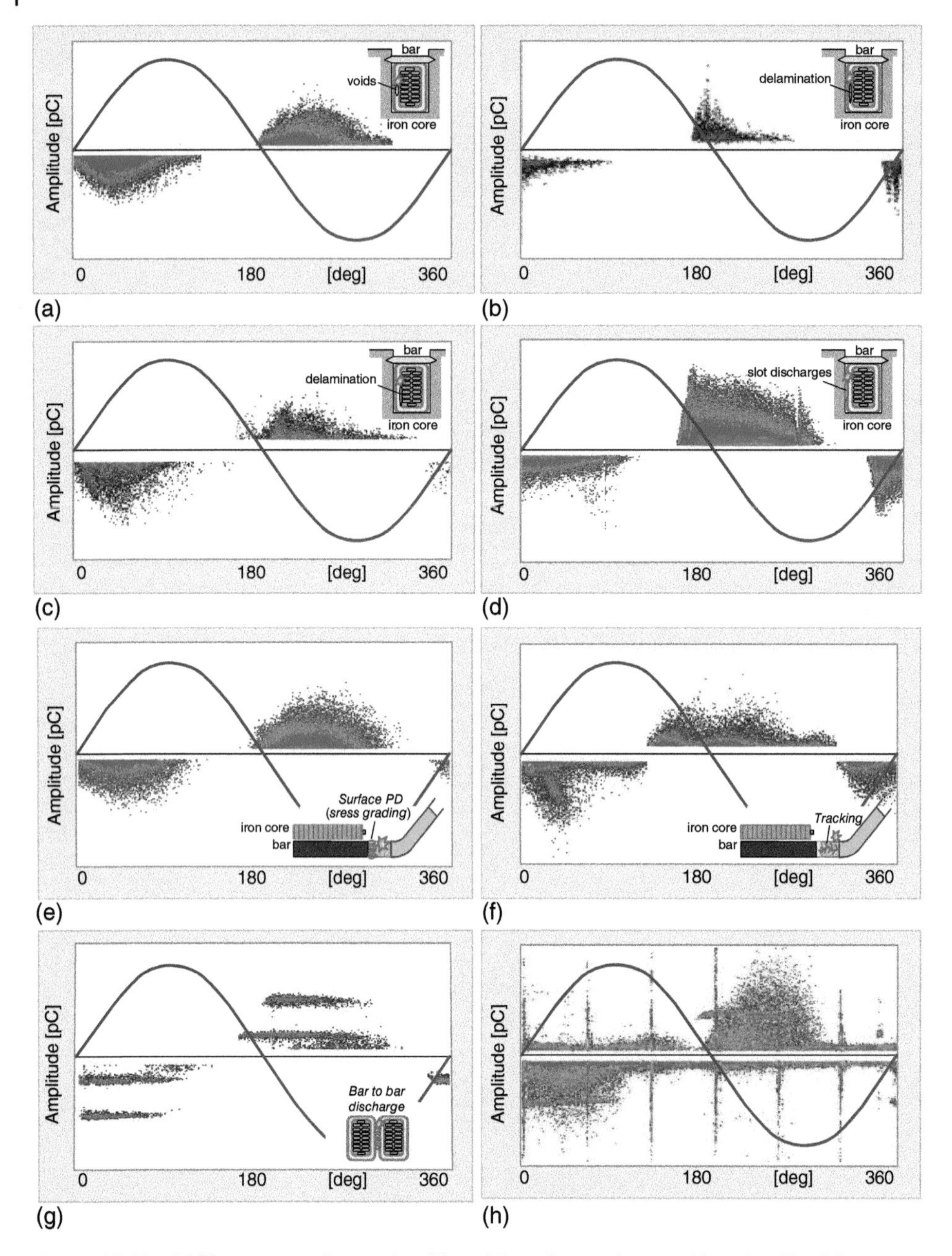

Figure 10.12 PRPD patterns of stator bar PD problems in rotating machines (on-line PD measurement) [4]: (a) Internal void discharge PRPD pattern; (b) Internal delamination PRPD pattern; (c) Delamination between conductor and insulation PRPD pattern; (d) Slot partial discharge activity; (e) Corona activity at the S/C and stress grading coating; (f) Surface tracking activity along the end arm; (g) PD activity between two bars; and (h) Multiple PD source.

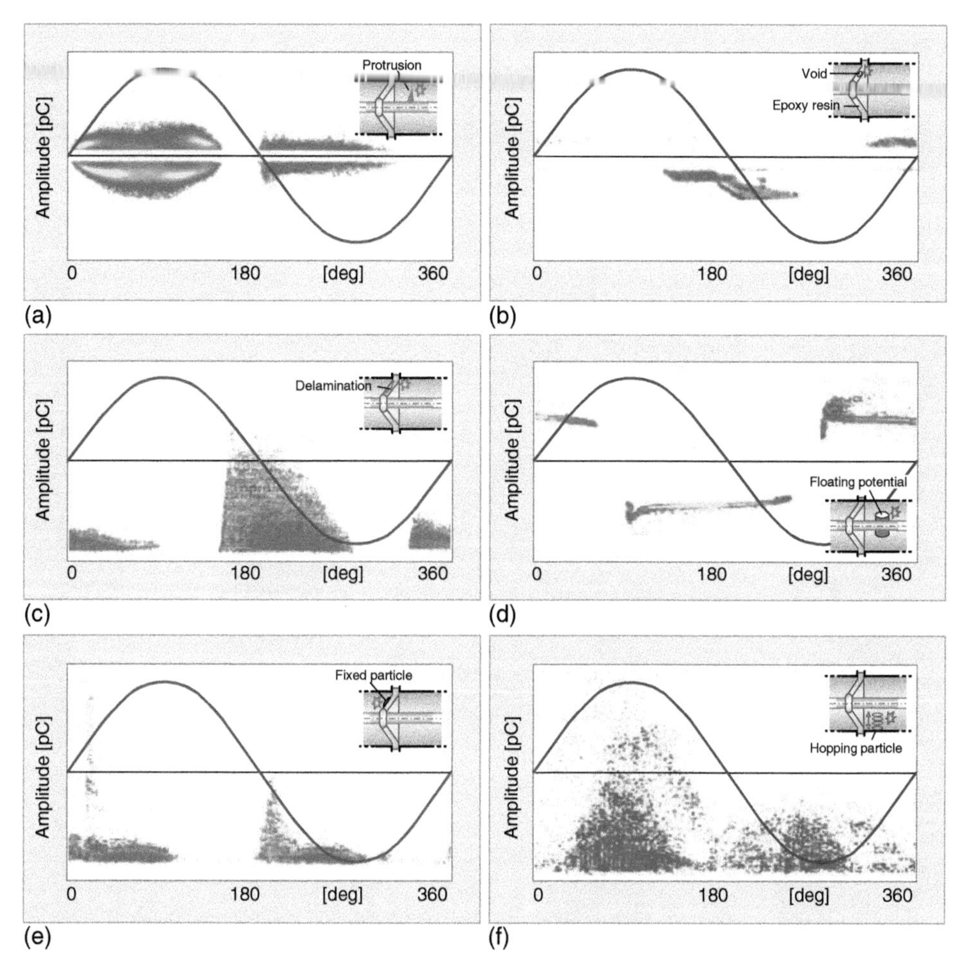

Figure 10.13 Typical PRPD patterns in GIS (off-line PD measurement) [10]: (a) Protrusion on inner conductor in GIS; (b) Void in epoxy resin; (c) Floating compound in GIS; (d) Fixed particle on GIS spacer; and (e) Hopping particle in GIS.

10.8 PD Signal in Time Domain and Frequency Domain Analysis

The time-domain PD signal (e.g. the PD pulse current) can be analyzed to determine the severity of problems caused by PD [1]. The parameters that characterize time-resolution of PD signals are as follows:

1) Maximum amplitude of the PD signal in mV
2) Rise-time of the PD current signal
3) PD signal oscillation
4) Reproducibility of the PD signal

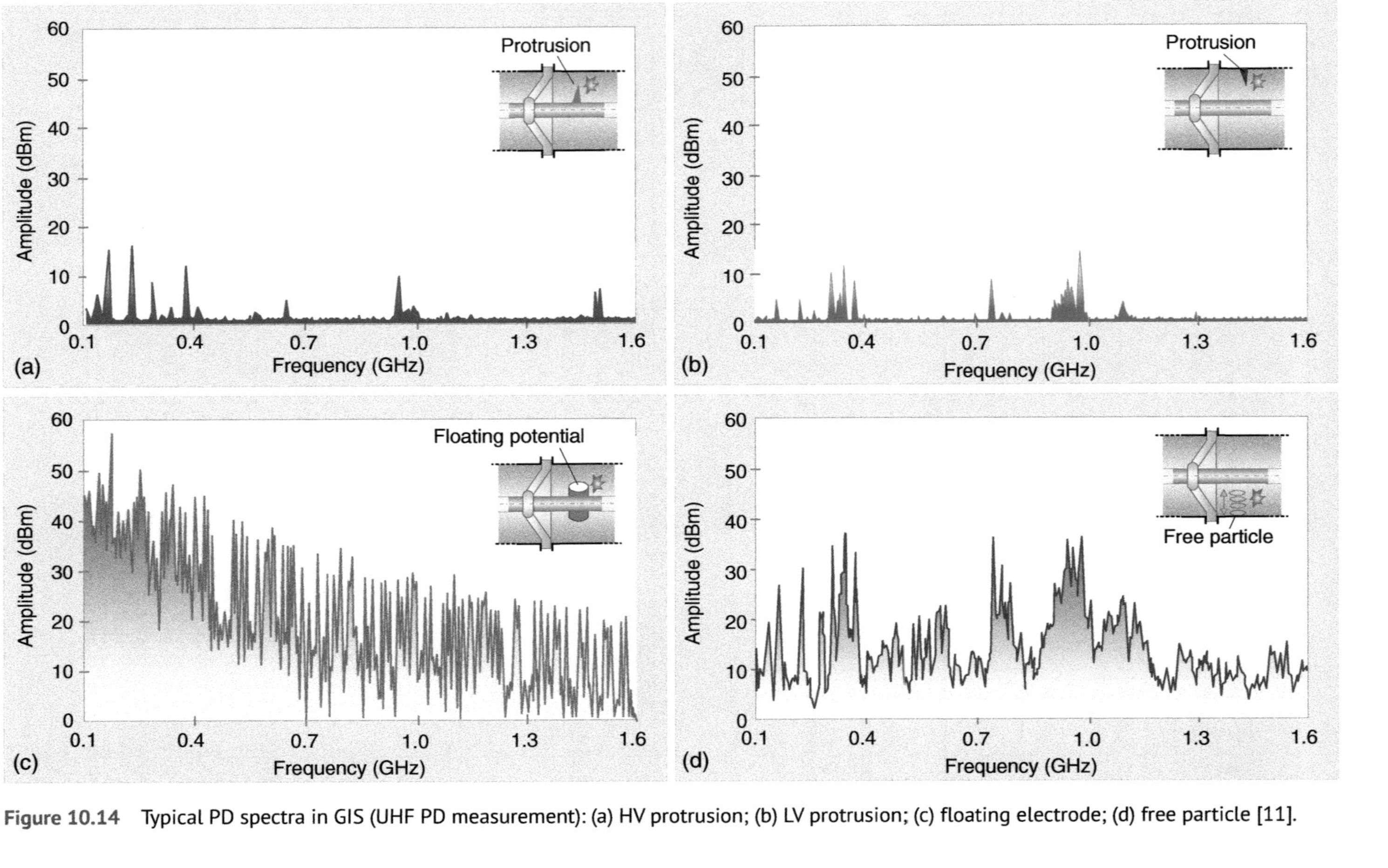

Figure 10.14 Typical PD spectra in GIS (UHF PD measurement): (a) HV protrusion; (b) LV protrusion; (c) floating electrode; (d) free particle [11].

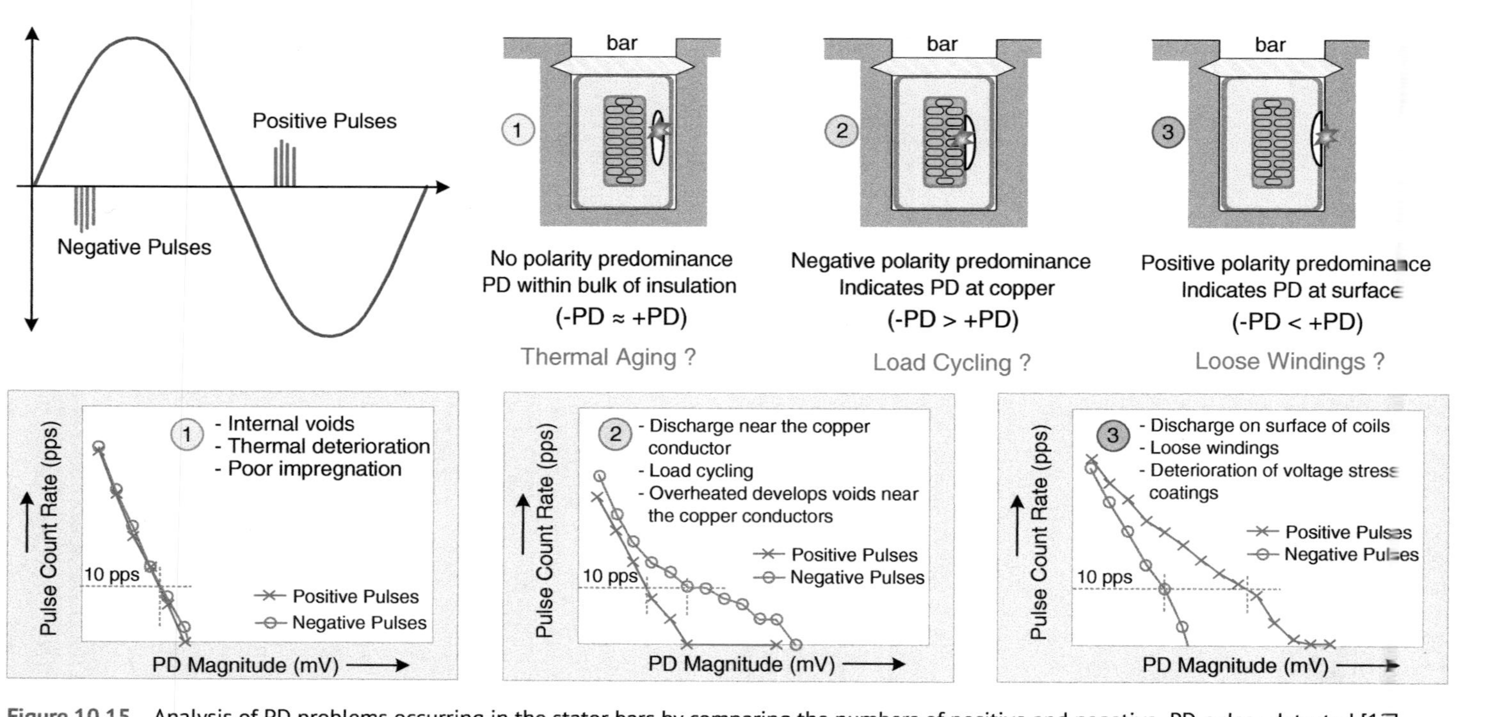

Figure 10.15 Analysis of PD problems occurring in the stator bars by comparing the numbers of positive and negative PD pulses detected [12].

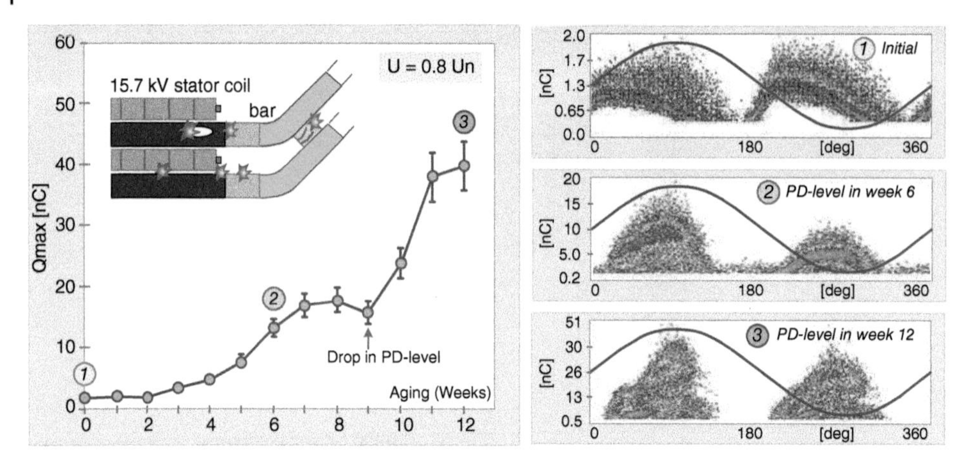

Figure 10.16 PD development before breakdown of a stator bar of the rotating machine [8].

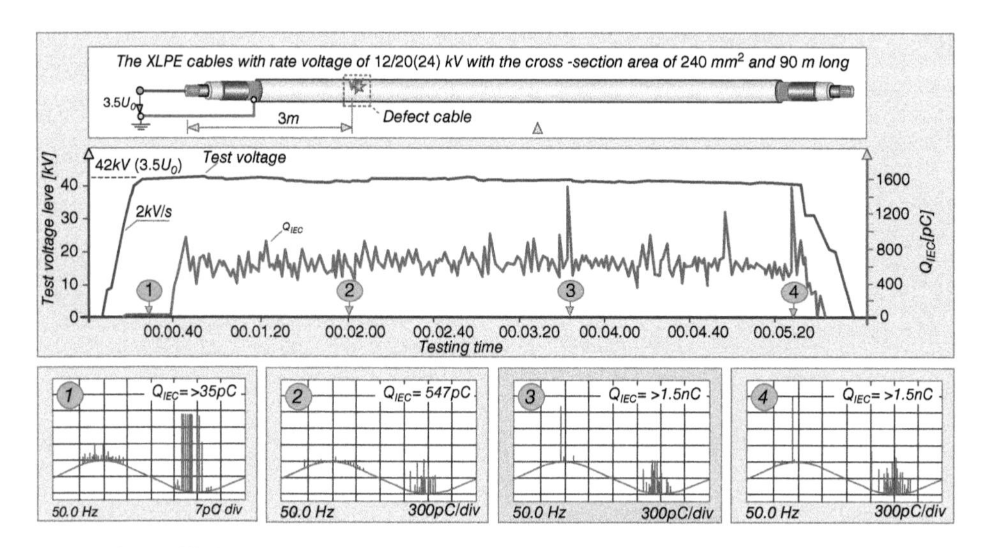

Figure 10.17 PD development in a defect cable prior to breakdown [13].

Amplitude and rise time of the PD signal can also be useful in determining PD location and PD-noise separation. PD signal monitoring in the time-domain (e.g. the PD pulse obtained from high-voltage bushing and neutral bushing) is shown in Figure 10.19a.

By contrast, the basic spectral characteristics of frequency-domain PD measurement are as follows:

1) Amplitude of the power spectrum in dBm
2) Frequency range of the power spectrum
3) Typical resonances
4) Reproducibility of the power spectra

Table 10.4 PRPD patterns and PD spectra of the PD problems due to foreign particles in GIS [8].

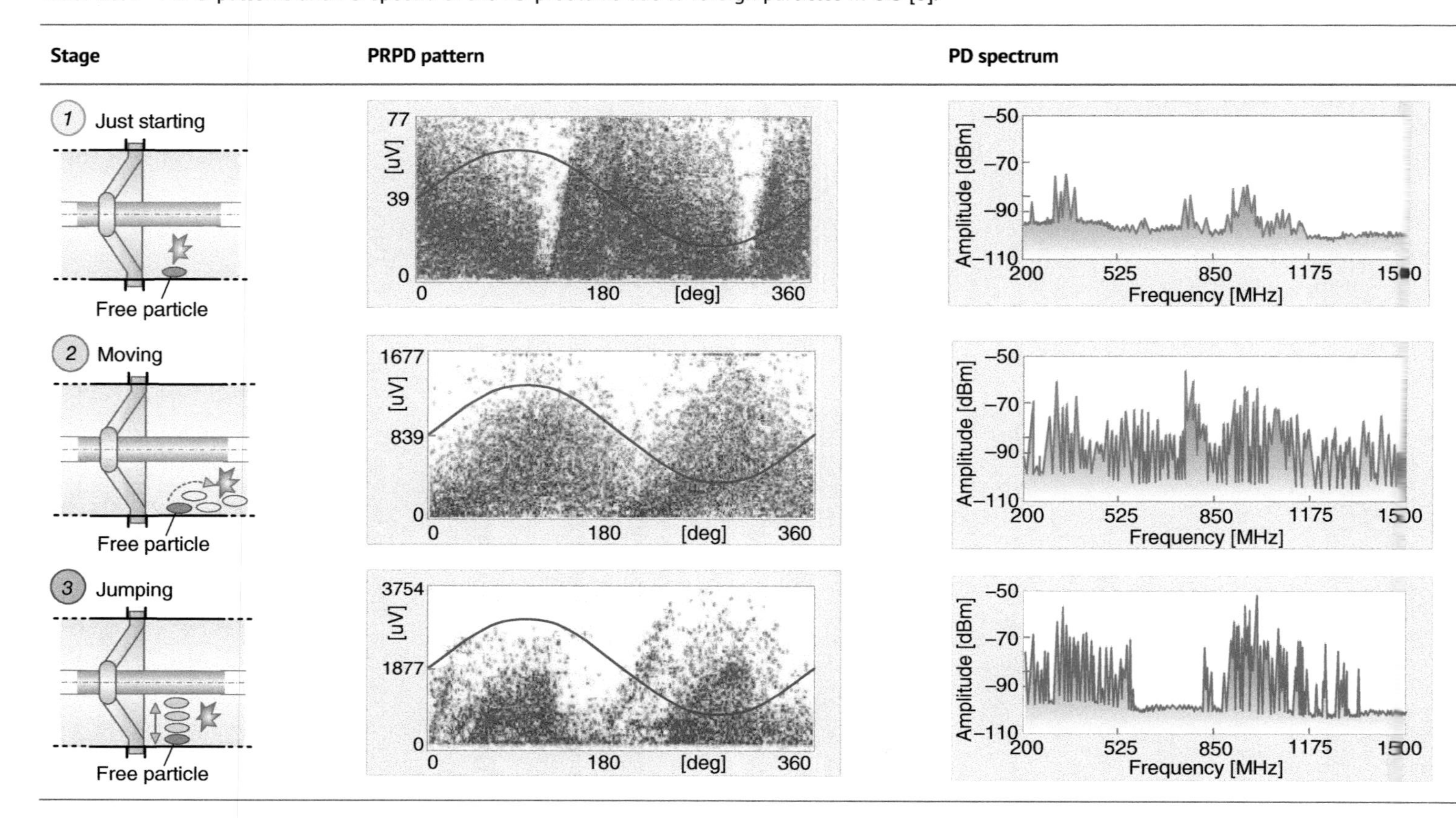

Table 10.5 Characteristics of PRPD patterns and PD spectra due to foreign particles in GIS [8].

	PRPD		PD spectrum	
Stage	**Magnitude**	**Intensity**	**Magnitude**	**Intensity**
1) Just starting	low	high	low	low selective
2) Moving	high	less	high	high and full
3) Jumping	high	less	high	high and full

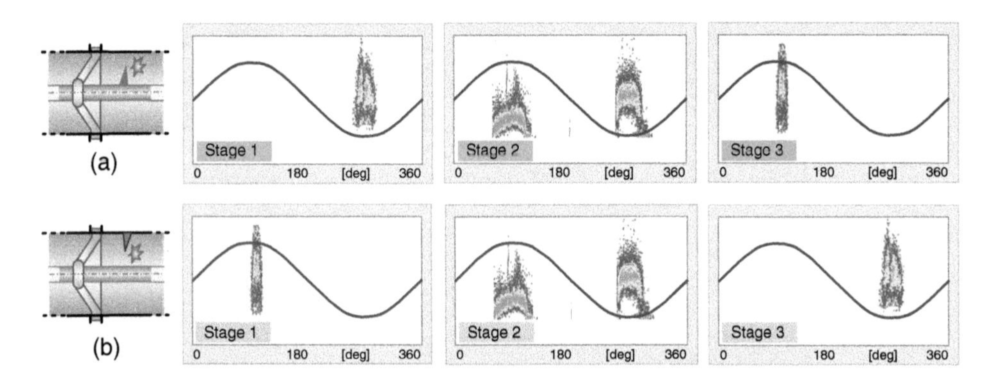

Figure 10.18 PD pattern development form the inception stage (stage 1) to close to breakdown stage (stage 3) of the GIS insulation system caused by the protrusion (a) the protrusion was fixed to HV conductor or to the enclosure (b) the protrusion was fixed to the enclosure [8].

For PD problem analysis, generally PD signal spectra are clearly influenced by external noise signals, as illustrated in Figure 10.19b. However, analysis of PD signals in the frequency-domain are only applicable to repeated PD activity.

10.9 PD Source as Criteria for Evaluation of Insulation Condition

Once PD sources are identified, the degree of severity of the PD insulation problems may be determined by taking into account the PD source, as shown in Tables 10.6–10.7.

10.10 Noise Patterns and Noise Reduction

10.10.1 Noise Patterns

Usually when PD measurement is performed, two types of background noise (i.e. conducted noise and/or radiated noise) are able to couple with the PD measuring circuit. Conducted noise may be generated in several ways, such as by a poorly setup high-voltage test circuit, by switching operations in the main circuit or proximity circuits, and by the

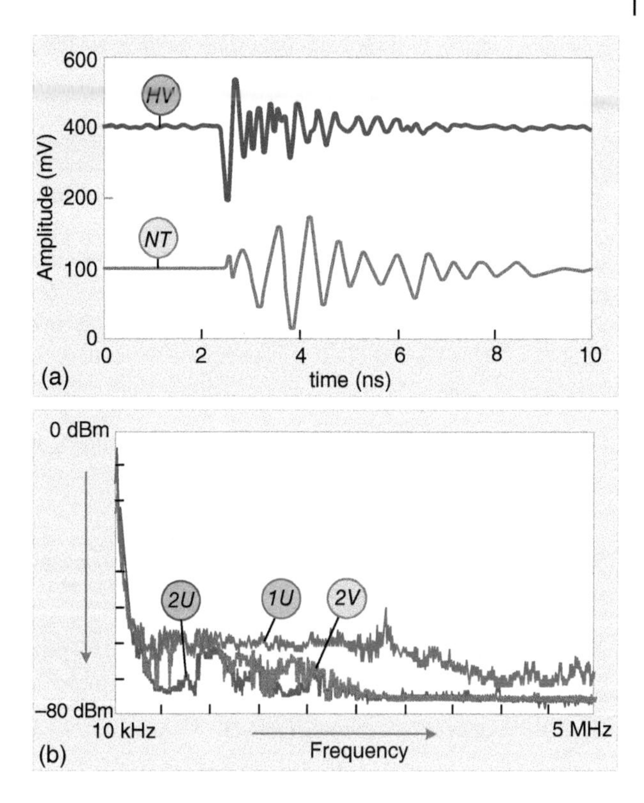

Figure 10.19 PD signals: (a) time domain; (b) frequency domain where (1U) is the recorded PD signal, (2V) is PD signal due to cross coupling from phase U, and (2U) is background noise [1].

Table 10.6 Degree of severity for the PD problems pertaining to various PD sources in transformers [1].

PD source	Degree of severity
PD sources in solid insulating material (e.g. PD in pressboard barriers of the main insulation, PD in the laminated wood, PD voids in glued plates, and PD caused by the presence of metallic particles in solid insulating material)	high
PD source located outside the active parts (e.g. corona discharge due to bubbles in liquid insulation, and PD caused by metallic particles lying on the surface of solid insulation)	low

noise signal from the grounding system and/or power circuit – even though from PD measuring device. Radiated noise may also be generated in several ways, such as by radio and television signals, by mobile phone basement, or by transient signals from nearby high-voltage equipment in substations. Examples of noise signature are shown in Figure 10.20.

10.10.2 Noise Reduction

As previously mentioned, the sensitivity of conventional PD measurement is usually impaired by background noise level – i.e. higher background noise tends to lower PD sensitivity. Furthermore, the mixing of background noise with PD signals increases the

Table 10.7 Degree of severity for the PD problems pertaining to various PD sources in rotating machines [4].

PD source	Degree of severity
Internal voids – i.e. PD in air-filled or gas-filled pockets embedded in the main insulation	low
Internal delamination – i.e. PD in air-filled or gas-filled elongated pockets oriented in a longitudinal direction	moderate
De-bonding between conductor and insulation – i.e. PD in air- or gas-filled elongated pockets (in longitudinal direction) that are imbedded between the main insulation and the field grading material	moderate
Slot discharge – i.e. PD generated by poor contact or missing between the conductive slot coating and the stator slot wall	Moderate if the winding is tight/high if the winding is loose and can vibrate
End-winding gap and surface discharge – i.e. PD generated on the surface of the insulation material due to conductive contaminants, abrasion, or damaged field-grading material	moderate
Foreign conductive materials and contamination – i.e. PD generated on the surface of the insulation material, which may cause by conductive contamination (carbon, oily dust, abrasion, etc.) or separated regions of field grading material	moderate

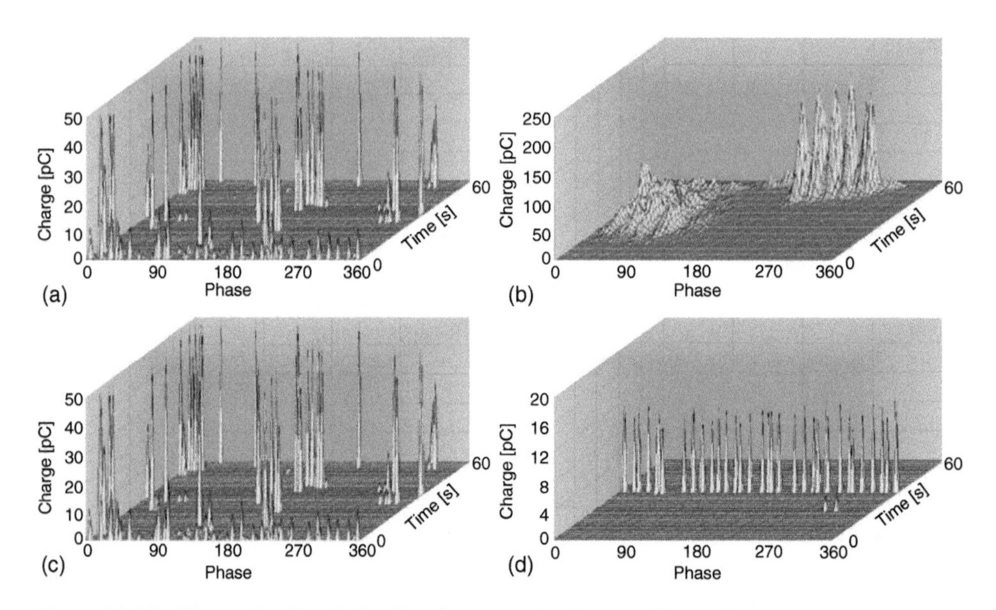

Figure 10.20 3D-graphs of typical noise signatures (*x*: phase angle, *y*: pulse number, *z*: apparent charge): (a) discharges from protrusions of HV shielding electrodes; (b) discharges between floating metallic parts; (c) maintenance work (drill); (d) signals form a cell phone call [2].

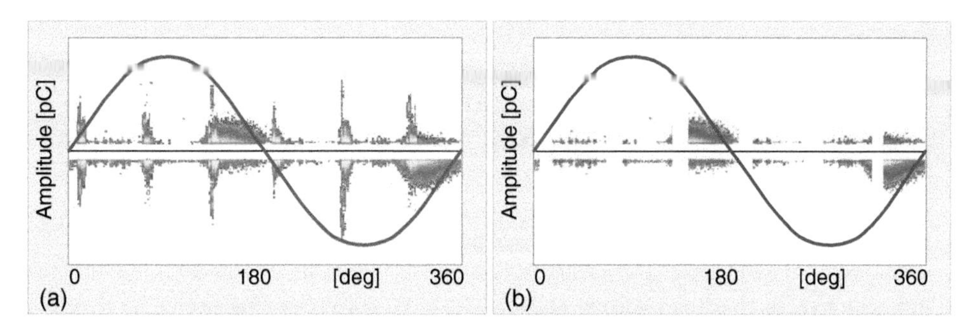

Figure 10.21 Noise reduction by window masking technique: (a) PD pattern mixed with noise signal; (b) PD pattern after denoising by window masking technique [14].

difficulty of PD pattern analysis. Therefore, conventional PD measurement should be performed in a properly shielded laboratory. Furthermore, noise-reduction techniques are required when conventional PD measurement must be performed in a high background noise area. As already mentioned, the PD test circuit must also be set up properly, with the appropriate filters in the electrical supply to the test circuit. In addition, narrow-band rather than wide-band PD monitoring may be conducted. In recent times, digital noise suppression software has become increasingly available for both conventional and nonconventional PD measurement. Examples of digital techniques applied for noise reduction and/or PD-noise separation are window masking, cluster separation, and time–frequency mapping. Figure 10.21 shows the application of window masking for noise reduction [4].

10.11 Effective Evaluation of PD Phenomena

In order to conduct effective evaluation of PD phenomena in high-voltage equipment, the person conducting the analysis of PD test data should have a proper understanding of the following:

1) Operational design limits pertaining to all the equipment involved.
2) Specific properties of all material used in the high-voltage equipment.
3) Operating stresses on the equipment: i.e. thermal, mechanical, electrical, and ambient. For example, the main stresses on transformers in service are electrical and thermal, whereas the main stresses on rotating machines are mechanical and electrical. Furthermore, the degree of thermal stress on rotating machines is determined by the load factor or working load.
4) Nature of the PD mechanism: i.e. the process by which PD develops, which itself is mainly determined by PD type and insulation type.
5) Principles of PD measurement as they relate to various PD phenomena: e.g. the nature of the PD phenomena should determine the frequency range for taking measurements.
6) Operating principles of the various PD measurement devices available and their limitations.
7) Principles of PD data evaluation: e.g. PD quantities and PD patterns.

References

1 Conseil International des Grands Réseaux Électriques. (2017). Partial discharges in transformers. *Technical Brochure. Ref. 676.*

2 Conseil International des Grands Réseaux Électriques. (2012) High-voltage on-site testing with partial discharge measurement. *Technical Brochure. Ref. 502.*

3 Kimura, K. (1993). Progress of insulation ageing and diagnostics of high voltage rotating machine windings in Japan. *IEEE Electrical Insulation Magazine* 9 (3): 13–20.

4 International Electrotechnical Commission. (2012). IEC 60034: rotating electrical machines – Part 27-2: on-line partial discharge measurements on the stator winding insulation of rotating electrical machines. Geneva, Switzerland: IEC.

5 Conseil International des Grands Réseaux Électriques. (2013) Risk assessment on defects in GIS based on PD diagnostics. *Technical Brochure. Ref. 525.*

6 International Electrotechnical Commission. (2015) IEC 60270: high-voltage test techniques – partial discharge measurements. Geneva, Switzerland: IEC.

7 König, D. and Rao, Y.N. (1993). *Partial Discharges in Electrical Power Apparatus*. Berlin: VDE-Verlag Gmbh.

8 Conseil International des Grands Réseaux Électriques. (2003). Knowledge rules for partial discharge diagnosis in service. *Technical Brochure. Ref. 226.*

9 Conseil International des Grands Réseaux Électriques (2018). On-site partial discharge assessment of HV and EHV cable systems. *Technical Brochure. Ref. 728.*

10 Litkhitsupin, S. (2013). UHF partial discharge measurement on GIS. Lecture note, Electricity Generating Authority of Thailand (26 July 2013).

11 Meijer, S. (2001). Partial discharge diagnosis of high-voltage gas-insulated systems. PhD thesis. Delft University of Technology.

12 Kankaanpaa, J. (2016). On-line partial discharge testing for stator winding of motors & generators. *Keynote speech for Proceedings of the 2nd International Symposium on Lightning Protection and High Voltage Engineering 2016 at King Mongkut's Institute of Technology Ladkrabang*, Bangkok, Thailand (9 March 2016).

13 Pattanadech, N. (2018). PD testing for underground cables. Lecture note, Chaophya park hotel in Bangkok (28 February 2018).

14 International Electrotechnical Commission. (2017). *IEC 60034: Rotating Electrical Machines – Part 27-1: Off-line Partial Discharge Measurements on the Winding Insulation*. Geneva, Switzerland: IEC.

11

Standards

11.1 Standards

PD tests according to the standards are to prove the quality and condition of insulation systems applied for high-voltage (HV) apparatus and their components. The test procedures for quality assurance after manufacturing and repair, as well as the test voltages and their limits of tolerated PD magnitude, are deduced from long-time experience. The limits are specified for every HV apparatus. The horizontal standard that applies to all areas is IEC 60270 "High-Voltage Test Techniques – Partial Discharge Measurement" [1]. But there are also associated standards for the various apparatus that include regulations for PD testing.

As the standard IEC 60270 has been used for a long time, it is now the task in IEC to revise and upgrade this document. Also, the task is to look forward to include expected research results into standards regarding PD testing with direct voltages. There is some progress in this work, but there is a lot to do. The revised standard has a new title of IEC 60270: "Charge-Based Partial Discharge Measurements" [2]. The reason was a deduction to the other possibilities of PD measurements as the electromagnetic (EM) methods. While IEC 60270 specifies the measuring value in pC, the EM PD detection methods measure the value in mV. Furthermore, the different frequency ranges must also be taken into account by these methods. However, it should be noted that until now, there was no correlation between the measured values in pC and the measured values in mV. For this reason, at this moment, only the revised IEC 60270 is valid for acceptance tests and quality control, while EM methods are used for diagnostics, research, and development. But the EM techniques have become very important, especially in on-site and on-line measurements. As there is a special need to recommend nonconventional PD detection methods (EM and acoustic methods), a new document has been worked out, which is currently a technical specification (TS). The title of this document is IEC TS 62478, "High-Voltage Test Techniques – Measurement of Partial Discharges by EM and Acoustic Methods" [3].

Partial Discharges (PD) - Detection, Identification, and Localization, First Edition. Norasage Pattanadech, Rainer Haller, Stefan Kornhuber, and Michael Muhr.

TS is often published when the subject under question is still under development or when sufficient consensus for approval of standard is available. Technical specifications provide the details and completeness of standards, but they have not yet passed through approval stages. TS must be checked every three years to evaluate whether it will be withdrawn, continued, or converted into a standard. Therefore, TS has more of an informative character.

IEC TS 62478 deals with a large variety of applications, sensors of different frequency ranges, and different sensitivities. The tasks of PD location, measuring system calibration, and sensitivity checks are also taken into account. Since the development of these nonconventional methods has not yet been completed, it was decided not to publish all results in the document as a standard but as a TS, scheduled for review in 2023. The ideas and steps that have been discussed in the working group to fulfill the requirements for a good document are the following: physical phenomena, sensors, location, transmission aspects, reading and derived quantities, PD sources, performance and system checks, pC correlation, and new quantities.

Besides the IEC standards, there are many international standards and documents regarding PD behavior, detection, measurement, and interpretation. Especially IEEE has a lot of standards regarding this subject. The basic document is "P454 – Guide for the Detection, Measurement, and Interpretation of Partial Discharges" [4]. This guide applies to the detection, measurement, and interpretation of PDs occurring in or around the insulation of electrical apparatus or their components. Topics included in this guide are:

- Basic discharge physics and material related discharge mechanisms
- Principles of PD calibration, detection, and measurement using apparent charge method
- Explanation and interpretation of phase-resolved PD pattern associated with common insulation defects
- Explanation and interpretation of the effects of time and the voltage level on PD behavior
- Information regarding the internal propagation of high-frequency signals resulting from PDs occurring in HV equipment, resulting impact on measurement sensitivity

Furthermore, IEEE working groups are now preparing the PD measurement of high-voltage apparatus standard, as follows:

- IEEE PC57.160, "IEEE Draft Guide for the Electrical Measurement of Partial Discharges in High Voltage Bushings and Instrument Transformers" [5]
- IEEE P2465 "Recommended Practice for Pulse-Type Partial Discharge Measurements on Individual Stator Coils and Bars" [6]
- IEEE P400.3 "Guide for Partial Discharge Field Diagnostic Testing of Shielded Power Cable Systems" [7]
- IEEE PC57.124, "Recommended Practice for the Detection of Partial Discharge and the Measurement of Apparent Charge in Dry-Type Transformers" [8]

Tables 11.1 and 11.2 summarize the IEC and IEEE PD testing standards, respectively.

Table 11.1 IEC PD testing standards.

No.	Name	Year
IEC60270 :2000 + AMD1 :2015 CSV	High-Voltage Test Techniques – Partial Discharge Measurements (Consolidated version)	2015
IEC TS 62478	High voltage test techniques – Measurement of Partial Discharges by EM and Acoustic Methods	2016
IEC TR 61294	Insulating Liquids – Determination of the Partial Discharge Inception Voltage (PDIV) – Test Procedure	1993
IEC 60076–3 :2013 + AMD1 :2018 CSV	Power Transformers – Part 3: Insulation Levels, Dielectric Tests, and External Clearances in Air (Consolidated Version)	2018
IEC 60034–27-1	Rotating Electrical Machines – Part 27-1: Off-line Partial Discharge Measurements on the Winding Insulation	2017
IEC 60034-27-2	Rotating Electrical Machines – Part 27–2: On-line Partial Discharge Measurements on the Stator Winding Insulation of Rotating Electrical Machines	2012
IEC/TS 61934	Electrical Insulating Materials and Systems – Electrical Measurement of Partial Discharges (PD) under Short Rise Time and Repetitive Voltage Impulses	2011
IEC 60034-18-41 :2014 + AMD1 :2019 CSV	Rotating Electrical Machines – Part 18–41: Partial Discharge Free Electrical Insulation Systems (Type I) Used in Rotating Electrical Machines Fed from Voltage Converters – Qualification and Quality Control Tests (Consolidated Version)	2019
IEC 60885-2	Electrical Test Methods for Electric Cables. Part 2: Partial Discharge Tests	1987
IEC 60885-3	Electrical Test Methods for Electric Cables – Part 3: Test methods for Partial Discharge Measurements on Lengths of Extruded Power Cables	2015
IEC 62271-203	High-Voltage Switchgear and Controlgear – Part 203: Gas-Insulated Metal-Enclosed Switchgear for Rated Voltages above 52 kV	2011

11.2 Technical Brochures

Beyond the standards, there are many important documents and publications regarding PDs. Especially, CIGRE has published a lot of technical brochures (TB) referring PD behavior in HV apparatus. Based on an agreement between IEC and CIGRE, CIGRE supports the work of IEC with studies and reports and thus provides the basis for developing and writing standards. Three CIGRE brochures specifically provide a better understanding of the standards IEC 60270 and IEC 62478: TB 366 – "Guide for Electrical Partial Discharge Measurements in Compliance to IEC 60270" [9]; TB 662 – "Guidelines for Partial Discharge Detection Using Conventional (IEC 60270) and

Table 11.2 IEEE PD testing standards.

No.	Name	Year
IEEE C57.113	IEEE Recommended Practice for Partial Discharge Measurement in Liquid-Filled Power Transformers and Shunt Reactors	2010
IEEE C57.127	IEEE Draft Guide for the Detection, Location and Interpretation of Sources of Acoustic Emissions from Electrical Discharges in Power Transformers and Power Reactors	2018
IEEE 436	IEEE Guide for Making Corona (Partial Discharge) Measurements on Electronics Transformers	1991
IEEE C57.124	IEEE Recommended Practice for the Detection of Partial Discharge and the Measurement of Apparent Charge in Dry-Type Transformers	1991
IEEE C57.160	IEEE Guide for the Electrical Measurement of Partial Discharges in High-Voltage Bushings and Instrument Transformers	2018
IEEE 1434	IEEE Guide for the Measurement of Partial Discharges in AC Electric Machinery	2014
IEEE 2862	IEEE Approved Draft Recommended Practice for Partial Discharge Measurements under AC Voltage with VHF/UHF Sensors during Routine Tests on Factory and Pre-molded Joints of HVDC Extruded Cable Systems up to 800 kV	2020
IEEE 400.4	IEEE Guide for Field Testing of Shielded Power Cable Systems Rated 5 kV and Above with Damped Alternating Current (DAC) Voltage	2015
IEEE 400.2	IEEE Guide for Field Testing of Shielded Power Cable Systems Using Very Low Frequency (VLF)(less than 1 Hz)	2013
IEEE 400.3	IEEE Guide for Partial Discharge Testing of Shielded Power Cable Systems in a Field Environment	2006
IEEE C37.301	IEEE Standard for High-Voltage Switchgear (Above 1000 V) Test Techniques – Partial Discharge Measurements	2009
IEEE C37.122.1	IEEE Guide for Gas-Insulated Substations Rated Above 52 kV	2014
IEEE 1291	IEEE Guide for Partial Discharge Measurement in Power Switchgear	1993

Unconventional Methods" [10]; and TB 444 – "Guidelines for Unconventional Partial Discharge Measurements" [11].

Besides these technical brochures, there are a lot of other CIGRE papers regarding PDs. Examples include TB 676 – "Partial Discharges in Transformers" [12]; TB 226 – "Knowledge Rules for Partial Discharge Diagnosis in Service" [13]; TB 502 – "High-Voltage On-Site Testing with Partial Discharge Measurement" [14]; and TB 674 – "Benefits of PD Diagnosis on GIS Condition Monitoring Assessment" [15]. Table 11.3 summarizes CIGRE technical brochures for PD testing.

Table 11.3 CIGRE technical brochures for PD testing.

No.	Name	Year
662	Guidelines for Partial Discharge Detection Using Conventional (IEC 60270) and Unconventional Methods	2016
502	High-Voltage On-Site Testing with Partial Discharge Measurement	2012
444	Guidelines for Unconventional Partial Discharge Measurements	2010
366	Guide for Electrical Partial Discharge Measurements in Compliance to IEC 60270	2008
226	Knowledge Rules for Partial Discharge Diagnosis in Service	2003
676	Partial Discharges in Transformers	2017
581	Guide – Corona Electromagnetic Probe Tests (TVA)	2014
728	On-Site Partial Discharge Assessment of HV and EHV Cable Systems	2018
297	Practical Aspects of the Detection and Location of PD in Power Cables	2006
674	Benefits of PD Diagnosis on GIS Condition Assessment	2017
654	UHF Partial Discharge Detection System for GIS: Application Guide for Sensitivity Verification	2016
525	Risk Assessment on Defects in GIS Based on PD Diagnostics	2013
703	Insulation Degradation under Fast Repetitive Pulses	2017

11.3 Books

PD books are those that deal specifically with PD only. The content of these books should include: phenomena, basic physics, behavior, measurement technique, detection and localization, diagnostics, evaluation, and impact to insulation systems. But there are also a lot of books dealing with PD phenomena and measurement but only in chapters of these books, including the following:

Kreuger, F. H. (1964). *Discharge Detection in High Voltage Equipment.* London: Temple Press Books Ltd [16].

König, D., and Rao, Y. N. (1993). *Partial Discharges in Electrical Power Apparatus.* Berlin: VDE-Verlag [17].

Ramu, T. S. and Nagamani, H. N. (2010). *Partial Based Condition Monitoring in High Voltage Equipment.* New Delhi: New Age International Publishers [18].

References

1 International Electrotechnical Commission (2000). *IEC 60270: High-Voltage Test Techniques – Partial Discharge Measurements.* Geneva, Switzerland: IEC.

2 International Electrotechnical Commission (2022). Standards Development TC-42 High-Voltage Test Techniques – Charge-Based Partial Discharge Measurements, https://www.iec.ch/dyn/www/f?p=103:38:2071137231035::::FSP_ORG_ID,FSP_APEX_PAGE,FSP_PROJECT_ID:1243,23,103245 (accessed 29 January 2021).

3 International Electrotechnical Commission (2016). *IEC TS 62478: High-voltage Test Techniques – Measurement of Partial Discharges by Electromagnetic and Acoustic Methods*. Switzerland: IEC.

4 IEEE Project (2019).P454 – guide for the detection, measurement and interpretation of partial discharges, https://standards.ieee.org/project/454.html (accessed 29 January 2021).

5 IEEE Project(2017). PC57.160 – IEEE draft guide for the electrical measurement of partial discharges in high voltage bushings and instrument transformers, https://standards.ieee.org/project/C57_160.html (accessed 29 January 2021).

6 IEEE Project(2017). P2465 – recommended practice for pulse-type partial discharge measurements on individual stator coils and bars, https://standards.ieee.org/project/2465.html (accessed 29 January 2021).

7 IEEE Project(2022). P400.3 – guide for partial discharge field diagnostic testing of shielded power cable systems, https://standards.ieee.org/project/400_3.html (accessed 29 January 2021)

8 IEEE Project(2017). PC57.124 – recommended practice for the detection of partial discharge and the measurement of apparent charge in dry-type transformers, https://standards.ieee.org/project/C57_124.html (accessed 29 January 2021)

9 Conseil International des Grands Réseaux Électriques (2008). Guide for Electrical Partial Discharge Measurements in Compliance to IEC 60270, *Technical Brochure. Ref. 366.*

10 Conseil International des Grands Réseaux Électriques. (2016). Guidelines for Partial Discharge Detection Using Conventional (IEC 60270) and Unconventional Methods. *Technical Brochure. Ref. 662.*

11 Conseil International des Grands Réseaux Électriques (2010). Guidelines for Unconventional Partial Discharge Measurements. *Technical Brochure. Ref. 444.*

12 Conseil International des Grands Réseaux Électriques (2017). Partial Discharges in Transformers. *Technical Brochure. Ref. 676.*

13 Conseil International des Grands Réseaux Électriques (2003). Knowledge Rules for Partial Discharge Diagnosis in Service. *Technical Brochure. Ref. 226.*

14 Conseil International des Grands Réseaux Électriques (2012). High-Voltage Onsite Testing with Partial Discharge Measurement. *Technical Brochure. Ref. 502.*

15 Conseil International des Grands Réseaux Électriques (2017). Benefits of PD Diagnosis on GIS Condition Assessment. *Technical Brochure. Ref. 674.*

16 Kreuger, F.H. (1964). *Discharge Detection in High-Voltage Equipment*. London: Temple Press Books Ltd.

17 König, D. and Rao, Y.N. (1993). *Partial Discharges in Electrical Power Apparatus*. Berlin: VDE-Verlag Gmbh.

18 Ramu, T.S. and Nagamani, H. N. (2010). *Partial Discharge Based Condition Monitoring of High Voltage Equipment*. New Delhi: New Age International Publishers.

12

Conclusions and Outlook

Electrical energy generation, transmission, distribution, and conversion utilize the conduction of electrical current as the basis for such activities. Therefore, potential separation becomes a necessary prerequisite. This potential separation is the task of electrical insulation technology. Consequently, it is very important to research, test, and use appropriate insulating materials to guarantee safety and reliability.

In addition to current and voltage stress, other factors such as temperature, humidity, environmental influences, mechanical stress, and aging play an essential role in the insulation material. There are a number of investigations and tests to determine the quality and reliability of the insulation material. Crucial criteria in these tasks are the detection, localization, and evaluation of partial discharges (PDs).

PDs are partial breakdowns in insulation materials, surface discharges on insulators, or discharges in gases on tips or edges of conductive parts. PDs have different effects depending on the type of insulating material (solid, liquid, gaseous). In the worst case, the insulation breaks down, and consequently, the device finally fails. Therefore, PD investigations are essential in the quality control of electrical power devices.

The importance of PD testing has been realized since the 1960s. Simultaneously, it was also defined in an IEC standard. With the advancing development of electronics and computer technology, PD acquisition has significantly been improved. Recently, PD behavior under alternating voltage stress has been researched and applied, and PD behavior under stress of direct voltage has become the focus of interest due to the increasing use of direct current (DC) transmission.

Since the importance of detection and localization of PDs is of such great significance in the quality assurance of electrical systems and apparatus, there has likewise been great interest in books and publications on this subject. Thus, there have been a large number of publications, reports, book chapters, as well as standards on PDs, but hardly any books devoted exclusively to PDs, and especially with testing and measurement techniques in high-voltage engineering and for insulating materials in electrical energy technology.

Accordingly, the authors have tried to deal with this topic comprehensively. This book reported the physical basics to the modeling of PD processes and measurement technology, along with the current status of PD technology. Other vital chapters dealt with electromagnetic

Partial Discharges (PD) - Detection, Identification, and Localization, First Edition. Norasage Pattanadech, Rainer Haller, Stefan Kornhuber, and Michael Muhr.

methods of PD detection, which are mostly used in diagnostics, and all nonelectrical methods of PD detection.

The localization of PDs is crucial because determining their origin allows for conclusions about the effect of PDs in insulating material; this was also delineated in this book. In addition, the evaluation and monitoring of the PD behavior, as well as the different effects by stress with alternating voltage or direct voltage, are essential, and they have been portrayed as well.

Due to the greater need for transmission and use of direct voltage, the effect of such voltage stress on systems and apparatus is of substantial interest. The direct voltage (DC) PD phenomena must be clearly understood since these phenomena are entirely different from the alternating voltage (AC) PDs. Unfortunately, the direct voltage PD phenomena have not yet been fundamentally researched.

The book concludes with a description of the intensive standardization on the use of PD behavior and applications.

The authors have tried to make PD topics as understandable as possible for all readers (users, students, and interested people) so that the PD mechanism and effects are better understood and easier to access. This book should serve as a good technical guide and textbook and represent sound support for the PD topics. The authors are also aware that some things may be missing, but this book should be a general and comprehensive presentation of the current status of PD detection, identification, and localization, without addressing the specific problems that are mostly detailed in publications.

PD detection, identification, localization, and evaluation have been dramatically developed and immensely employed with new devices. Plentiful PD patterns have been observed and recorded in recent years, so it is not easy to understand all of these. Correspondingly, this situation makes it increasingly difficult to assess whether detected PD is dangerous and therefore requires a very good knowledge of PD behavior and its effects.

To achieve technical development in areas where fundamental knowledge about the PD behavior is not available, especially the PD phenomena in the DC system, from the academic point of view, there is always light at the end of the tunnel. The ocean of knowledge for the direct voltage PDs will be explored finally, due to the increased use of direct voltage, and more research will be carried out in this field. Consequently, the knowledge gained will be useful to the understanding of the PD mechanism, enabling users to utilize electrical equipment with the highest efficiency and reliability throughout the expected lifetime.

The authors would therefore be happy if this book about PD was met with interest and approval and would give the reader a better understanding of the field of PD.

Index

Partial Discharges (PD) - Detection, Identification, and Localization, First Edition. Norasage Pattanadech, Rainer Haller, Stefan Kornhuber, and Michael Muhr.

b

d

f

O

p

q